RETHINKING
THE OZONE PROBLEM
IN URBAN AND REGIONAL
AIR POLLUTION

Committee on Tropospheric Ozone
Formation and Measurement

Board on Environmental Studies
and Toxicology

Board on Atmospheric Sciences
and Climate

Commission on Geosciences,
Environment, and Resources

National Research Council

NATIONAL ACADEMY PRESS
Washington, D.C. 1991

National Academy Press   2101 Constitution Ave., N.W.   Washington, D.C.   20418

Library of Congress Catalog No. 91-68142
International Standard Book Number 0-309-04631-9

Cover photo: M. Cerone/Superstock, Inc.   First Printing, January 1992
                                                  Second Printing, March 1994
Printed in the United States of America

Committee on Tropospheric Ozone Formation and Measurement

Board on Environmental Studies and Toxicology

Board on Atmospheric Sciences and Climate

v

Commission on Geosciences, Environment, and Resources

vi

Preface

Ambient ozone in urban and regional air pollution represents one of this country's most pervasive and stubborn environmental problems. Despite more than two decades of massive and costly efforts to bring this problem under control, the lack of ozone abatement progress in many areas of the country has been disappointing and perplexing.

It is encouraging to note that the U.S. Environmental Protection Agency recognized a need for this independent assessment from the National Research Council and agreed to co-sponsor the study in 1989, even before it was mandated in Section 185B of the Clean Air Act Amendments of 1990. It is further encouraging to note the additional support for this study by the U.S. Department of Energy, the American Petroleum Institute, and the Motor Vehicle Manufacturers Association of the United States. The authors of this report have undertaken an effort to re-think the problem of ambient ozone and to suggest steps by which the nation can begin to address this problem on a more rigorous scientific basis.

The Committee on Tropospheric Ozone Formation and Measurement was established by the National Research Council to evaluate scientific information relevant to precursors and tropospheric formation of ozone and to recommend strategies and priorities for addressing the critical gaps in scientific information necessary to help address the problem of high ozone concentrations in the lower atmosphere. The committee was specifically charged to address emissions of volatile organic compounds (anthropogenic and biogenic) and oxides of nitrogen; significant photochemical reactions that form ozone, including differences in various geographic regions; precursor emission effects on daily patterns of ozone concentration; ambient monitoring techniques; input data and performance evaluations of air quality models; regional source-receptor relationships; statistical approaches in tracking ozone abatement progress; and patterns of concentration, time, and interactions with other atmospheric pollutants.

During the course of the committee's deliberations, we solicited information from many federal, state, academic, and industrial experts. We also reviewed the scientific literature, government agency reports, and unpublished data bases. The committee benefitted from having earlier National Research Council and Congressional Office of Technology Assessment reports as a starting base. Gregory Whetstone of the House Energy and Commerce Committee staff, John Bachmann and John Calcagni of the Environmental Protection Agency, and representatives of the other sponsors kindly provided useful information and perspectives to the committee. The committee's efforts were also greatly aided by information provided by David Chock of Ford Motor Company's Research and Engineering Division, Brian Lamb of Washington State University, Douglas Lawson of the California Air Resources Board, S. T. Rao of the New York State Department of Environmental Conservation, and Donald Stedman of the University of Denver.

We wish especially to thank Raymond Wassel, the National Research Council project director, who assisted the committee all along the way, and was particularly valuable in the final stages of preparation of the report. We are also grateful to James Reisa, director of the Board on Environmental Studies and Toxicology, for his guidance and contributions throughout the study. Kate Kelly did an excellent job as editor. Other staff who contributed greatly to the effort were research assistant William Lipscomb, who helped in the final stages; Lee Paulson and Tania Williams, who prepared the document for publication; Felita Buckner, the project secretary; information specialist Anne Sprague; and other dedicated staff of BEST's Technical Information Center.

John H. Seinfeld
Chairman

viii

Dedication

The committee dedicates this report
to our late colleague and committee member,
Dr. Bernard J. Steigerwald,
whose three decades of distinguished public service
with the United States Public Health Service
the National Air Pollution Control Administration,
and the Environmental Protection Agency
contributed significantly to
scientific knowledge and protection of
the nation's air quality.

Contents

Tables

xvi

Figures

Executive Summary

INTRODUCTION

Of the six major air pollutants for which National Ambient Air Quality Standards (NAAQS) have been designated under the Clean Air Act, the most pervasive problem continues to be ozone,[1] the most prevalent photochemical oxidant and an important component of "smog." The most critical aspect of this problem is the formation of ozone in and downwind of large urban areas where, under certain meteorological conditions, emissions of nitric oxide and nitrogen dioxide (known together as NO_x) and volatile organic compounds (VOCs) can result in ambient ozone concentrations up to three times the concentration considered protective of public health by the U.S. Environmental Protection Agency (EPA).

Major sources of VOCs in the atmosphere include motor vehicle exhaust, emissions from the use of solvents, and emissions from the chemical and petroleum industries. In addition, there is now a heightened appreciation of the importance of reactive VOCs emitted by vegetation. NO_x comes mainly from the combustion of fossil fuels; major sources include motor vehicles and electricity generating stations.

The occurrence of ozone concentrations that exceed the NAAQS in various

---

[1]The scientific community now has strong reason to believe that, unlike stratospheric (i.e., high-altitude) ozone concentrations, which are declining, concentrations of tropospheric (i.e., near-ground) ozone are generally increasing over large regions of the United States.

1

regions of the United States indicates that many people may be exposed to concentrations of ozone that EPA has determined to be potentially harmful. EPA reported that in 1989, about 67 million people lived in areas where the second-highest ozone concentration, a principal measure of compliance, exceeded the NAAQS concentration. Despite considerable regulatory and pollution control efforts over the past 20 years, high ozone concentrations in urban, suburban, and rural areas of the United States continue to be a major environmental and health concern.

The nationwide extent of the problem, coupled with increased public attention resulting from the high concentrations of ozone over the eastern United States during the summer of 1988, adds to the urgency for developing effective control measures. As the ozone attainment strategy presented in the 1990 amendments to the Clean Air Act is put into effect, the success of efforts to control the precursors of ozone will be of vital concern to Congress, to governmental regulatory agencies, to industry, and to the public.

THE CHARGE TO THE COMMITTEE

The Committee on Tropospheric Ozone Formation and Measurement was established in 1989 by the Board on Environmental Studies and Toxicology of the National Research Council (NRC) in collaboration with the NRC's Board on Atmospheric Sciences and Climate to evaluate scientific information and data bases relevant to precursors and tropospheric formation of ozone, and to recommend strategies and priorities for filling critical scientific and technical gaps in the information and data bases. The committee's members had expertise that included atmospheric chemistry, measurement, mathematical modeling, pollution trends monitoring, transport meteorology, exposure assessment, air-pollution engineering, and environmental policy. The committee was specifically charged to address

- Emissions of VOCs (anthropogenic and biogenic[2]) and NO_x;
- Significant photochemical reactions that form ozone, including differences in various geographic regions;
- Effects of precursor emissions on daily patterns of ozone concentration;
- Ambient monitoring techniques;

---

[2]"Anthropogenic emissions" refers to emissions resulting from the actions of human society. "Biogenic emissions" refers to natural emissions, mainly from trees and other vegetation.

- Input data and performance evaluations of air-quality models;
- Regional source-receptor relationships;
- Statistical approaches in tracking ozone abatement progress;
- Patterns of concentration, time, and interactions with other atmospheric pollutants.

The committee was not charged to evaluate and did not address the adequacy of the NAAQS for protecting human health and welfare, nor the technologic, economic, or sociologic implications of current or potential ozone precursor control strategies.

The committee's work was sponsored principally by the U.S. Environmental Protection Agency and Department of Energy. Additional funding was provided by the American Petroleum Institute and the Motor Vehicle Manufacturers Association. EPA is expected to provide the committee's report to Congress as partial fulfillment of Section 185B of the Clean Air Act amendments of 1990, which requires EPA to conduct a study in conjunction with the National Academy of Sciences on the role of ozone precursors in tropospheric ozone formation and control. The study required in Section 185B is more extensive than the committee's charge, and EPA is separately addressing the portions of the study required in Section 185B that are beyond the scope of this NRC study.

THE COMMITTEE'S APPROACH TO ITS CHARGE

In this report, the committee examines trends in tropospheric ozone concentrations in the United States; reviews current approaches to control ozone precursors; and assesses current understanding of the chemical, physical, and meteorological influences on tropospheric ozone. Based on this understanding, the committee provides a critique of the scientific basis for current regulatory strategies and offers recommendations for improving the scientific basis for future regulatory strategies. It also recommends an integrated research program to further clarify the factors that affect tropospheric ozone formation within the context of changing regional and global environmental conditions. The committee presents data and arguments to address these issues and to help resolve the continuing national debate over devising an effective program to achieve the NAAQS for ozone.

The major findings and recommendations of the committee are discussed in the remainder of this summary.

OZONE IN THE UNITED STATES

FINDING: Despite the major regulatory and pollution-control programs of the past 20 years, efforts to attain the National Ambient Air Quality Standard for ozone largely have failed.

DISCUSSION: Since passage of the 1970 Clean Air Act amendments, extensive efforts to control ozone have failed three times to meet legislated deadlines for complying with the ozone NAAQS. Congress set 1975 as the first deadline, but 2 years after this deadline, many areas were still in violation of the NAAQS. The 1977 amendments to the Clean Air Act extended the deadline for compliance until 1982 and allowed certain areas that could not meet the 1982 deadline until 1987. For 1987, however, more than 60 areas still exceeded the NAAQS; the following year, the number of areas exceeding the NAAQS jumped to 101. In 1990, 98 areas were in violation of the NAAQS.

EPA has reported a trend toward lower nationwide average ozone concentrations from 1980 through 1989, with anomalously high concentrations in 1983 and 1988. Ozone concentrations were much lower in 1989 than in 1988, possibly the lowest of the decade. However, since the trend analysis covers only a 10-year period, the high concentrations in 1983 and 1988 cannot be assumed to be true anomalies, nor can the lower concentrations in 1989 be assumed to be evidence of progress. It is likely that meteorological fluctuations are largely responsible for the highs in 1983 and 1988 and the low in 1989. Meteorological variability and its effect on ozone make it difficult to determine from year to year whether changes in ozone concentrations result from fluctuations in the weather or from reductions in the emissions of precursors of ozone. However, it is clear that progress toward nationwide attainment of the ozone NAAQS has been extremely slow at best, in spite of the substantial regulatory programs and control efforts of the past 20 years.

OZONE TRENDS

FINDING: The principal measure currently used to assess ozone trends (i.e., the second-highest daily maximum 1-hour concentration in a given year) is highly sensitive to meteorological fluctuations and is not a reliable measure of progress in reducing ozone over several years for a given area.

RECOMMENDATION: More statistically robust methods should be developed to assist in tracking progress in reducing ozone. Such methods should account for the effects of meteorological fluctuations and other relevant factors.

DISCUSSION: Year-to-year meteorological fluctuations might mask downward ozone trends in cases where precursor emission controls are having the desired effect. Such fluctuations might also mask trends of increasing ozone or evidence that progress in reducing ozone has been slower than expected. Alternative statistical measures, discussed in Chapter 2, should be developed. These measures should not be mere statistical entities; they should bear some relation to the range of ozone concentrations considered harmful to human health and welfare. Support should be given to the development of methods to normalize ozone trends for meteorological variation. Several techniques that have shown promise for individual cities and regions could be useful on a national scale.

STATE IMPLEMENTATION PLANNING

FINDING: The State Implementation Plan (SIP) process, outlined in the Clean Air Act for developing and implementing ozone reduction strategies, is fundamentally sound in principle but is seriously flawed in practice because of the lack of adequate verification programs.

RECOMMENDATION: Reliable methods for monitoring progress in reducing emissions of VOCs and NO_x must be established to verify directly regulatory compliance and the effectiveness associated with mandated emission controls.

DISCUSSION: The Clean Air Act places the responsibility for attaining the NAAQS on a federal-state partnership. EPA develops uniform NAAQS, and the states formulate and implement emission reduction strategies to bring each area exceeding the NAAQS into compliance.

The SIP process for air-quality management is based on the premise that emission reductions can be inferred directly from observed improvements in air quality. For some pollutants, such as sulfur dioxide (SO_2) and carbon monoxide (CO), it has generally been possible to infer progress by tracking emission reductions through measurements of ambient concentrations of the pollutants. However, in the case of ozone, which is formed by highly complex and nonlinear reactions that involve VOC and NO_x precursors, it is extremely questionable in most cases to conclude that emissions reductions have occurred solely on the basis of observed trends in ambient ozone concentrations.

EPA's approach to ozone control, originally developed in 1971, has relied largely upon unverified estimates of reductions in precursor emissions; EPA has not required systematic measurements of ambient precursor concentrations. Systematic measurements of NO_x and VOCs are needed in addition to ozone measurements to determine the extent to which precursor emissions must be controlled and to verify the effectiveness of the control measures undertaken. Over the past two decades, the substantial reductions in ozone concentrations predicted to result from the VOC emission reductions in major urban centers have not occurred. Moreover, the limited data available on ambient concentrations of VOCs suggest that the actual VOC emission reductions have been smaller than estimated in the SIP process. The reasons for this failure are largely unknown.

Designing a strategy to control precursor emissions, and tracking that strategy's effectiveness in controlling a specific VOC or NO_x, requires systematic VOC and NOx measurements in strategic locations. Until verification programs are incorporated into the SIP process, the use of unverified emission inventories[3] in air-quality models will continue to involve considerable uncertainties in predicting changes in ozone concentrations resulting from emission controls.

ANTHROPOGENIC VOC EMISSIONS

FINDING: Current emissions inventories significantly underestimate anthropogenic emissions of VOCs. As a result, past ozone control strategies may have been misdirected.

RECOMMENDATION: The methods and protocols used to develop inventories of ozone precursor emissions must be reviewed and revised. Independent tests, including monitoring of ambient VOCs, should be used by government agencies to assess whether emissions are indeed as they are represented by emissions inventories.

DISCUSSION: As discussed in Chapter 9, there is substantial evidence that the methods and protocols used to develop anthropogenic VOC inventories are flawed and do not account adequately for all types of sources, nor for

---

[3]An emission inventory is a data base containing estimated emissions from various sources in a specific area and period (for example, nationwide emissions of VOCs per year).

the magnitude of VOC emissions. Ambient measurements of VOC/NO_x ratios, for instance, consistently yield ratios that are larger than would be expected from emissions inventories. In addition, measurements of VOC concentrations near roadways and in tunnels, as well as ambient measurements of specific VOCs in urban areas, indicate that VOC emissions from mobile sources have been underestimated in these inventories by a factor of two to four. The discrepancy in mobile-source emissions is probably a result of several factors in current emission models. It is likely that the fleets used in dynamometer testing to determine emission factors are not representative of on-road vehicles, that speed correction factors and estimates of evaporative emissions are inaccurate, and that the Federal Test Procedure does not adequately simulate actual driving behavior. Moreover, current Inspection and Maintenance (I/M) programs do not appear to be leading to the emission reductions anticipated.

The underestimation of VOC emission inventories presents two problems. One problem involves the limited effectiveness of older emissions-control technology. It now seems likely that mandated emissions controls in past years have not been as effective as EPA had estimated. Even if the reductions claimed are correct, their effect as a percentage of the total VOC inventory is clearly smaller than expected, because total anthropogenic VOC emissions were underestimated. Although VOC reductions of 25-50% have been claimed for some areas, it is likely that the actual reductions were in the 10-25% range.

A more profound problem resulting from underestimating VOC emissions is the implication for future emission controls. As discussed in Chapters 6 and 11, the relative effectiveness of VOC and NO_x controls for reducing ozone in a particular area depends on the ambient VOC/NO_x ratios in that area. At VOC/NO_x ratios of about 10 or less, VOC control is generally more effective, and NO_x control may actually be counterproductive. At VOC/NO_x ratios of 20 or more, NO_x control is generally more effective. The nation's ozone reduction strategy has been based largely on the premise that VOC/NO_x ratios in most polluted urban areas fall in the less-than-10 range, where VOC control is more effective than NO_x control. Hence, a major upward correction in VOC emission inventories could indicate the need for a fundamental change in the strategy used to abate ozone in many geographic areas.

The fact that a fundamental change in the nation's ozone reduction strategy could be necessary after two decades of costly efforts indicates that there could be an even more fundamental flaw in the overall design of the strategy. The current design depends too much on the assumption that emissions inventories are accurate, and not enough on adequate checks and tests of these inventories. Independent tests of the inventories and alternative approaches, such as tunnel studies and remote sensing, need to be added to the design so that errors in the inventories can be uncovered and corrections can be made. The checks should include thorough monitoring of precursor concentrations in nonattainment areas to verify the effectiveness of emission controls. In addition, air-quality models that use ambient data instead of emissions inventories to identify important precursor sources, to verify emissions algorithms, and to determine ozone precursor relationships could serve as useful checks of models that require the entry of accurate data from emissions inventories.

BIOGENIC VOC EMISSIONS

FINDING: The combination of biogenic VOCs with anthropogenic NO_x can have a significant effect on photochemical ozone formation in urban and rural regions of the United States.

RECOMMENDATION: In the future, emissions of biogenic VOCs must be more adequately assessed to provide a baseline from which the effectiveness of ozone control strategies can be estimated before such strategies are applied for a specific urban core or larger regions. Ambient measurements of concentrations and emission rates are needed to improve the accuracy of biogenic VOC inventories.

DISCUSSION: Measurements of the ambient concentrations of isoprene and other VOCs known to be emitted by vegetation, as well as estimates of total emissions from biogenic VOC sources, suggest that these compounds help foster episodes of high concentrations of ozone in urban cores and other areas affected by anthropogenic NO_x. Moreover, in many rural areas, especially those in the eastern and southern United States, more biogenic VOCs than anthropogenic VOCs are oxidized in the troposphere. However, quantifying the role of biogenic VOCs in specific episodes of high ozone concentrations is difficult because of large uncertainties in inventories.

The biogenic VOC contribution is a background concentration that cannot be removed from the atmosphere by emission controls. If anthropogenic VOC emissions are reduced, this background concentration will become a larger and more significant fraction of total VOCs. The committee's analysis (see Chapter 8) suggests that in many urban cores and their environs, even if anthropogenic VOC emissions are totally eliminated, a high background concentration of reactive biogenic VOCs will remain; for example, on hot summer days this can be the equivalent of a propylene concentration of 10-30 parts per billion (ppb) (see Chapter 8). In the presence of anthropogenic NO_x and under favorable meteorological conditions, these background biogenic VOCs can contribute to summertime ozone concentrations exceeding the NAAQS concentration of 120 ppb.

AMBIENT AIR QUALITY MEASUREMENTS

FINDING: Ambient air quality measurements now being performed are inadequate to elucidate the chemistry of atmospheric VOCs or to assess the contributions of different sources to individual concentrations of these compounds.

RECOMMENDATION: New measurement strategies that incorporate more accurate and precise measurements of the individual trace compounds involved in ozone chemistry should be developed to advance understanding of the formation of high concentrations of ozone in the United States and to verify estimates of VOC and NO_x emissions.

DISCUSSION: There have been major advances in the ability to measure ambient VOCs, NO_x, and other reactive nitrogen compounds (see Chapter 7). These more accurate and precise measurements should be used in coordinated programs to test the reliability of estimated VOC and NO_x emissions from major anthropogenic and natural sources and to study the distribution of these sources to characterize ozone accumulation and destruction. Trends in atmospheric measurements of emissions can provide a valuable check of the reductions estimated by using emissions inventories.

In addition, despite significant progress in the past 2 decades, the state of knowledge of many of the fundamental processes that govern the formation and distribution of ozone in the atmosphere needs improvement. To increase understanding of the processes of ozone formation and removal, reliable measurements of ozone and its precursors are needed. These measurements can be obtained only by intensive field studies that use validated methods to provide data for evaluating the representation of physical and chemical processes in air-quality models designed to predict the effects of future emissions controls. Past field studies have offered limited spatial coverage; sites in a wide variety of areas must be studied. Future studies must consider winter as well as summer conditions. The role of aerosol particles and cloud chemistry in ozone formation must also be investigated.

AIR-QUALITY MODELS

FINDING: Although three-dimensional or grid-based ozone air-quality models are currently the best available for representing the chemical and physical processes of ozone formation, the models contain important uncertainties about chemical mechanisms, wind-field modeling, and removal processes. Moreover, important uncertainties in input data, such as emissions inventory data, must be considered when using such models to project the effects of future emissions controls.

RECOMMENDATION: Air-quality models are essential in predicting the anticipated effects of proposed emissions controls on ambient ozone concentrations. Therefore, the effects of uncertainties on model predictions, such as uncertainties in the emissions inventory and in the chemistry incorporated in the models, must be elucidated as completely as possible. Predictions of the effects of future VOC and NO_x controls should be accompanied by carefully designed studies of the sensitivity of model results to these uncertainties.

DISCUSSION: Most future SIPs will rely on trajectory and grid-based photochemical air-quality models to estimate the effects of strategies to control emissions of ozone precursors. Air-quality models are designed to represent the complex physics and chemistry of the atmosphere and require a number of important types of input data, such as initial and boundary conditions, meteorological fields, and emissions inventories (see Chapter 10). Models are evaluated for use in control strategy assessment by first establishing their ability to simulate one or more past episodes of high concentrations of ozone. Grid-based models generally have been able to simulate observed 1-hour average ozone concentrations with a gross error of 30% or less, but known biases in certain inputs, particularly in emissions inventories, have raised the concern that there are uncertainties as yet unknown in other areas such as meteorology. Thus, good predictions of ozone concentrations can result from offsetting errors in inputs. Predictions of the effect of future emission controls should be accompanied by estimates of the uncertainty about ozone concentrations that stems from uncertainties in input variables.

Uncertainties in air-quality models include those in the chemical mechanism, those in the treatment of physical processes, and those that result from the choice of numerical algorithms. One measure of the uncertainty arising from a chemical mechanism can be obtained by performing simulations with different chemical mechanisms. By necessity, simplified equations are used and constant values are assigned to represent physical processes in air photochemistry, such as atmospheric transport including dry deposition. The best approach to quantifying the effects of uncertainties in a model's representation of physical processes is to conduct simulations with a different model for the same set of input data.

Uncertainties in model input data include those involving meteorological factors (e.g., winds and mixing heights), initial and boundary conditions, base case emissions, and projected future emissions. The data that describe the wind field and mixing height contain uncertainties of magnitudes that are difficult to estimate. One way to quantify these uncertainties is to simulate different episodes for the same city or region. The overall influence of initial and boundary conditions on predictions of ozone concentrations can be assessed by performing simulations with initial conditions set to zero and boundary conditions set to zero. Similar calculations can be carried out to assess the uncertainties in ozone predictions that arise from uncertainties in base-year emissions inventories and projected future emissions.

Although considerable effort has gone into developing and applying air-quality models, the lack of ambient data that can be used to evaluate the models comprehensively has impeded progress in their development and use. To obtain such data will require intensive field programs, designed to obtain the data needed to evaluate models. The Southern California Air Quality Study (SCAQS), the San Joaquin Valley Air Quality Study (SJVAQS)/Atmospheric Utility Signatures, Predictions, and Experiments (AUSPEX), the Southern Oxidants Study, and the Lake Michigan Ozone Study are examples of such programs. More such programs should be developed, especially in the eastern and southern United States.

VOC VERSUS NO_x CONTROL

FINDING: State-of-the-art air-quality models and improved knowledge of the ambient concentrations of VOCs and NO_x indicate that NO_x control is necessary for effective reduction of ozone in many areas of the United States.

RECOMMENDATION: To substantially reduce ozone concentrations in many urban, suburban, and rural areas of the United States, the control of NO_x emissions will probably be necessary in addition to, or instead of, the control of VOCs.

DISCUSSION: Application of grid-based air-quality models to various cities in the United States shows that the relative effectiveness of VOC and NO_x controls in ozone abatement varies widely. NO_x reductions can have either a beneficial or detrimental effect on ozone concentrations, depending on the locations and emission rates of VOC and NO_x sources in a region. The effect of NO_x reductions depends on the local VOC/NO_x ratio and a variety of other factors. Modeling studies show that ozone should decrease in response to NO_x reductions in many urban areas. However, some modeling and field studies show that ozone concentrations can increase in the near field in response to NO_x reductions, but decrease in the far field. Thus, NO_x controls could reduce ozone under some conditions, but under different conditions might lead to smaller ozone decreases than if VOCs alone are reduced. NO_x controls should be evaluated not only on the basis of reducing peak ozone concentrations, but also for their effects on other nitrogen-containing species (see Chapter 6). In any event, the ramifications of NO_x control on ozone concentrations are complex and must be considered carefully in devising ozone abatement strategies.

A decrease in emissions of NO_x should lower ozone concentrations in many parts of the United States. When rural measurements of VOCs and NO_x are used in regional air-quality modeling, the resulting ozone concentrations can exceed 100 ppb—as is consistent with observed concentrations. In some areas, concentrations of the biogenic VOC isoprene by itself are high enough to generate this much ozone in the presence of ambient NO_x. In rural areas, formation of ozone appears to be insensitive to changes in concentrations of anthropogenic VOCs because of the generally high VOC/NO_x ratios, but this insensitivity depends strongly on how much NO_x is present. Much of the NO_x in rural areas is generated by mobile and stationary sources in the urban cores and their major connecting regions; such NO_x sources appear to contribute to the pervasive high ozone concentrations found in the eastern United States. Simulations with the Regional Oxidant Model (ROM) have shown that ozone concentrations above 80 ppb can be generated in the synoptic-scale transport region, including the Ohio River Valley and the entire Northeast corridor, from the prevailing NO_x and biogenic VOCs alone.Ozone concentrations were greater than 100 ppb downwind of the major urban areas.

Ozone is predicted to decrease in response to NO_x reductions in most urban locations. Models show that ozone concentrations rise in some urban cores, such as New York City and Los Angeles, in response to NO_x reductions but decrease in downwind areas, where maximum amounts of ozone are found. Choosing not to reduce NO_x in those urban centers, while ameliorating a local problem, could exacerbate the ozone problem in downwind regions. Moreover, total population exposure to ozone (and other harmful pollutants) might not respond in the same manner as peak ozone to control techniques designed to reduce peak ozone; population exposure should be considered in the design of future control strategies.

Many simulations conducted to date have relied on emissions inventories that did not include biogenic emissions and are strongly suspected of significantly underestimating anthropogenic VOC emissions. The result is an overestimate of the effectiveness of VOC controls and an underestimate of the efficacy of NO_x controls. Faulty inventories have likely led to underprediction of ozone concentrations in central urban areas (see Chapters 10 and 11). An increase by a factor of two to three in mobile-source VOCs, as suggested by recent studies discussed in Chapter 9, leads to predicted ozone concentrations closer to those observed. If the anthropogenic VOC inventory is as badly underestimated as recent studies indicate, areas that were previously believed to be adversely affected by NO_x controls might actually benefit from them.

ALTERNATIVE FUELS FOR MOTOR VEHICLES

FINDING: The use of alternative fuels has the potential to improve air quality, especially in urban areas. However, the extent of the improvement that might result is uncertain and will vary depending on the location and on the fuels used. Alternative fuel use, alone, will not solve ozone problems nationwide. Moreover, it will not necessarily alleviate the most critical problem associated with motor vehicle emissions—increased emissions as in-use vehicles age.

RECOMMENDATION: Because there is uncertainty about the degree to which alternative fuels would reduce ozone, requiring the widespread use of any specific fuel would be premature. An exception may be electric vehicles, which can lead to substantial reductions in all ozone precursor emissions. Coordinated emissions measurement and modeling studies should be used to determine which fuels will work best to control formation of ozone.

DISCUSSION: The possible widespread use of alternative fuels in the next several years would change the emission characteristics of motor vehicles. Therefore it is important to assess the potential improvement in air quality resulting from such use. Alternative fuels are viewed as a means to improve air quality by reducing the mass emission rates from motor vehicles or by reducing the ozone-forming potential (or reactivity) of those emissions (see Chapter 12). Candidate alternative fuels include natural gas, methanol, ethanol, hydrogen, and electricity. Another candidate fuel is reformulated gasoline[4], whose composition has been altered to make exhaust products less photochemically reactive and toxic and to lower total emissions of VOCs, CO, or NO_x. Reformulated gasoline has the advantage that it may be used immediately in exist-

---

[4]Reformulated gasoline is not considered a true alternative fuel, but it is discussed here because its use could potentially improve air quality.

ing vehicles.

Vehicles that run on electricity or on hydrogen in fuel cells would emit virtually no precursors to ozone, although the production of the electricity or hydrogen can contribute to ozone formation depending on the feedstock and location of the generation facility. Modeling studies show that at relatively low ambient VOC/NO_x ratios (in the range of 4 to 6), the use of natural gas as a motor fuel could reduce ozone formation on a mass basis (grams of ozone formed per gram of VOC emitted) by as much as 75% compared with the use of conventional gasoline vehicles. Emissions from methanol-fueled vehicles are largely methanol and formaldehyde. On a per-mass-emitted basis, methanol-fueled vehicles are predicted to reduce ozone formation by 15-40% relative to conventionally fueled vehicles if formaldehyde emissions are controlled. Ethanol would provide less benefit, especially if an increased vapor pressure is allowed. At higher VOC/NO_x ratios, less improvement is expected from VOC reactivity reduction.

When considered for their effect on ozone concentrations in an urban area, alternative fuels are predicted to provide considerably less benefit to air quality than projected by the numbers cited above because motor vehicle emissions do not constitute the total of VOC emissions. For example, use of methanol in Los Angeles is predicted to lead to no more than a 10-15% reduction in ozone exposure, and little decrease in peak ozone. A major uncertainty is the effect of alternative fuels on the in-use emissions of the motor vehicle fleet, which is dominated by a relatively small fraction of vehicles having high VOC emissions. Under NO_x-limited conditions, typified by high VOC/NO_x ratios such as those found in Houston or Atlanta, reducing the reactivity of emissions probably would have little benefit. In high NO_x regions, reducing reactivities would complement NO_x control that might otherwise lead to local ozone increases.

Alone, no alternative fuel will solve the air pollution problems that face most large cities. Each fuel must be considered in conjunction with other controls. Specific fuels could work effectively in some regions, but would provide little benefit if used in others.

A RESEARCH PROGRAM ON TROPOSPHERIC OZONE

FINDING: Progress toward reducing ozone concentrations in the United States has been severely hampered by the lack of a coordinated national research program directed at elucidating the chemical, physical, and meteorological processes that control ozone formation and concentrations over North America.

RECOMMENDATION: A coherent and focused national program should be established for the study of tropospheric ozone and related aspects of air quality in North America. This program should include coordinated field measurements, laboratory studies, and numerical modeling that will lead to a better predictive capability. In particular, the program should elucidate the response of ambient ozone concentrations to possible regulatory actions or to natural changes in atmospheric composition or climate. To avoid conflict between the long-term planning essential for scientific research and the immediacy of requirements imposed on regulatory agencies, the research program should be managed independently from the EPA office that develops regulations under the Clean Air Act and from other government offices that develop regulations. The research program must have a long-term commitment to fund research on tropospheric ozone. The direction and goals of this fundamental research program should not be subjected to short-term perturbations or other influences arising from ongoing debates over policy strategies and regulatory issues. The program should also be broadly based to draw on the best atmospheric scientists available in the nation's academic, government, industrial, and contract research laboratories. Further, the national program should foster international exchange and scientific evaluations of global tropospheric ozone and its importance in atmospheric chemistry and climate change. The recommended tropospheric ozone research program should be carefully coordinated with the Global Tropospheric Chemistry Program currently funded and coordinated by the National Science Foundation (NSF) and with corresponding global change programs in the National Aeronautics and Space Administration (NASA), the National Oceanic and Atmospheric Administration (NOAA), the Department of Energy (DoE), and other agencies.

DISCUSSION: A good analogy for the research program needed is the U.S. effort to address depletion of the stratospheric ozone layer by chlorofluorocarbons.[5] For this program, EPA is the relevant regulatory agency, but NASA's Upper Atmosphere Research Program was directed in the Clean Air Act amendments of 1977 to "continue programs of research, technology, and monitoring of the phenomena of the stratosphere for the purpose of understanding the physics and chemistry of the stratosphere and for the early detection of potentially harmful changes in the ozone of the stratosphere." The partnership has worked well, and the basic research program has prepared the

---

[5]This program is discussed as an example because it has many features that would be desirable in a tropospheric ozone research program. The committee does not recommend which agency should direct such a program.

scientific foundation for international assessment and for the Montreal Protocol on Substances that Deplete the Ozone Layer (1987). NASA has developed a basic research program of laboratory and field measurements, satellite data analysis, and theoretical modeling. The particular strengths of the program have been its broad participation base, which draws on academic, government, industrial, and contract research groups, and its careful coordination with other federal and industrial programs and non-U.S. research efforts. The results of this comprehensive and coordinated research effort have been reported to Congress and to EPA. Its scientific assessments often include specific modeling studies that meet the regulatory and policy needs of EPA. A similar partnership that meets the needs of the research community and those of regulatory agencies will be necessary to establish a reliable scientific basis for the improvement of the nation's air quality.

Rethinking
the Ozone Problem
in Urban and Regional
Air Pollution

1

What Is the Problem?

NATURAL ATMOSPHERIC OZONE

Ozone (O_3) is a reactive oxidant gas produced naturally in trace amounts in the earth's atmosphere. The ozone molecule is composed of three oxygen atoms, in contrast to normal molecular oxygen (O_2), which makes up roughly 21% of our air. Ozone forms when an atom of oxygen, O, usually produced in the troposphere by solar photodissociation of nitrogen dioxide (NO_2), combines with molecular oxygen to form ozone. Ozone was discovered by C.F. Schönbein in the middle of the last century; he also was first to detect ozone in air (Schönbein, 1840; 1854).

Most of the earth's atmospheric ozone is found in the stratosphere—the portion of the atmosphere between about 10 and 50 kilometers (km) altitude—where it plays a critical role in absorbing ultraviolet radiation emitted by the sun. A plot of ozone partial pressures and atmospheric mixing ratios as a function of altitude is shown in Figure 1-1 for a typical latitude profile (AFGL, 1985). The bulge shown in Figure 1-1, the stratospheric ozone layer, peaks in ozone concentration (partial pressure) at 20-30 km and prevents most solar radiation in the wavelength range of 200-300 nanometers (nm) from reaching the lower atmosphere and the earth's surface, where it would damage plant and animal life. Any significant weakening of the stratospheric ozone layer is predicted to lead to greatly increased instances of skin cancer in humans (NRC, 1982, 1984).

The lowest part of the atmosphere is called the troposphere; it begins at the earth's surface and extends upward to about 10 km, cooling with altitude

19

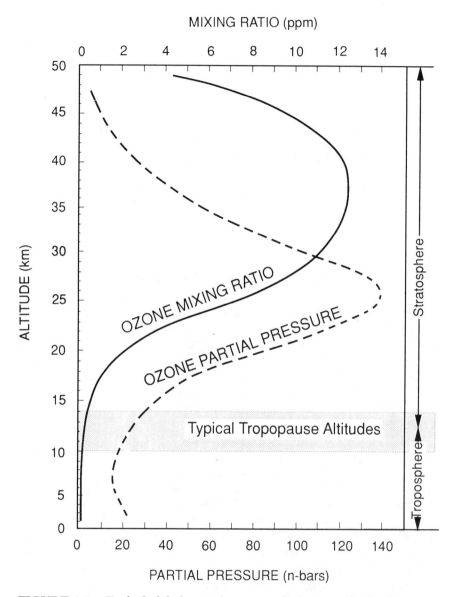

FIGURE 1-1 Typical global annual mean vertical ozone distribution.
Source: Adapted from AFGL, 1985.

at the rate of 6-8 kelvins (K) per km. The stratosphere is much colder (mid-latitude temperatures are typically 210-270K [-82—26°F]) and much lower in pressure (typically 0.25-0.0005 atmospheres) than the troposphere. The juncture of the troposphere and the stratosphere is called the tropopause, the upper limit of turbulent mixing for atmospheric gases. The tropopause is typically the coldest part of the lower atmosphere, and at midlatitudes, it is 10-14 km in altitude, as shown in Figure 1-1. The tropopause can be as low as 8 km at high latitudes, near the poles, and as high as 18 km in the tropics. The region above the tropopause is very stable and impedes the exchange of chemical species between the troposphere and stratosphere.

A much smaller portion of naturally occurring ozone is found in the troposphere. The data plotted in Figure 1-1 show that natural concentrations of tropospheric ozone are very small—usually a few tens of parts per billion (ppb) in mixing ratio[1] (molecules of O_3/molecules of air; 10 ppb = 2.5×10^{11} molecules/cm^3 at sea level and 298 K) compared with more than 10,000 ppb (10 parts per million (ppm)) typically found at peak stratospheric mixing ratios. However, the atmosphere thins out exponentially with altitude, which is why the peak in ozone mixing ratio occurs at a higher altitude than does its peak in partial pressure (concentration), as shown in Figure 1-1. Nevertheless, a significant amount of naturally occurring ozone, about 10-15% of the atmospheric total, is found in the troposphere (Chatfield and Harrison, 1977; Fishman et al., 1990). Ozone was first accurately measured in the lower atmosphere by Strutt (1918), who used its ability to absorb ultraviolet light to quantify ozone at ground level over long atmospheric paths.

For the discussions presented in this report it is important to remember that ozone is truly a trace atmospheric species; if the entire atmospheric ozone volume, as plotted in Figure 1-1, were collapsed to a pressure of one atmosphere, it would form a layer only ≈ 3 millimeters (mm) thick. Therefore, it is not surprising that naturally or anthropogenically induced changes in trace chemical emissions might cause changes in atmospheric chemistry which, in turn, could have a significant effect on atmospheric ozone concentrations.

Changes in Stratospheric Ozone

Over the past 2 decades, our improved ability to monitor atmospheric

---

[1]Mixing ratios are reported on a volume basis in this report. The units of the mixing ratio will not show the volume designation (ppb instead of ppbv). "Concentration" is used throughout the report as a synonym for "mixing ratio," although in strict usage, "mixing ratio" is the correct term.

ozone with ground-sited, aircraft-mounted, and satellite-borne instruments has led to several concerns about changes in the amounts of ozone found in the atmosphere. One highly publicized issue has been the catalytic destruction of significant portions of the stratospheric ozone layer, most dramatically seen each spring in the Antarctic ozone hole, which is the result of stratospheric halogen-induced photochemistry due largely to the decomposition of anthropogenic halocarbon compounds in the stratosphere (WMO, 1986, 1988, 1990). The current and projected loss of stratospheric ozone has stimulated a major, continuing global research program in stratospheric chemistry and an international treaty that restricts the release of chlorofluorocarbons into the atmosphere.

Changes in Tropospheric Ozone

This report is concerned with the problem of elevated tropospheric ozone concentrations, particularly in densely populated urban and suburban areas. Such elevated ozone causes damage to exposed people, plants, and animals. The scientific community has strong reason to believe that concentrations of tropospheric ozone generally are increasing over large regions of the country—extending over the northern midlatitudes. During the past decade, an increase of approximately 10% (1%/year) in ozone throughout the height of the troposphere has been demonstrated over Europe (WMO, 1986, 1990). If stratospheric ozone concentrations remained constant, the 10% increase in tropospheric ozone would increase the total column abundance of ozone by about 1%.

However, stratospheric ozone concentrations are declining, and increasing amounts of ultraviolet solar radiation are leaking through a thinning stratospheric ozone layer. An average decrease of 5% in the total column abundance of ozone at 50° N has been observed over the past decade (Stolarski et al., 1991). Thus, the additional tropospheric ozone is believed to have counteracted only a small fraction of the stratospheric loss, even if the trends observed over Europe are representative of the entire northern midlatitude region.

The build-up in tropospheric ozone has broad implications for atmospheric chemistry. Ozone and associated atmospheric oxidants play a significant role in controlling the chemical lifetimes and reaction products of many atmospheric species and also influence organic aerosol formation. In addition, tropospheric ozone is a greenhouse gas that traps radiation emitted by the earth and an increase in tropospheric ozone might contribute to a warming of the earth's surface.

Some of the evidence for increased baseline levels of tropospheric ozone comes from Europe, where during the late 1800s there was much interest in atmospheric ozone. Because ozone was known to be a disinfectant, it was believed to promote health (Warneck, 1988). Measurements of atmospheric ozone made at Montsouris, near Paris, from 1876 to 1910 have been reanalyzed by Volz and Kley (1988), who recalibrated the original measurement technique. Their analysis showed that surface ozone concentrations near Paris 100 years ago averaged about 10 ppb; current concentrations in the most unpolluted parts of Europe average between 20 and 45 ppb (Volz and Kley, 1988; Janach, 1989). An analysis of ozone measurements made in relatively remote European sites indicates a 1-2% annual increase in average concentrations over the past 30 years (Janach, 1989).

The presence of tropospheric ozone is generally attributed to a combination of its in situ photochemical production and destruction coupled with regular incursions of ozone-rich stratospheric air (Logan, 1985). Worldwide expansions in agriculture, transportation, and industry are producing a growing burden of waste gases, most particularly oxides of nitrogen (especially NO and NO_2, designated as NO_x) and volatile organic compounds (including hydrocarbon and oxyhydrocarbon compounds, designated as VOCs), which enter the atmosphere and exacerbate the photochemical production of ozone. Computer models have been used to extrapolate the response of tropospheric ozone production as a function of atmospheric concentrations of VOCs and NO_x, both backward and forward in time; these models estimate the low concentrations of tropospheric ozone of the past century and forecast increasing concentrations for the future, unless projected emissions of precursor trace gases are curbed (Hough and Derwent, 1990; Thompson et al., 1990).

The most critical aspect of the tropospheric ozone problem is its formation in and downwind of large urban areas, where, under certain meteorological conditions, emissions of NO_x and VOCs can result in ozone concentrations as high as 200-400 ppb. Such production of ozone and related oxidant species is called photochemical air pollution; it was first recognized in the Los Angeles basin in the 1940s, when vegetable crops began to show damage. Work in the 1950s by Haagen-Smit and co-workers established the photochemical nature of the agents that were causing plant damage. This work elucidated the key roles of NO_x and VOCs in ozone formation (Haagen-Smit et al., 1951, 1953; Haagen-Smit, 1952; Haagen-Smit and Fox, 1954, 1955, 1956). By 1961 the topic was well enough established to be presented comprehensively in a classic monograph by Leighton (1961).

UNDERSTANDING TROPOSPHERIC OZONE AND
PHOTOCHEMICAL AIR POLLUTION

By the 1960s, major anthropogenic sources of VOCs that react in the atmosphere had been identified: motor vehicle exhaust, emissions from the commercial and industrial use of solvents, and fugitive emissions from the chemical and petroleum industries. Most of the original research was directed toward clarifying the nature of the problem using the Los Angeles basin as the classic example. Motor vehicle exhausts and stationary combustion system exhausts were identified as important NO_x sources (OTA, 1989). Oxides of nitrogen are emitted into the atmosphere as nitric oxide (NO) and nitrogen dioxide (NO_2) and cycled within the atmosphere through nitrate radical (NO_3), organic nitrates, and dinitrogen pentoxide (N_2O_5), eventually forming nitric acid (HNO_3); the sum of these atmospheric oxides of nitrogen is often designated as NO_y. There is now a heightened appreciation of the importance of biogenic VOC emissions, whose reaction products react with NO_x from anthropogenic and natural sources for regional and urban ozone production (Trainer et al., 1987; Chameides et al., 1988; Cardelino and Chameides, 1990).

Research throughout the 1960s, 1970s, and 1980s focused on several critical aspects of photochemical air pollution (Finlayson-Pitts and Pitts, 1977, 1986; Seinfeld, 1986, 1989) including: a detailed elucidation of photochemical mechanisms and rates (Hough, 1988; Dodge, 1989); the development of monitoring networks and measurement of ozone in urban centers; the measurement of ozone in suburban and rural settings (after the findings in Glens Falls, New York, by Stasiuk and Coffey [1974]), where ozone was observed in a relatively isolated rural community at concentrations similar to those observed in New York City); transport experiments for distances beyond 100 km; the creation of photochemical and transport models to predict the evolution of air pollution episodes (Seinfeld, 1988); and the development of field measurement techniques for monitoring photochemical reactants, intermediates, and products to test results from models.

The atmospheric chemistry of tropospheric ozone formation is complex and is presented, in detail, in Chapter 5 of this report. Briefly, reactive VOCs, represented as RH, react with hydroxyl radicals (OH) to form organic radicals (R):

$$RH + OH \rightarrow R\cdot + H_2O \tag{1.1}$$

(Additional reactions of some RH species with ozone and the nitrate radical, NO_3, also could be significant.) Organic radicals combine with molecular oxygen to form peroxy radicals (RO_2), a process that usually requires an inert third body, M (e.g., N_2 or O_2):

$$R \cdot + O_2 \xrightarrow{M} RO_2 \cdot \qquad (1.2)$$

Peroxy radicals react with nitric oxide (NO) to form nitrogen dioxide (NO_2):

$$RO_2 \cdot + NO \rightarrow NO_2 + RO \cdot \qquad (1.3)$$

Nitrogen dioxide is photodissociated by solar radiation to release ground state oxygen atoms, $O(^3P)$, and reform nitric oxide:

$$NO_2 + h\nu \rightarrow NO + O(^3P) \qquad (1.4)$$

Energy from solar radiation is represented by $h\nu$, the product of Planck's constant, h, and the frequency, ν, of the electromagnetic wave of solar radiation. Finally, oxygen atoms combine with molecular oxygen, in the presence of a third body, to form ozone:

$$O(^3P) + O_2 + \xrightarrow{M} O_3 + M \qquad (1.5)$$

The process is a chain reaction: Ozone is photodissociated by near-ultraviolet solar radiation to form an excited oxygen atom, $O(^1D)$:

$$O_3 + h\nu \rightarrow O_2 + O(^1D) \qquad (1.6)$$

which, in turn, can react with water vapor (H_2O) to form two OH radicals:

$$O(^1D) + H_2O \rightarrow 2OH \qquad (1.7)$$

The resulting OH radicals drive the chain process. Furthermore, reactions initiated by the RO radicals formed in Reaction 1.3 can, in the presence of NO, lead to further production of OH. With enough VOCs and NO_x in the atmosphere, the chain reactions represented above can, in the presence of sunlight, lead to unhealthful concentrations of tropospheric ozone.

Because of the importance of precursors in ozone photochemistry, a comprehensive understanding of tropospheric ozone chemistry also requires knowledge of the atmospheric sources and sinks of VOCs and NO_x. The sources include VOC emissions from vegetation, industrial and commercial facilities, and motor vehicles and NO_x emissions from motor vehicles, power plants, industrial facilities, and to a lesser degree, biomass burning, soil, and lightning. Sinks include dry deposition of VOCs and NO_y to vegetation, land, and water surfaces and wet (cloud, fog, and rain droplet) scavenging of oxygenated VOCs and NO_y. Heterogeneous chemical transformations of precursors, which occur on aerosol particles and in clouds and fogs, are also important. The resulting chemical products can reenter the gas phase as cloud droplets evaporate, or can be deposited to the ground in precipitation. Figure 1-2 diagrams relevant photochemical and transport processes.

Understanding the ways in which ozone is formed, accumulates, and moves though space and time requires a conceptual framework. The basic processes of atmospheric chemistry that lead to ozone formation, the effects of local emissions of precursor compounds on the processes, and the transport of ozone and other species (precursors, radicals, and products) can be described using the hypothesis that there are different types of canonical regions where elevated concentrations of tropospheric ozone or its precursors can be found within the United States. Such canonical regions can be identified for a specific part of the country by determining the ambient concentrations of VOCs and NO_x and how their reactivity affects ozone formation. However, the extent of ozone reduction depends on the effectiveness of strategies to control the VOCs and NO_x both within a region and in regions which are upwind sources of ozone. These canonical regions are important to recognize and

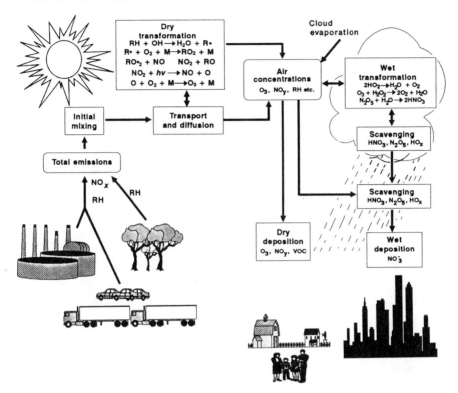

FIGURE 1-2   Photochemical air pollution, from emission to deposition.

evaluate because, as the relationships between source emissions and their atmospheric chemistry are better understood, certain consistencies in those relationships can provide opportunities for tailoring ozone precursor control strategies to specific cities or canonical regions.

Five types of canonical regions of ozone accumulation are relevant to the discussion in this report (see Figure 1-3). A brief discussion of each is presented below:

Central Urban Core. These areas generally contain major and minor point sources and area sources of ozone precursors. In many instances there are low VOC/NO$_x$ ratios that lead to depressed ozone concentrations within the central core. These regions often include the densely populated residential or commercial sectors of large metropolitan areas.

Urban Perimeter. The urban perimeter is marked by major transportation arteries that move people to and from the central urban core. There are

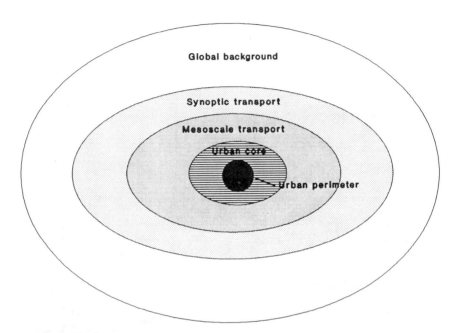

FIGURE 1-3   Conceptual canonical regions for evaluating tropospheric ozone formation and control.

lower population densities, but the people living in these areas rely heavily on motor vehicles for most personal and business transportation. Major point sources of precursors can be present. The perimeter would be considered the first region to receive ozone produced from precursor-laden air that is photolyzed as it moves away from the central core.

Mesoscale Transport Region. This is a much more diverse kind of area affected by ozone. It is defined by major geographical and topographical boundaries, specific precursor sources, and specific meteorological conditions. A mesoscale transport region could include a valley, a corridor of urban areas, an ocean shoreline, forests or large parks, agricultural land, and isolated major point sources, such as power stations and factories. The definition of boundaries is difficult, but these regions must be considered because the ozone present in a specific location can either influence or be affected by other canonical regions. Sustained patterns in the chemistry and meteorology of the region might affect the ozone concentration in such areas for several days.

Synoptic-Scale Transport Region. Processes associated with large-scale weather systems control the characteristics and intensity of ozone chemistry and transport in this kind of region. Such weather processes include frontal activity, subsidence, and the anticyclonic and cyclonic flows associated with high- and low-pressure systems, respectively. As discussed in Chapter 4, high-pressure systems allow ozone to accumulate, and storms associated with low-pressure systems can disperse or scavenge ozone within a radius of approximately 400 km. The occurrence and duration of such weather systems provide the basis for each photochemical smog episode.

Background. The tropospheric ozone concentrations in specific regions of the globe are affected by several processes, including methane and carbon monoxide oxidation, movement of ozone from the stratosphere, and other persistent natural activities that produce or scavenge ozone. The apparent rate of increase in continental background ozone concentrations of 1-2% each year will not help ozone reduction efforts in urban areas.

Each description of a canonical region includes the geographic and land-use characteristics that can contribute to severe episodes of ozone pollution. Because the sources of precursors are not uniform, each area must be examined to determine the most important contributors to high concentrations of ozone. The canonical regions framework does not provide a rigid format for examining tropospheric ozone, but it does provide necessary building blocks to elucidate both the situations described throughout this report and areas that could be affected by increased ozone concentrations in the future.

OZONE AND AIR-QUALITY REGULATIONS

The high concentrations of ozone, coupled with lesser, but still serious, concentrations of other smog products, including NO_2, nitrate radical (NO_3), N_2O_5, peroxyacetyl nitrate (PAN), and nitric acid (HNO_3), created during serious urban and regional photochemical pollution episodes, pose significant threats to human health. Accordingly, ozone has been identified in the Clean Air Act as a common and widespread air pollutant. In 1971, as required by the Clean Air Act, the U.S. Environmental Protection Agency (EPA) established primary (human health) and secondary (welfare) national ambient air-quality standards (NAAQS) for oxidants. Oxidants (e.g., hydrogen peroxide (H_2O_2)) are defined as those compounds giving a positive response using the iodide oxidation technique. For the primary standard the concentration of oxidants not to be exceeded was 80 ppb for any 1-hour period in a year. In 1978, the standard was changed to a standard for ozone—an indicator for all oxidants—and the concentration limit was raised to 120 ppb for any 1-hour period, which was not to be exceeded more than once each year. This relaxation of the primary standard temporarily reduced the number of regions that were out of compliance, without necessarily improving air quality. The secondary standard for welfare effects was also changed in 1978; it is identical to the primary standard.

An area is said to be in attainment of the ozone NAAQS if the expected number of days per year with a maximum 1-hour average concentration of ozone exceeding 120 ppb is less than or equal to one. The expected number of days per year with ozone concentrations above 120 ppb is calculated by averaging over 3 years of monitoring data. Thus, any area with 4 or more days having 1-hour ozone concentrations above 120 ppb during 1987 through 1989 would exceed the NAAQS in 1989, regardless of when those 4 days occurred during the 3-year period.

The Clean Air Science Advisory Committee (CASAC) of the EPA Science Advisory Board has recommended to the EPA administrator that the 1-hour averaging time for the ozone NAAQS be retained and that the maximum allowed concentration be set between 80 and 120 ppb. However, there is significant support in the scientific community and the CASAC for EPA to continue research on the criteria for two primary (health-based) standards: an acute-exposure standard and a chronic-exposure standard—both with averaging times of more than 1 hour. The Committee on Tropospheric Ozone Formation and Measurement has recognized the legal requirement to meet the current NAAQS, but its members also are aware that different standards could be adopted or at least considered that would achieve the same end, and that careful consideration is required of the types of precursor control strategies necessary to achieve different standards.

The Clean Air Act states that the responsibility for attaining the NAAQS is to be borne by a federal-state partnership. EPA develops the NAAQS, and the states formulate and enforce cleanup strategies to bring each nonattaining area into compliance. Some states, notably California, set stricter emission and air-quality standards for ozone than those mandated nationally.

NATIONAL TRENDS IN OZONE

Two indicators that current practices do not achieve their goals in controlling the production of ozone are the number of areas that exceed the NAAQS and the frequency and magnitude of high concentrations in each location. EPA has summarized the trends in second-highest daily maximum ozone concentrations from 1978 through 1989. (See Chapters 2 and 3 for a description of the methods used to determine and track compliance with the ozone NAAQS.) EPA (1991a) has reported a trend toward lower concentrations for the period from 1980 through 1989, with what have been described as anomalous years in 1983 and 1988. Because the trend analysis covers only a 10-year period, however, the increases in 1983 and 1988 cannot be considered true anomalies.

The number of areas not meeting the NAAQS from 1982 through 1989 is shown in Table 1-1. The table is based on an analysis using consistent area boundaries. In 1988, the number of areas not meeting the NAAQS jumped to 101 from 63 the year before. Photochemical pollution conditions in 1988 were extensive enough to cause violations in locations where the NAAQS had not been exceeded in the past. EPA reported that, in 1990, based on revised designations of area boundaries, 98 areas were not in attainment of the NAAQS (EPA/OAQPS, 1991).

The trends in various sections of the country have not necessarily been the same. For example, in Figure 1-4 it is apparent that the 26 sites in the Los Angeles basin have shown a consistent, although slight, downward trend. The two sites in Atlanta and 11 sites in the Washington, D.C., metropolitan area show an upward trend since 1985. For the metropolitan New York-New Jersey area, the number of ozone excursions above the NAAQS increased substantially in 1988 because of the summer heat wave. In contrast, in 1989 the number of days on which maximum 1-hour ozone concentrations were above 120 ppb fell to a low for the decade.

Although generally reliable data on urban ozone concentrations have been collected, corollary data on ozone precursors (VOCs and NO_x) are not generally available. Thus, we do not know whether the factors that suggest increased precursor emissions, such as higher population density, automobile use, and industrial activity, have been successfully offset by emission controls. We know when ozone concentrations go up or down, but not necessarily why.

DETRIMENTAL EFFECTS OF OZONE

Although the harm caused by tropospheric ozone to humans, animals, and plants is not a focus of this report, a brief summary of the current understanding of its effects on human health and plant viability is presented below to put the tropospheric ozone problem into perspective.

Health Effects of Atmospheric Ozone

Concentrations of ozone above and below the maximum allowed NAAQS concentration of 120 ppb have been shown to be associated with transient

TABLE 1-1   Number of Areas Not Meeting the Ozone NAAQS (1982-
1989)[a,b,c]

Year	Number of areas[c]
1982	96
1983	90
1984	84
1985	77
1986	64
1987	63
1988	101
1989	96

[a]The number of areas not meeting the NAAQS in a given year is determined by averaging over three years of monitoring data. For example, the
determination of the number of areas not meeting the NAAQS in 1989 is
based on monitoring data from 1987, 1988, and 1989.

[b]The monitoring network for ozone has expanded significantly during
the past 10 years. According to EPA, the estimates of the number of areas
not meeting the ozone NAAQS for 1985 and earlier have not been subject
to the same level of quality assurance and monitoring network review as
the estimates for more recent years.

[c]Areas, whose boundaries are designated by EPA, have consisted of
Metropolitan Statistical Areas (MSAs), Consolidated Metropolitan Statistical Areas (CMSAs), and counties. The table is based on an analysis using
consistent area boundaries (W. Freas, pers. comm., EPA, November 1991).
Boundary designations were changed in 1991 (56 Federal Register 56694).
Based on these new designations, 98 areas did not meet the NAAQS in
1990 (EPA/OAQPS, 1991).

Source: EPA, Aerometric Information Retrieval System, 1987, 1991.

effects on the human respiratory system. Of those documented, the most
significant are the dose-response relationships established for decrements in
pulmonary function of individuals while they participate in light to heavy
exercise. The more important findings from human studies were summarized

in a review by Lippmann (1989). The review discussed field health studies that showed equivalent or greater decrements in pulmonary function per exposure concentration of ozone than were observed in the control population. Possible reasons are the longer durations of exposure and the increased effect of ozone due to the presence of other pollutants in smog-laden ambient air. Each reason should be considered in examining current modeling results and measurement data when defining the transport and transformation processes that can place populations at risk to health damage induced by high concentrations of ozone.

For the entire suite of human experiments, acute effects have been observed after exposure that lasted from less than 1 hour up to several days and at concentrations above and below 120 ppb. Changes have occurred in functional lung capacity, lung flow rate, epithelial permeability, and reactivity to bronchial challenges. In some cases, the effects persisted for many hours or days after exposure during exercise stopped.

Chronic effects that have resulted from recurrent seasonal exposure to ozone have been studied only to a limited degree. Most of the current evidence is derived from animal responses to chronic ozone exposures. The most revealing evidence to date is from chronic toxicologic studies in rats and monkeys at exposures of ~1,000 ppb that showed persistent functional and morphologic changes in the gas exchange region (terminal bronchioles and alveoli) of the lung. These changes must be studied further, but they suggest that the scientific research community must formulate measurement and modeling approaches that can be used in conjunction with health indices to detect or predict annual increments of improvement or degradation of human health.

Human Exposure Issues

The occurrence of ozone concentrations that exceed 120 ppb in various regions of the United States indicates that many people could be exposed to potentially harmful concentrations of ozone. Although the daily maximum concentration usually occurs between 12 noon and 5 p.m. in most central or downtown urban settings, areas downwind of those settings have experienced occasional excursions above 120 ppb, and these can occur well into the evening. This is illustrated for a multiday episode of high concentrations of ozone in an isolated small city, Montague, Massachusetts, that was affected by transported ozone (Figure 1-5).

EPA estimated that about 67 million people lived in areas with second-highest daily maximum 1-hour ozone concentrations above 120 ppb in 1989

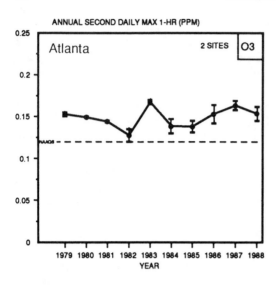

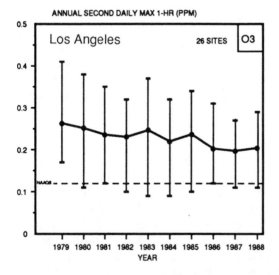

FIGURE 1-4   Trends in the annual second-highest daily maximum 1-hour concentration of ozone in Atlanta, Los Angeles, and Washington, D.C., metropolitan areas (Source: EPA, 1990a).

(EPA, 1991a). Using the National Exposure Model (NEM), EPA has estimated that 13 million moderately exercising adults are at risk from exposure to ozone in excess of 120 ppb for at least 1 hour per week during the summer (Paul et al., 1987).   Because ozone gradually builds and decreases over the course of the day in urban and nonurban areas, varying degrees of population

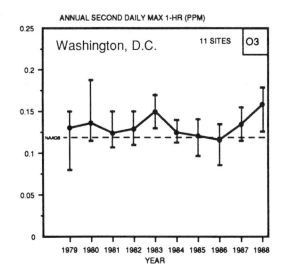

FIGURE 1-4 (continued)

exposure are associated with the 1-hour maximum. In addition, there are multiple opportunities for individuals to be exposed to a 1-hour value above 120 ppb (Figure 1-6). This fact alone is enough to warrant concern, but as shown in the brief summary on the potential human health effects of ozone, concerns have been raised about public health risks due to exposures to ozone concentrations that are in compliance with the NAAQS.

Lioy and Dyba (1989) and Berglund et al. (1988) have demonstrated that there are places in the United States, especially in the Northeast, where at times the NAAQS of 120 ppb for 1 hour is not violated, but the workplace

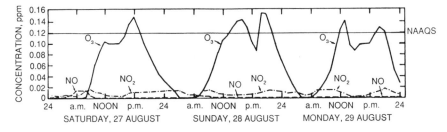

FIGURE 1-5   Three-day sequence of hourly ozone concentration at Montague, Massachusetts. Sulfate Regional Experiment (SURE) station showing locally generated midday peaks and transported late peaks. Source: Martinez and Singh, 1979.

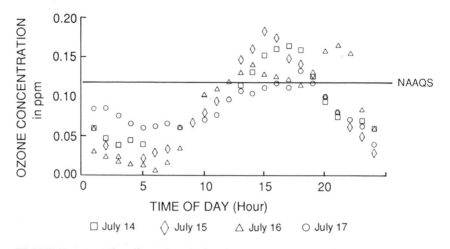

FIGURE 1-6   The diurnal variation in ozone concentration during the
summer 1982 ozone episode at Mendham, New Jersey, associated with
the health effects study conducted by Lioy et al., 1985. (Source: Lioy
and Dyba, 1989).

permissible exposure limit (PEL) of 100 ppb for 8 hours is exceeded in outdoor air. Their analyses indicated that in 25% of the cases where the PEL
was violated, the NAAQS was not violated. Controlled human studies and
field health studies have indicated that the potential for producing transitory
pulmonary effects is greater than originally projected and have indicated the
need to consider a demonstration period longer than 1 hour for exposure to
ozone.

The discussion of ozone's effects must also consider the relative importance
of indoor ozone concentrations and the potential duration of exposure. Total
exposure to ozone is made up of contact in the indoor and outdoor environments. It must be remembered, however, that a single hour or several hours
of exposure to ozone in the outdoor environment is sufficient to cause respiratory effects. Indoor exposures are important for many of the other five criteria pollutants listed in the Clean Air Act and for toxic air pollutants, but
generally not for ozone. Therefore concern for possible effects will remain
associated with outdoor air. The areas of greatest concern will be open spac-

es within the urban, suburban, and rural regions in and around the major metropolitan areas of the United States.

Effects on Vegetation

Elevated ozone exposures affect agricultural crops (Heck et al., 1982; EPA, 1986a) and trees (EPA, 1986a). Short-term, high-concentration exposures are identified by many researchers as being more important than long-term, low-concentration exposures (Heck et al., 1966; Heck and Tingey, 1971; Bicak, 1978; Henderson and Reinert, 1979; Nouchi and Aoki, 1979; Reinert and Nelson, 1979; Bennett, 1979; Stan et al., 1981; Musselman et al., 1983, 1986; Ashmore, 1984; Amiro et al., 1984; Tonneijck, 1984; Hogsett et al., 1985a). Although little is known about the ozone distribution patterns that affect trees, support for the hypothesis that peak concentrations are an important factor in determining the effects of ozone on trees comes from the work of   Hayes and Skelly (1977), who reported injury to white pine in rural Virginia, and from Mann et al. (1980), who described oxidant injury in the Cumberland Plateau. Work by Hogsett et al. (1985b) with two varieties of slash pine seedlings suggested that exposures to peak ozone concentrations elicit a greater response than do exposures to mostly lower concentrations over similar time periods. Miller et al. (1989) showed that better air quality in the central valley of California has led to improved viability in stands of ponderosa and Jeffrey pine.

The search for a vegetative exposure index for plant response has been the subject of intensive discussion in the research community (EPA, 1986a; Lefohn and Runeckles, 1987; Hogsett et al., 1988; Tingey et al., 1989). Both the magnitude of a pollutant concentration and the length of exposure are important; however, there is evidence that the magnitude of vegetation responses to air pollution depends more on the magnitude of the concentration than on the length of the exposure (EPA, 1986a). Several exposure indexes have been proposed. The 7-hour (0900 and 1559 hours) mean, calculated over an experimental period, was adopted as the statistic of choice by EPA's National Crop Loss Assessment Network (NCLAN) program (Heck et al., 1982). Toward the end of the program, NCLAN redesigned its experimental protocol and applied proportional additions of ozone to its crops for 12-hour periods. The results of the NCLAN experiments have been used to estimate agricultural crop losses (Adams et al., 1985, 1989).

Recently, attention has turned from long-term seasonal means to cumulative indexes (exposure measurements that sum the products of concentrations multiplied by time over an exposure period). Oshima (1975) and Lefohn and

Benedict (1982) proposed similar cumulative indexes. NCLAN data have been used to test the usefulness of cumulative indexes to describe ozone exposure (Lee et al., 1988; Lefohn et al., 1988; Lee et al., 1989).

PURPOSE OF THIS REPORT

In this report the committee systematically examines trends in tropospheric ozone measurements within the United States; reviews current approaches to control ozone by regulating ozone precursors; and assesses the present understanding of the chemical, physical, and meteorological influences on tropospheric ozone. Based on this current understanding, the committee provides a critique of the scientific basis of current regulatory strategies and lists recommendations for improving the scientific basis of future regulatory strategies. The committee determined that a coherent, integrated research program is needed to clarify further the factors that control tropospheric ozone formation within the context of changing regional and global environmental conditions. The committee also recommends increased air-quality measurement and modeling to systematically devise and check future regulatory strategies.

In this report, the committee does not critique the current NAAQS for ozone or other criteria pollutants, but is sensitive to issues associated with the magnitude and form of current and anticipated standards. It also does not address the technologic, economic, or sociologic implications of current or potential ozone precursor control strategies.

Subsequent chapters in this report address the following questions:

Chapter

2   What are the trends in tropospheric concentrations of ozone in the United States?

3   What criteria should be used to design and evaluate ozone reduction strategies?

4   What are the effects of meteorology on tropospheric ozone?

5   What is the atmospheric chemistry of ozone and its precursors?

6   What is the maximum amount of ozone that can form from a given initial mixture of VOCs and NO_x?

7   How well can we measure tropospheric ozone and its chemical precursors?

8   What VOC/NO_x ratios are found in the atmosphere?

9   What are the emissions that result in ambient concentrations of ozone?

10 What is the role of air-quality models in determining ozone reduction strategies?

11 What is the trade-off between control of VOCs and of NO_x?

12 Can alternative fuels for transportation improve air quality?

13 What is the interaction between tropospheric ozone concentrations and global change?

14 Is a comprehensive, long-term research program on tropospheric ozone formation and measurement necessary?

The committee presents data and arguments to address these questions and help focus the continuing national debate over devising an effective program to provide healthful air over the United States.

2

Trends in Tropospheric Concentrations of Ozone

INTRODUCTION

Various levels of government monitor the concentration of ozone in the ambient air (the troposphere) to document the severity of the ozone problem and to measure progress in reducing ozone concentrations. The U.S. Environmental Protection Agency (EPA) has developed a National Ambient Air Quality Standard (NAAQS) for ozone that is intended to provide a margin of safety for the protection of public health and welfare, and its attainment has been pursued in many areas of the country.

Despite comprehensive local, state, and national regulatory initiatives over the past 20 years, ambient ozone concentrations in urban, suburban, and rural areas of the United States continue to be a major environmental and health concern. Sixty-three areas did not meet the ozone NAAQS in 1987; 101 areas were out of compliance in 1988, 96 areas in 1989, and 98 areas in 1990 (see Table 1-1 and Figure 2-1) (EPA, 1990a; EPA, 1991a; EPA/OAQPS, 1991). The nationwide extent of the problem, coupled with the increased public attention resulting from the high concentrations of ozone over the eastern United States during the summer of 1988, add to the urgency for developing effective approaches to reduce the risk to public health and welfare (as measured by the effects of pollution on vegetation, materials, and visibility). The 1990 amendments to the Clean Air Act present a broad ozone attainment strategy that will require areas violating the ozone NAAQS to demonstrate reasonable further progress in reducing precursor emissions (see Chapter 3). The success of these efforts to control the precursors to ozone will be of vital

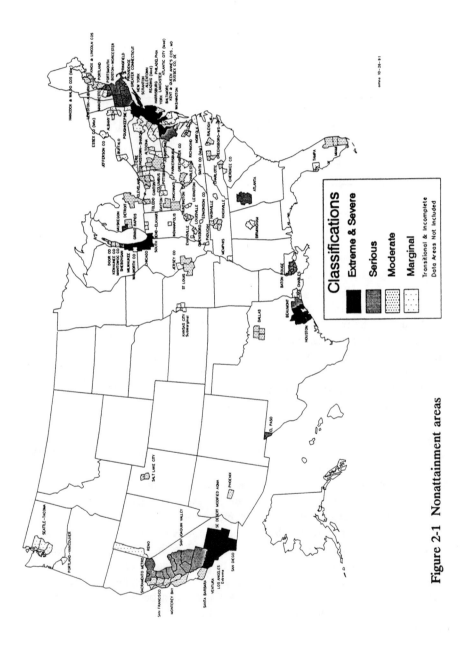

Figure 2-1 Nonattainment areas

concern to Congress; to federal, state, and local regulatory agencies; to industry; and to the public.

It is important that areas not in attainment of the NAAQS be able to track their progress in reducing ozone. A downward trend in ozone concentrations over several years might suggest that emission controls are having the desired effect, whereas the absence of such a trend might suggest the need for additional controls. However, analyses of ambient-air-monitoring data for ozone in major U.S. cities indicate that the number of days on which ozone concentrations exceed the NAAQS 1-hour average concentration of 120 parts per billion (ppb) varies widely from one year to the next (Stoeckenius, 1990). EPA's principal statistical measure of ozone trends, the composite average of second-highest daily maximum 1-hour concentrations in a given year, also varies considerably from year to year. Furthermore, there is no discernible downward trend (EPA, 1991a). A substantial portion of this variability is attributable not to year-to-year changes in precursor emissions, but to natural fluctuations in the weather. Meteorologically induced variability makes it difficult to identify underlying trends in ozone concentrations that could result from changes in the amount, type, and geographical distribution of precursor emissions. This chapter outlines the existing form of the ozone NAAQS, summarizes studies on ozone trends, evaluates recent efforts to screen meteorological influences on ozone trends, and highlights research needs for more robust indicators of progress in reducing ozone concentrations.

NATIONAL AMBIENT AIR QUALITY STANDARD FOR OZONE

To guide in monitoring and controlling ambient concentrations of ozone, EPA has developed ambient air quality standards for ozone. The national form of the ozone NAAQS is explicitly defined in the U.S. Code of Federal Regulations (40 CFR 50.9(a)), in part:

The [ozone] standard is attained when the expected number of days per calendar year with a maximum hourly average concentration above 0.12 part[s] per million (235 $\mu g/m^3$) is equal to or less than 1, as determined by Appendix H [of the regulations].

Appendix H provides a method to account for incomplete monitoring data and allows the number of days exceeding the standard each year to be averaged over the past 3 calendar years. An area is in compliance as long as this average remains less than or equal to one (McCurdy and Atherton, 1990).

A complete standard requires at least eight attributes, as shown in Table 2-1 for ozone (McCurdy, 1990). Although not all of them are formally specified for many NAAQS or for other air pollutant standards, each can be important in developing a standard to protect the public health, in assessing the attainment status of a region, or in determining progress in reducing exposure concentrations.

The first, second, and fifth attributes shown in Table 2-1 are those most commonly considered in relation to the ozone NAAQS. For the existing ozone NAAQS, the averaging time is 1 hour, but ozone trends have been examined using averaging times of 7, 12, and 24 hours for a given day, and the Occupational Safety and Health Administration exposure standard for ozone is based on an 8-hour averaging period. The second attribute does not have a good name, but was described by McCurdy (1990) as the period over which short-term monitoring data are aggregated for purposes of comparison with the standard. Only the highest hourly value in a day is compared with the NAAQS concentration evaluation level (CEL), the fifth attribute, to determine whether the CEL of 120 ppb is exceeded. In some cases, monthly or quarterly averages of the daily maximum value have been used as a basis for determining a trend.

The third attribute of a NAAQS, the frequency of repeated peaks allowed before a violation is declared, has not been applied to the ozone NAAQS. The fourth attribute, the epoch, is the period over which ozone exceedances can occur and in which ozone data can be collected. This attribute is important when percentile rank statistics are used to evaluate trends. (A percentile rank is the percent of measurements out of the total that fall at or below that rank.) For example, Korsog and Wolff (1991) have reported that the 95th percentile statistic used in California (yearly monitoring), would correspond to the 80th percentile in areas where only 3 months' data were collected, because the top five percentile points for yearly monitoring correspond to the top twenty percentile points for 3 months' data.

The sixth and seventh attributes shown in Table 2-1 are the standard statistics and the allowed violation rate. For the ozone NAAQS, these correspond to a maximum of 1 expected day per year exceeding the daily CEL, averaged over a 3-year period. Thus, 4 days exceeding the CEL over 3 years constitute a violation of the NAAQS. The last attribute is a combination of data handling and analyses methods that are used to determine NAAQS violations (and design values as discussed in Chapter 3). Decisions needed for this determination might be whether to use the highest monitor reading in a region or an average of all readings; what analytical method to use to monitor for ozone; and how to record, average, or round off the data.

As discussed in Chapter 1, alternatives to the current 1-hour averaging time

Table 2-1   Attributes of an Ozone NAAQS

Attribute	Current NAAQS	Alternative NAAQS formulations		
		SUMO6	8-hour mean	
Sampling averaging time	Clock hour	Clock hour	Clock hour	
Temporal aggregation period	Maximum hour in each day (1-hour daily maximum)	None	Maximum 8-hour average in each day (8-hour daily maximum)	
Frequency of repeated peaks	a	a	a	
Epoch	Ozone season[b]	Three consecutive months	April-October	
Concentration evaluation level (CEL)	120 ppb	60 ppb	a	
NAAQS standard statistic	Expected exceedances of CEL	Number of hours ≥ CEL	Mean of all 8-hour daily maximums	
Allowed violation rate	Three-year mean of annual expected exceedances ≤1.0	a	a	
Data-handling and analysis conventions	Defined by Appendix H of 40 CFR 50 and EPA guide-line documents[c]	a	a	

aThe attribute is not used or is not yet defined; bOzone season is based on calendar months and varies by state; season length varies from 5 to 12 months; cIncludes such items as what constitutes a valid day of data, how missing values are handled, and what rounding conventions are to be used.
Source: McCurdy, 1990

Table 2-1   Attributes of an Ozone NAAQS

45

for the ozone NAAQS are being assessed as new health and ecological effects studies are completed and as new standards are considered. McCurdy (1990) reported that when all the attributes of a NAAQS are considered over a 3-year period, most short-term standards (those that use averaging times of less than 1 day) show a good correlation to the existing 1-hour standard. However, the 1-hour maximum is not as well correlated to longer averaging times (monthly to seasonal).

NATIONAL TRENDS IN
TROPOSPHERIC CONCENTRATIONS OF OZONE

As discussed above, an area is out of attainment with the ozone NAAQS when the expected number of days per year exceeding the allowed ozone concentration, averaged over the past 3 years, is greater than one anywhere in that area. Areas are defined by EPA and may consist of counties, Metropolitan Statistical Areas (MSAs), or Consolidated Metropolitan Statistical Areas (CMSAs). According to EPA, the number of areas in violation of the current NAAQS showed a general downward trend from 1984 until 1987, then increased substantially in 1988. Table 1-1 in Chapter 1 lists the number of areas not meeting the NAAQS from 1982 through 1989.

Little national progress is apparent in the number of areas reaching attainment between 1982 and 1989. Because attainment is based on an average of 3 years of data, high ozone concentrations in 1 year, such as 1988, will increase the number of areas exceeding the NAAQS for the following 2 years (1989 and 1990). Although the definition of areas was modified in 1987, the number and definitions of base areas remained fairly constant from 1987 to 1990.

Two kinds of measures are commonly used to monitor year-to-year trends: concentration indicators (such as the second highest 1-hour maximum ozone concentration) and threshold indicators (such as the number of days on which the 1-hour daily maximum of 120 ppb is exceeded).

To complement its annual report on the number of areas out of attainment, EPA issues an annual analysis of trends for ozone and other criteria pollutants over the preceding decade. In 1991, the trends report included the 1989 data and a partial analysis of the 1990 data (EPA, 1991a). EPA's principal statistical measure of ozone concentrations is the composite nationwide average of second highest 1-hour daily maximum concentrations in a given year. EPA reported that this composite average decreased by 14% between 1980 and 1989 (Figure 2-2). However, according to the previous year's trends report (EPA, 1990a), ozone concentrations increased by 1% between 1979 and 1988. EPA also reported a decrease during the 1980s in the composite average of

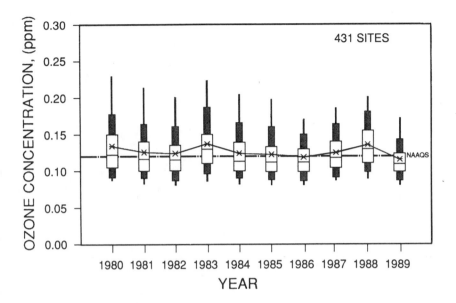

FIGURE 2-2.   Boxplot comparisons of trends in annual second highest
daily maximum 1-hour ozone concentration at 431 monitoring sites,
1980-1989. The trends in ambient air quality, presented as boxplots,
display the 5th, 10th, 25th, 50th (median), 75th, 90th, and 95th percentiles of the data, as well as the composite average. The 5th, 10th, and
25th percentiles depict the "cleaner" sites; the 75th, 90th, and 95th depict
the "higher" sites; and the median and average describe the "typical"
sites. For example, 90% of the sites would have concentrations equal to
or lower than the 90th percentile. Source: EPA, 1991a.

the expected number of days exceeding the maximum allowed 1-hour concentration of 120 ppb. From 1980 to 1989, the expected number of days
exceeding the standard for the 431 ozone monitoring sites in the analysis fell
by 53% (Figure 2-3). However, the number of days exceeding the standard
increased by 38% between 1987 and 1988.

EPA has acknowledged that the trend has never been smooth. There are
year-to-year fluctuations, and in some years (1983 and 1988, for example),
ozone concentrations are much higher than average. Ozone concentrations
in 1989 were much lower than in 1988 and were possibly the lowest of the
decade. It has been suggested that meteorological variations can explain the
high in 1988 and the low in 1989 (EPA, 1991a).

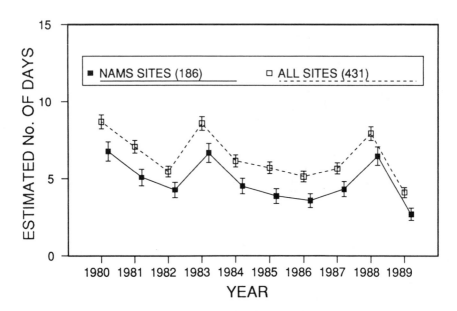

FIGURE 2-3.   National trend in the composite average of the estimated number of days exceeding the ozone NAAQS concentration during the ozone season at monitoring sites, with 95% confidence intervals, 1980-1989. The National Air Monitoring Stations (NAMS) sites were established through monitoring regulations promulgated by EPA (44 Federal Register 27558, May 10, 1979). The NAMS are located in areas with high pollutant concentrations and high population exposures. These stations must meet EPA's criteria for siting, quality assurance, equivalent analytical methodology, sampling intervals, and instrument selection. "All sites" includes sites having complete data for at least 8 of the 10 years under consideration. Source: EPA, 1991a.

TRENDS IN PRECURSOR EMISSIONS

EPA has annually presented a trends analysis of emissions of oxides of nitrogen (NO_x) and of VOCs (EPA, 1991a) to accompany its analysis of ozone. Between 1980 and 1989, annual VOC emissions were reported to have decreased by 19% (Figure 2-4), while annual NO_x emissions were reported to have decreased by 5% (Figure 2-5). As discussed later in the report, however, the methods EPA uses to estimate precursor emissions are flawed (see Chapter 9). In particular, it is believed that emissions of anthropogenic and biogenic VOCs have been significantly underestimated.

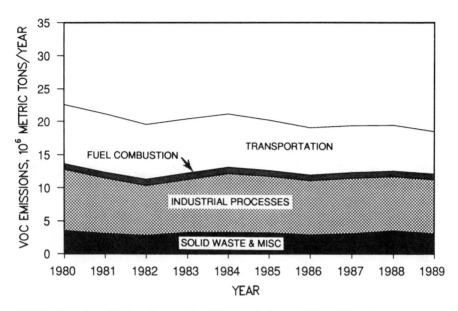

FIGURE 2-4.   National trend in VOC emissions, 1980-1989. Source:
EPA, 1991a.

Although nationwide reductions in precursor emissions and ozone concentrations between 1980 and 1989 have been reported, little relationship has
been demonstrated between year-to-year variations in ozone concentrations
and changes in estimated precursor emissions. For example, there was no
increase in precursor emissions to accompany the large increase in ozone
concentrations observed between 1987 and 1988. This increase, as discussed
above, is believed to have been primarily the result of unusual weather in
1988.

There also has been a lack of correlation between changes in ambient
concentrations of precursors and in emissions inventories. For example, in the
California South Coast air basin, Kuntasal and Chang (1987) estimated that
NO_x emissions decreased by only 1% between 1970 and 1985, whereas ambient concentrations of NO_x decreased by 30%. For VOCs, measured as nonmethane hydrocarbons (NMHCs), ambient concentrations decreased by 33%
and estimated NMHC emissions decreased by 40% during this period. The
ambient concentrations of NMHCs decreased continuously, whereas estimated
emissions of NMHCs have decreased much faster since the late 1970s. The
discrepancy between measured precursor concentrations and emissions inventories is discussed further in Chapter 9.

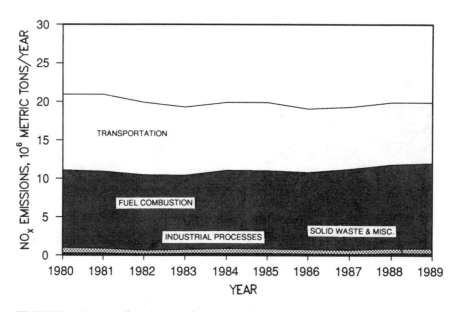

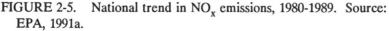

FIGURE 2-5.   National trend in NO_x emissions, 1980-1989. Source:
EPA, 1991a.

OZONE TRENDS NORMALIZED
FOR METEOROLOGICAL VARIATION

The formation of ozone is heavily dependent on meteorological conditions
(see Chapter 4). In fact, in most areas of the country, monitoring for ozone
is not required during the 6 to 9 months when atmospheric conditions are not
conducive to the formation of ozone.

The effect of weather on ozone formation has been studied a good deal
over the past 20 years (e.g., Ludwig et al., 1977; Zeldin and Meisel, 1978).
The most frequently examined relationship has been that between ozone
formation and atmospheric temperature. For example, this relationship has
been studied for the South Coast Air basin (Kuntasal and Chang, 1987), New
England (Wolff and Lioy, 1978; Atwater, 1984; Wackter and Bayly, 1988), and
Philadelphia (Pollack, 1986). Generally, the relationship of daily ozone concentrations to temperature is nonlinear. Ozone concentrations appear to show
no dependence on temperature below 70-80°F, but they become strongly
dependent on temperature above 90°F (Figure 2-6).

As discussed above, the current principal statistical measure of ozone concentrations, the composite average of second highest daily maximum 1-hour

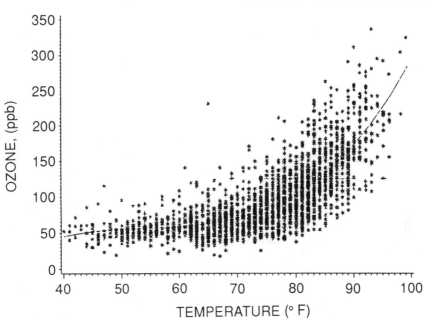

FIGURE 2-6   Connecticut daily maximum ozone vs. daily maximum temperature, 1976-1986. Temperatures represent the average daily maxima
at three inland sites. Source: Wackter and Bayly, 1988.

concentrations, is highly sensitive to meteorological fluctuations. This measure
masks underlying trends and is therefore not a reliable indicator of an area's
progress in reducing ozone over several years. The following sections discuss
various approaches to obtaining a measure of ozone concentrations that accounts for variations in meteorology. Three general approaches are discussed:

   • *Measurement*—selecting an ozone indicator that is less sensitive to variations in meteorology
   • *Classification*—classifying ozone measurements by meteorological conditions
   • *Regression*—correlating ozone measurements to meteorological conditions.

Measurement Approaches:
A More Robust Ozone Indicator

One way to account for weather is to use a more robust indicator of ozone trends, defined as an indicator that is not as strongly dominated by a single unusual meteorological event (Chock, 1988; see also Walker, 1985). Chock (1991) has emphasized that the meteorological fluctuations that result in large year-to-year variations in ozone concentrations are beyond human control. He has proposed that the number-of-exceedances statistic be replaced by a percentile rank order statistic (such as the 95th percentile) to determine the attainment status of a region, essentially replacing the more variable extreme-value statistic by a less variable, less extreme statistic. Reducing the influence of random fluctuations in the weather could assist in the establishment of consistent emission reduction strategies.

Several other potentially robust indicators of ozone trends have been proposed (Figure 2-7). These include the 95th percentile of the daily maxima (Chock, 1988) if data for a full year are collected or the 80th percentile (Korsog and Wolff, 1991) if data for only 3 months are collected. In companion studies to the EPA trends report, Niemann (1988) looked at 9-year trends (1979 to 1987) for selected urban and rural regions of the United States using monthly averages of a 1-hour maximum and 7-, 12-, and 24-hour averaging periods. McCurdy (1990) reported that nationwide, on the basis of the number of areas meeting the NAAQS, an 8-hour maximum concentration of 95 ppb would correspond to a 1-hour maximum of 120 ppb. Curran and Frank (1990) examined trends from 1979 to 1988 using a different set of alternative indicators, including the 99th, 95th, and 90th percentiles as well as the seasonal mean and the median. The trends for the seasonal mean were flat, but the trends for the three percentile indicators were similar to those observed for the second highest maximum ozone concentration (Figure 2-7). Little reduction in the overall year-to-year variation would be observed by using one of these percentile indicators.

One comprehensive study (Larsen et al., 1990), chartered under the California Clean Air Act of 1988, characterized the intrinsic uncertainty or "native variability" of 27 air-quality-related indicators in 43 areas within the state and determined the uncertainty intrinsic to each indicator in the absence of reduction of ozone concentrations. From this characterization, a list of approved indicators could be used to estimate progress in the attainment of air quality standards. About two-thirds of the measures examined were "concentration indicators" (95th-percentile concentration, annual average, mean daily maximum 1-hour concentration, etc.). The remainder were "threshold indicators" (number of days or hours exceeding a concentration threshold). As expected,

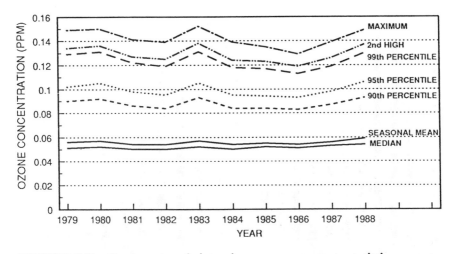

FIGURE 2-7   Ten-year trends in various ozone summary statistics.
Ozone concentrations represent composite nationwide averages for each
summary statistic. Source: Chock, 1988.

the most robust indicators were those that depend less on extreme values.
The threshold indicators (with a native variability of 7-150% of the average
number of exceedances) were much more variable than were the concentration indicators (with a variability of 1-20% of the average concentration). The
indicator judged to be most promising was based on percentile-centered
means (an average of 30 points centered around a percentile value of the
distribution of ambient ozone concentrations).

In the South Coast air basin of California, progress has been evaluated by
the South Coast Air Quality Management District (SCAQMD, 1989) using
exposure measurements such as hours at or above 130 ppb (and above 200 or
350 ppb) for the average resident (Figure 2-8), and per capita ozone exposure
in the basin (Figure 2-9). Exposure was presented in units of concentration
multiplied by time (ppm-hours). In the former analysis (Figure 2-8), significant reductions in the number of hours above all concentrations were reported, with the number above 350 ppb reaching zero during 1986 and 1987 (from
a high of 1.9 for 1978 and 1979). In the latter analysis (Figure 2-9), for the
period from 1975 to 1987, it was reported that the basin's per capita exposure
of ozone concentrations exceeding 0.12 ppm (120 ppb) showed an overall
decrease of 38% (from 42.1 ppm-hours to 26.1 ppm-hours, respectively).

The robustness of an ozone trends indicator depends on its separation from
the extreme values associated with the NAAQS. However, extreme ozone

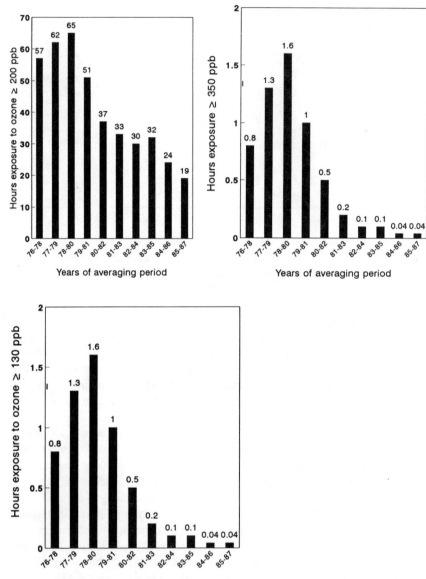

FIGURE 2-8   Three-year running mean of South Coast basin population-weighted ozone exposure hours for the average resident.   Source: SCAQMD, 1989.

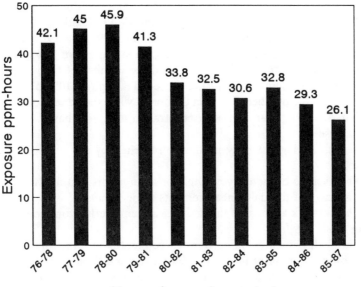

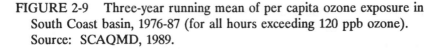

Years of averaging period

FIGURE 2-9   Three-year running mean of per capita ozone exposure in South Coast basin, 1976-87 (for all hours exceeding 120 ppb ozone). Source: SCAQMD, 1989.

concentrations threaten human health and welfare. Hence, it is a challenge to find a robust indicator that is not a mere statistical entity. To be a useful measure of an area's progress in reducing harmful concentrations of ozone, an indicator must bear some relation to those harmful concentrations.

Classification Techniques

Classification techniques move a step beyond the measurement approaches described above in that they attempt to account more explicitly for meteorological variability.

The first step necessary to account for meteorological effects is to determine what ozone data should be used in an analysis. Data often are selected to reduce the dependence of the analysis on the extreme values generally associated with the existing ozone NAAQS. Thus, trends are developed from violations of standards based on lower concentration cutoffs (105 ppb or 80 ppb) or using percentile distributions. Most of these measures have some association with the existing ozone NAAQS, in the form of either threshold violations or ozone concentrations. For threshold violations, Jones et al. (1989), Kolaz and Swinford (1990), and Wakim (1990) used the number of days on which the maximum ozone concentration was above 120 ppb (on any monitor in a city); Stoeckenius (1990) used the number of times during the year that the daily summary statistics exceeded 80 ppb or 105 ppb; Zeldin et al. (1990) used the number of days in California when the ozone concentration exceeded 200 ppb. Many ozone concentration measures have been used in trend analyses:

• The single highest daily maximum recorded by any monitor in a city (Wakim, 1989)

• The average of the daily maxima for all monitors in a network, and the annual average of the daily measurements (Stoeckenius, 1990)

• The network average of the daily maximum hourly ozone concentration (Zeldin et al., 1990)

• The 75th percentile value from three months of daily 1-hour maxima from the highest monitor in a network (Korsog and Wolff, 1991)

• Three-month averages of the 1-hour daily maxima for 6, 9, and 13 stations in the South Coast Air Quality Management District (Kuntasal and Chang, 1987)

• The basinwide daily maximum oxidant concentrations (Chock et al., 1982; Kumar and Chock, 1984)

The simplest way to account for meteorological effects would be to determine the meteorological conditions associated with those days when there are high concentrations of ozone, then predict high- or nonhigh-ozone days, based on observed meteorological conditions. An ozone trend would be developed by normalizing the actual ozone data relative to the weather conditions that are conducive to high concentrations of ozone. The goal of this approach would be to determine the "ozone-conducive days" (Kolaz and Swinford, 1990) or high-ozone days (Chock et al. 1982; Pollack 1986). Kolaz and Swinford (1990) have classified days as ozone-conducive or not based on a meteorological index that requires several cutoffs to be exceeded. Chock et al. (1982) have calculated an ozone formation potential for the South Coast air basin that includes high, moderate, and low concentrations, then related this potential to five meteorological parameters. Korsog and Wolff (1991) have examined the significance of eight meteorological parameters for eight cities on the East Coast to select the high-ozone days for use in the trend analysis. Table 2-2 summarizes the weather conditions found to be important in classifying high-ozone days.

Where meteorological information is used to normalize ozone data collected on conducive days, the variability in the ozone trend can be reduced, but

Table 2-2   Parameters Affecting "High Ozone Days"

Meteorological Parameter	Chock et al. (1982)	Pollack (1986)	Korsog and Wolff (1991)	Kolaz and Swinford (1990)
Temperature	X	X	X	X
Upper air temperature			X	
Dew point temperature			X	
Wind speed	X	X	X	X
Solar radiation or cloud cover		X		X
Relative humidity or precipitation				X
Wind direction			X	

in some cases, significant variation remains (Pollack, 1986). In none of the cases were the high-ozone years (1983, 1987, and 1988) fully accounted for by normalizing for ozone-conducive days alone. For example, Kolaz and Swinford (1990) reported that in Chicago, June 1984 and June 1987 had about the same number of ozone-conducive days (10 and 12, respectively). In June 1984, however, there were no days with 1-hour concentrations above 120 ppb, and in June 1987, there were eight. Apparently, the ozone-conducive days of June 1987 were more prone to ozone formation than were the ozone-conducive days of June 1984. An important problem that impedes the routine prediction of ozone concentrations is that the ozone associated with transport cannot always be related to weather. The amounts and types of ozone precursors transported into an impacted area often change during a major episode of ozone pollution.

After an initial classification of the ozone and meteorological data, further subdivision of the ozone-conducive days is possible. One such approach is to develop a regression tree, such as the CART (classification and regression trees) methodology (Stoeckenius, 1989, 1990). Through a series of binary splits, a decision tree is grown by continuously splitting the ozone data into two groups based on the value of a single predictor variable. The particular meteorological variable and the cutoff value are selected to produce the best possible split of the data at that point. The binary splits continue until the data are subdivided such that the ozone concentrations on days within each final group are sufficiently uniform. While the regression tree does not reduce all the year-to-year variability, it does highlight the results from individual years that have unusual meteorological conditions, such as 1980, 1983, and 1988.

Another approach is based on the hypothesis that the probability of an ozone-conducive day exceeding the allowed threshold concentration increases as the magnitude of specific meteorological parameters exceeds the threshold values of those parameters. In this approach, a meteorological intensity index (MII) would measure the deviation of the observed meteorological parameters from the average conditions of an ozone-conducive day and relate these parameters to the distribution of ozone-conducive days and days exceeding the threshold. The higher the value of the MII, the greater the probability that an ozone-conducive day would exceed the threshold concentration. An MII of 1.0 would correspond to the average conditions of an ozone-conducive day. When the MII is above 1.65, it would be a virtual certainty that the day would exceed the threshold. By applying this index to all days in the year, the number of expected violations of the standard for the year can be calculated, and compared with the actual number of days exceeding the standard. This approach, when used by Kolaz and Swinford (1990) for the Chicago area, substantially reduced the variation in the 3-year running average (Figure 2-10) and showed that substantial improvement had been made in reducing ozone concentrations in the area in the period from 1979 to 1981, but that little progress had been made since. Only after the meteorological variability was removed did the earlier progress and later stabilization become apparent. These and similar modified classification techniques appear to have promise in normalizing ozone trends for meteorological variability and should be evaluated for application to other geographic areas.

Regression Techniques

A more refined way to account for meteorological effects on ozone concentrations is to correlate ozone air quality to one or more aspects of meteorology and use the subsequent association to estimate what change would occur

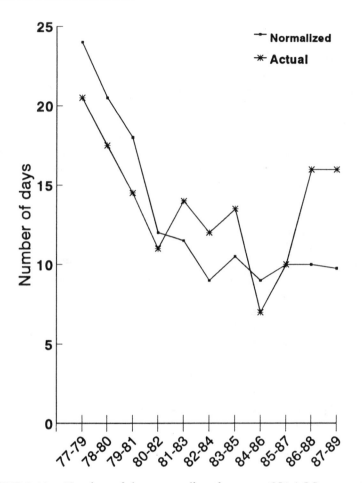

FIGURE 2-10   Number of days exceeding the ozone NAAQS concentration in the Chicago area, 1977-1989. Normalized data were obtained by removing meteorological variability. Source: Kolaz and Swinford, 1990.

in the ozone indicator given a change in meteorology. Such approaches were proposed shortly after trend analyses began. The most important variable used in these analyses is temperature. Several studies were conducted with temperature (in some form) being the only variable used to adjust the ozone trend. Jones et al. (1989) used such an approach to compare the days with ozone concentrations above 120 ppb in several cities in the United States to the days with temperatures above 90°F. By accounting for temperature using a simple ratio of the two measures, Jones reported that many areas of the

United States showed a downward trend in ozone concentrations from 1980 to 1988 (Figure 2-11). In one independent use of this approach for Philadelphia, Pollack et al. (1988) adjusted the number of days when the average maximum ozone concentration was above 105 ppb by the number of days when the maximum temperature was above 90°F, and the trend was even more variable. A direct linear dependence of high ozone concentrations on temperature could be difficult to develop because temperature is often related to other factors that influence ozone formation (Whitten and Gery, 1986), such as clear skies, light winds, mixing height, and thunderstorm activity.

Kuntasal and Chang (1987) performed a regression of ozone concentrations against temperature (at an atmospheric pressure of 850 mb) for the California South Coast air basin and reduced variability in the ozone concentration trend for the period from 1968 to 1985 (Figure 2-12). Wakim (1989) performed a regression analysis of high ozone concentrations against temperature for three cities: Houston, New York, and Washington. He explored 32 different configurations of temperature in his analysis, including the daily maximum temperature, the daily daylight temperature, and the daily average temperature. The best fit for the period from 1980 to 1987 was for ozone concentrations in New York, which, when normalized in this manner, showed a clear downward trend over the 7-year period. Generally, it can be observed that a high temperature is a necessary but not a sufficient condition for the occurrence of high ozone concentrations (Pollack et al., 1988), and other meteorological variables often need to be considered.

When other meteorological variables are considered, a regression equation (Chock et al., 1982; Kumar and Chock, 1984; Wakim, 1990; Zeldin et al., 1990; Korsog and Wolff, 1991) can be developed:

$$\text{Ozone indicator} = [\exp(a)](TEMP)^b(WS)^c(RH)^d(SKY)^e \quad (2.1)$$

Equation 2.1 considers temperature ($TEMP$), wind speed (WS), relative humidity (RH), and sky cover (SKY); other variables that could be accounted for are wind direction, dew point temperature, sea level pressure, and precipitation. In some cases, transport variables (wind direction, pressure gradient) also are important, but seldom are meteorological conditions associated with the previous day considered. As expected, the dominant variable aspect in this analysis is temperature (Korsog and Wolff, 1991).

In most cases, after correction for changes in the meteorological conditions, ozone trends showed reduced variability. In many cases (Wakim, 1990), downward trends in the corrected ozone data were observed.

One potential drawback associated with the use of a regression analysis to normalize a trend has been discussed in follow-up analyses (Pollack et al., 1988; Stoeckenius, 1990). It has been observed that high ozone concentrations are consistently underpredicted by the regression model. This is attributed to the least-squares fitting procedure used in linear regression, which is designed to limit the overall mean square error. Because high ozone values (100 ppb or more) are generally extreme values and occur only rarely, they might not be important in determining the regression coefficients. This was reported to be the case even when the meteorological data had first gone through an initial classification to remove the majority of the low ozone days (Pollack et al., 1988) (see Figure 2-13).

Without adjusting for meteorology, it is difficult to establish a relationship between ozone concentrations and precursor emissions (or ambient concentrations) over a period of a few years or less—for example, by comparing 1987 with 1988 or 1989. After adjustments for meteorology, consistent downward trends in ozone concentrations have been reported for some areas, supporting the view that meteorological conditions were more conducive during 1987 and 1988 than in other years to generating ozone. To make such adjustments, which could help areas reliably evaluate their progress in reducing ozone, additional meteorological data might need to be collected in some areas.

SUMMARY

Regulatory and pollution control efforts to attain the National Ambient Air Quality Standard (NAAQS) for ozone have not succeeded. In 1990, 98 areas exceeded the NAAQS. EPA (1991a) has reported a trend toward lower nationwide average ozone concentrations from 1980 through 1989, with anomalously high concentrations in 1983 and 1988. Concentrations in 1989 were possibly the lowest of the decade. However, since the trend analysis covers only a 10-year period, the high concentrations in 1983 and 1988 cannot be considered true anomalies. It is likely that meteorological fluctuations are largely responsible for the highs in 1983 and 1988 and the low in 1989. Meteorological variability and its effect on ozone make it difficult to determine from year to year whether changes in ozone concentrations result from changes in the weather or from precursor emission reductions.

The current principal statistical measure of ozone trends—the second-highest daily maximum 1-hour concentration—is highly sensitive to fluctuations in weather patterns and therefore is not a reliable measure of underlying trends. If the effectiveness of a program to reduce ozone concentrations in a particular area is to be tracked over a period of several years, then some way is needed to account for the effects of meteorological fluctuations. The condi-

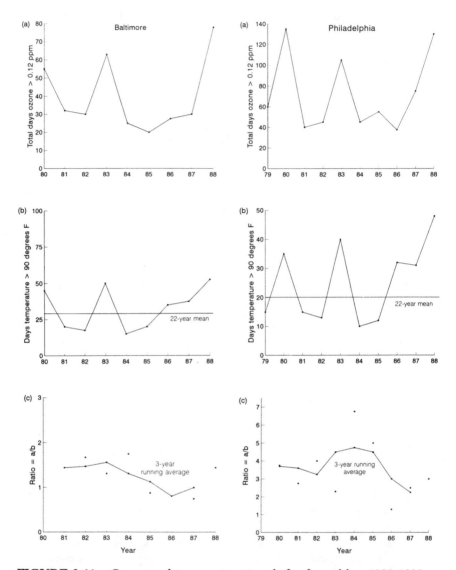

FIGURE 2-11   Ozone and temperature trends for four cities, 1980-1988. For each city, the number of days per year with ozone concentration greater than 0.12 ppm is shown in *a*. The number of days per year with temperatures greater than 90°F is shown in *b*. The ratio of data corresponding to the same year from *a* and *b* is shown in *c*. Source: Jones et al., 1989.

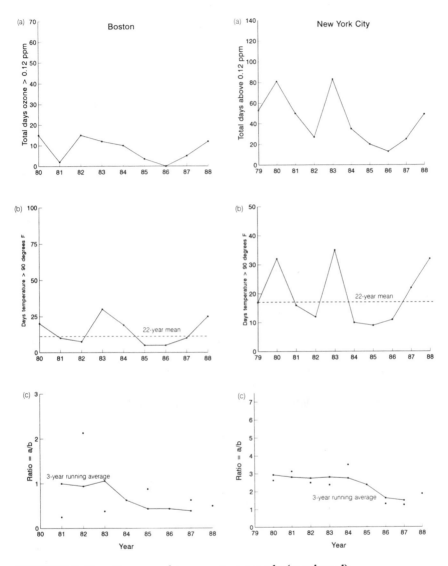

FIGURE 2-11   Ozone and temperature trends (continued).

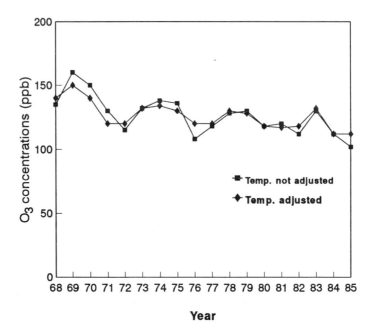

Year

FIGURE 2-12   Trends in ozone concentrations (temperature-adjusted and unadjusted) at nine sites in the California South Coast air basin, 1968-1985. Composite monthly averages of daily maximum 1-hour concentrations of ozone for July, August, and September were used. Atmospheric pressure was 850 millibars. Source: Kuntasal and Chang, 1987.

tions conducive to high ozone concentrations include high temperature, low wind speeds, intense solar radiation, and an absence of precipitation. The ozone-forming potential of a given day can generally be estimated using these factors as a starting point.

Support should be given to the development of methods to normalize ozone trends for meteorological variation. Several techniques that have shown promise for individual cities or regions could be useful on a national scale. In some cases, exposure indicators more robust than those associated with the NAAQS should be used to monitor ozone trends. One particularly promising approach in developing a relationship between ozone concentrations and meteorological

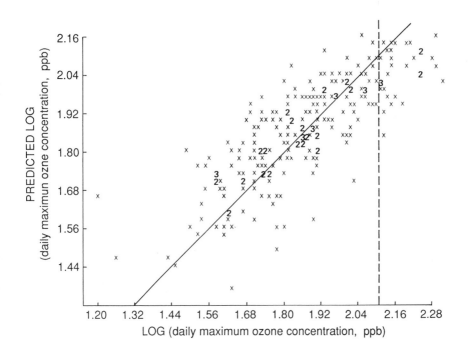

FIGURE 2-13   Predicted vs. actual maximum ozone concentration for days that passed the screening test at Bridgeport, Connecticut. The diagonal line is the line of equality between the predicted and actual values. The dashed vertical line indicates the log of the NAAQS concentration for ozone. Plotted numbers indicate the frequency of overlapping data points; no more than three points overlap at any one location. Source: Pollack et al., 1988.

variations is first to classify days as ozone-conducive or nonconducive and then further subdivide the conducive days using a binary decision tree or a regression equation to measure the degree to which the weather conditions of each conducive day deviate from the norm.

3

| Criteria for Designing |
| and Evaluating |
| Ozone Reduction Strategies |

INTRODUCTION

This chapter provides background information related to State Implementation Plans (SIPs), which the nation's states are required to use to attain the National Ambient Air Quality Standard (NAAQS) for ozone. The Clean Air Act, as amended in 1990, provides the legal foundation for this process. This chapter also discusses weaknesses in the existing SIPs' use of ambient air-quality data and emissions inventory data, the connections between air quality and emissions, new strategies for control, and the effectiveness of existing controls.

THE CLEAN AIR ACT

The Clean Air Act was the first modern environmental law enacted by Congress. The original act was signed into law in 1963, and major amendments were made in 1970, 1977, and 1990. The act establishes the federal-state relationship that requires the Environmental Protection Agency (EPA) to develop uniform air-quality standards (NAAQS) and empowers the states to implement and enforce regulations to attain them. The act also requires EPA to set NAAQS for common and widespread pollutants after preparing criteria documents summarizing scientific knowledge of their detrimental effects. EPA established NAAQS for each of six criteria pollutants: sulfur dioxide, particulate matter, nitrogen dioxide, carbon monoxide, ozone, and lead. These pollutants are emitted from numerous and diverse sources, and at certain concentrations and length of exposure they are anticipated to endanger public health or welfare. The NAAQS are threshold concentrations based on a detailed review of the scientific information contained in criteria documents prepared by EPA and peer reviewed. Pollution concentrations below the NAAQS are intended to expected have no adverse effects for humans and the environment. For each criteria pollutant, NAAQs comprise: a primary standard, which is intended to protect the public health with a margin of safety, and a secondary standard, which is intended to protect the public welfare as measured by the effects of the pollutant on vegetation, materials, and visibility.

Primary and secondary NAAQS for ozone, which were originally called NAAQS for oxidants, were established by EPA in 1971. Photochemical oxidants, a group of chemically related pollutants, are defined as those compounds giving a positive response using the iodide oxidation technique. In 1979 EPA revised the NAAQS for oxidants to the current NAAQS for ozone only. Both the primary and secondary NAAQS for ozone are now defined as a daily maximum 1-hour average concentration of 0.12 parts per million (ppm), or 120 parts per billion (ppb), not to be exceeded on average more than once each year. The average number of days exceeding the standard is calculated for a 3-year period. Based on recent health effects studies, the ozone NAAQS requires the shortest averaging time of any of the criteria pollutants' NAAQS. An area, whose boundaries are designated by EPA, is considered to exceed this threshold and is classified as being in "nonattainment" if a violation occurs anywhere within the area. For ozone, areas have consisted of Metropolitan Statistical Areas (MSAs), Consolidated Metropolitan Statistical Areas (CMSAs), and counties. The Clean Air Act requires that a SIP be developed for areas in nonattainment to reduce precursor emissions enough to bring air quality into compliance with the NAAQS. SIPs must be adopted by local and state governments and then approved by EPA. Once a SIP is fully approved, it is legally binding under both state and federal law.

In the Clean Air Act amendments of 1970, Congress set 1975 as the deadline for meeting the NAAQS. By 1977, 2 years after this deadline, many areas were still in violation of the ozone NAAQS. The 1977 amendments to the Clean Air Act delayed compliance with the ozone and carbon monoxide NAAQS until 1982, and areas that demonstrated they could not meet the 1982 deadline were given extensions until 1987. In 1990, 3 years after the final deadline, more than 133 million Americans were living in the 96 areas that were not in attainment of the ozone NAAQS the year before (EPA, 1990b).

The 1990 amendments classify nonattainment areas according to degree of noncompliance with the NAAQS. The classifications are extreme, severe, serious, moderate, or marginal, depending on the area's ozone design value and the percentage by which the value is greater than the NAAQS. Ozone design values are ozone concentrations that are statistically determined from air-quality measurements for each nonattainment area. If monitoring data for an area are complete, the design value is the fourth highest monitor reading over the past 3 years. The U.S. Code of Federal Regulations, Appendix H) provides a method to account for incomplete monitoring data (40 CFR 50.9(a). Design values are used to determine the extent of control needed for an area to reach attainment. These values and the target attainment years are shown in Table 3-1.

TABLE 3-1   Classification of Nonattainment Areas

Designation	% above 0.12 ppm ozone	Ozone design value range[a], ppm	Years allowed to attain ozone NAAQS[b]	Number of areas, 1989
Extreme	>133	>0.280	20	1
Severe	50-133	0.180-0.280	15	[c] 8
Serious	33-50	0.160-0.180	9	16
Moderate	15-33	0.138-0.160	6	35
Marginal	0-15	0.121-0.138	3	36

[a]Determined as the fourth highest value over 3 consecutive years.

[b]Number of years from November 15, 1990, allowed in the 1990 amendments to the Clean Air Act.

[c]Severe areas with design values between 0.19 and 0.28 ppm are allowed 17 years to attain the ozone NAAQS.

The 96 areas out of compliance in 1989, and their design values for 1983-1985, 1985-1987, and 1987-1989 are listed in Table 3-2. The areas classified as extreme or severe are in four major regions of the nation: the South Coast basin (Los Angeles) and San Diego, California; the greater Houston area; the Northeast Corridor (which extends from the Washington, D.C. area to Boston

TABLE 3-2 Classification of Nonattainment Areas for Ozone[a]

Area	Design value, ppm[b]		
	1983-1985	1985-1987	1987-1989
Extreme, design value 0.28 ppm or higher			
Los Angeles/Long Beach CMSA[c]	0.36	0.35	0.33
Severe, design value 0.18 to 0.28 ppm			
Baltimore	0.17	0.17	0.19
Chicago	0.20	0.17	0.19
Houston CMSA	0.25	0.20	0.22
Milwaukee	0.17	0.17	0.18
Muskegon, MI	0.14	0.17	0.18
New York City, NY/NJ/CT CMSA	0.22	0.19	0.20
Philadelphia, PA/NJ CMSA	0.18	0.16	0.19
San Diego	0.21	0.18	0.19
Serious, design value 0.16 to 0.18 ppm			
Atlanta	0.16	0.17	0.16
Bakersfield	0.16	0.16	0.17
Baton Rouge	0.16	0.14	0.16
Beaumont/Port Arthur, TX	0.16	0.13	0.16
Boston	0.16	0.14	0.17
El Paso	0.16	0.16	0.17
Fresno	0.17	0.17	0.17
Hartford	0.23	0.17	0.17
Huntington/Ashland, WV/KY/OH	0.14	0.14	0.16
Parkersburg/Marietta, WV/OH	—	0.13	0.17
Portsmouth/Dover/Rochester NH/MA	0.13	0.13	0.17
Providence CMSA	0.18	0.16	0.16

Area	Design value, ppm[b]		
	1983-1985	1985-1987	1987-1989
Serious, design value 0.16 to 0.18 ppm (continued)			
Sacramento	0.18	0.17	0.16
Sheboygan, WI	—	—	0.17
Springfield, MA	—	—	0.17
Washington, DC/MD/VA	0.16	0.15	0.17
Moderate, design value 0.138 to 0.16 ppm			
Atlantic City	0.19	0.14	0.15
Charleston, WV	0.13	—	0.14
Charlotte/Gastonia/Rock Hill, NC/SC	0.13	0.13	0.16
Cincinnati, OH/KY/IN	0.17	0.14	0.16
Cleveland	0.14	0.13	0.16
Dallas/Fort Worth	0.16	0.16	0.14
Dayton/Springfield	0.13	0.13	0.14
Detroit	0.13	0.13	0.14
Edmonson County, KY	—	—	0.14
Grand Rapids	0.13	0.13	0.14
Greensboro/Winston-Salem/ High Point, NC	—	—	0.15
Hancock County, ME	0.13	0.13	0.13
Jefferson County, NY	—	0.13	0.14
Knox County, ME	—	0.15	0.16
Louisville, KY/IN	0.15	0.16	0.15
Kewaunee County, WI	—	0.13	0.15
Knoxville	—	—	0.14
Memphis, TN/AR/MS	0.15	0.13	0.14

Area	Design value, ppm[b]		
	1983-1985	1985-1987	1987-1989
Moderate, design value 0.138 to			
0.16 ppm (continued)			
Miami/Hialeah	0.13	0.15	0.14
Modesto	0.15	0.15	0.14
Nashville	0.14	0.14	0.14
Pittsburgh	0.13	0.13	0.15
Portland, ME	0.16	0.14	0.16
Poughkeepsie	—	—	0.13
Raleigh/Durham	—	0.13	0.14
Reading, PA	0.13	—	0.14
Richmond/Petersburg, VA	0.13	0.13	0.14
Salt Lake City/Ogden	0.15	0.15	0.14
San Francisco CMSA	0.17	0.14	0.14
Santa Barbara/Santa Maria/			
Lompco, CA	0.16	0.14	0.14
St. Louis, MO/IL	0.16	0.16	0.16
Smyth County, VA	—	—	0.14
Visalie/Tulare/Porterville, CA	0.13	0.15	0.15
Worcester, MA	0.13	0.13	0.15
Marginal, design value 0.121 to			
0.138 ppm			
Albany/Schenectady/Troy	—	—	0.13
Allentown/Bethelehem, PA	0.14	0.13	0.14
Altoona, PA	—	—	0.13
Buffalo	—	—	0.13
Birmingham	0.13	0.15	0.13
Canton, OH	—	—	0.14
Columbus	—	—	0.13
Erie, PA	0.13	—	0.13
Essex County, NY	—	—	0.13
Evansville, IN/KY	—	—	0.13
Fayetteville	—	—	0.13

Area	Design value, ppm[b]		
	1983-1985	1985-1987	1987-1989
Marginal, design value 0.121 to			
0.138 ppm (continued)			
Greenbrier County, WV	—	—	0.13
Harrisburg/Lebanon/Carlisle	0.13	—	0.14
Indianapolis	0.13	—	0.13
Johnson City/Kingsport/			
Bristol, PA	—	—	0.13
Johnstown, PA	—	—	0.13
Kansas City, MO/KS	0.14	—	0.13
Lake Charles, LA	0.14	—	0.13
Lancaster, PA	0.13	—	0.13
Lexington, KY	—	0.13	0.13
Lewiston/Auburn, ME	—	—	0.14
Lincoln County, ME	—	0.13	0.13
Livingston County, KY	—	—	0.13
Manchester, NH	—	—	0.14
Montgomery, AL	—	0.14	0.14
Norfolk	—	0.13	0.13
Owensburg, KY	—	—	0.14
Scranton/Wilkes Barre	—	—	0.13
South Bend/Mishawaka, IN	—	—	0.12
Stockton	0.15	0.14	0.13
Sussex County, DE	—	—	0.13
Tampa/St. Petersburg/			
Clearwater	0.13	0.13	0.13
Waldo County, ME	—	—	0.13
York, PA	—	—	0.13
Youngstown, Warren, OH	—	—	0.13

[a]Based on data from 1987-1989; [b]Rounded to the nearest hundredth; [c]Consolidated Metropolitan Statistical Area. Source: OTA, 1989 and EPA, 1990b.

and beyond); and the Chicago area, including downwind areas in Wisconsin and Michigan.

The 1990 amendments to the Clean Air Act require EPA to designate the boundaries and classifications of the nonattainment areas. These designations are important in that they determine the SIP's geographic extent as well as the severity of the control program for these areas. From a scientific perspective they are no less important. Ozone formation is complex and time dependent. Given the time scales for the transport of pollutants from other areas, ozone maximum concentrations can be triggered by ozone precursor sources far upwind from the affected area. These atmospheric processes and the regional transport of ozone are described in Chapters 5 and 4, respectively.

Criteria for determining the size of the areas within a state have been established using jurisdictional boundaries (Metropolitan Statistical Areas or Consolidated Metropolitan Statistical Areas). In general, these areas should be as large as possible, and they should include sources and ozone-monitoring sites within the same area while allowing for expansion in accordance with population growth.

Section 184 of the 1990 amendments of the Clean Air Act establishes an interstate ozone transport region extending from the Washington, D.C. metropolitan area to Maine. In this densely populated region, ozone violations in one area might be caused, at least in part, by emissions in upwind areas. A transport commission is authorized to coordinate control measures within the interstate transport region and to recommend to EPA when additional control measures should be applied in all or part of the region in order to bring any area in the region into attainment. Hence areas within the transport region that are in attainment of the ozone NAAQS might become subject to the controls required for nonattainment areas in that region. EPA will likely establish other interstate ozone transport regions and transport commissions. How well these commissions carry out their responsibilities will be an early test of the effectiveness of the 1990 Clean Air Act amendments.

THE STATE IMPLEMENTATION PLAN

The State Implementation Plan is the technical and regulatory process for demonstrating attainment and maintenance of the requirements of the NAAQS. Once approved by EPA, the plan is legally enforceable under federal law and thus is a powerful tool for achieving the NAAQS. The current SIP mechanism is represented in Figure 3-1. Each side of the triangle represents a fundamental aspect of the SIP. The horizontal base (a) represents the time an area is allowed to achieve the NAAQS; this schedule is established by the

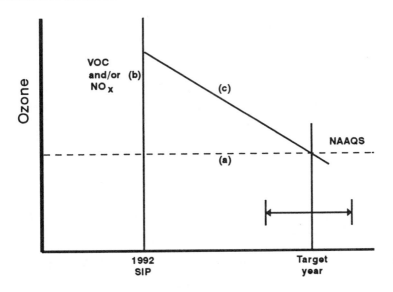

FIGURE 3-1   Conceptual diagram of SIP mechanism. (a) Time allowed to attain the ozone NAAQS; (b) required reduction in VOC and/or NO$_x$ emissions (c) reasonable further progress line.

Clean Air Act. The vertical ordinate (b) is the reduction in precursor emissions needed to reach attainment. The slope that completes the triangle is the rate of emission control progress needed to attain the NAAQS; it is called the reasonable further progress (RFP) line (c). The new RFP provision of the act will require a reduction in VOC emissions below the base year inventory by 15% over the first 6 years and 3% per year thereafter for all but "marginal" nonattainment areas. Areas classified as moderate, serious, or severe may choose an alternative to the 15% and 3% requirements, which includes, for example, the installation of all feasible controls on existing sources of emissions.   The 1990 amendments of the Clean Air Act also address the use of NO$_x$ controls in addition to, or instead of, VOC controls. Section 182(c) of the 1990 amendments allows states, with EPA guidance and approval, to supplement or replace VOC controls with NO$_x$ controls to an extent "that would result in a reduction in ozone concentrations at least equivalent to that which would result from the [required] amount of VOC emission reductions." In addition, Section 182(f) mandates that the control provisions required for major stationary sources of VOCs also apply to major stationary sources of NO$_x$ unless EPA determines that net air-quality benefits in an area are greater in the absence of NO$_x$ reductions, or that NO$_x$ reductions would not contribute to attainment of the ozone NAAQS in that area. Except in California, NO_x emission reductions have not previously been a major component of most SIPs.

SIPs will rely on enhanced monitoring of ozone, NO_x, and VOCs for the demonstration of attainment and maintenance of the NAAQS in serious, severe, and extreme areas. A less stringent process is allowed for moderate and marginal areas. Figure 3-2 diagrams the different phases of a SIP. The demonstration phase of the plan will take from 1 to 4 years after the plan begins, depending on an area's classification. The implementation phase ends when attainment is reached in accordance with the act's deadline or when EPA determines that the SIP is deficient and issues a recall or, as a last resort, develops a Federal Implementation Plan (FIP) for the area. The FIP could be a complete replacement of the SIP or a supplement to the SIP to correct its deficiencies.

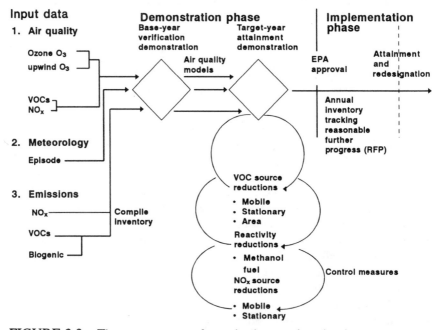

FIGURE 3-2   Three components of state implementation planning process.

In the demonstration phase, base-year modeling is performed to verify the air quality-emissions relationship for a specified meteorological episode. The base year is the year of the latest available emissions inventory. Target-year modeling, in which emissions are reduced according to the control measures selected by the area, is performed next. In general the meteorological episode is a typical episode of high ozone concentrations for which adequate meteorological data are available. Figure 3-2 illustrates the three SIP data bases: air quality, meteorology, and emissions. Because of the importance of these data bases to attainment, each is discussed in detail in subsequent chapters.

Once established by the base-year analysis, the same meteorological input data are applied to future years, and modeled VOC or NO_x emissions are incrementally reduced by applying estimates of reductions to be brought about by control measures proposed for mobile, stationary and area sources. If attainment is not demonstrated by the target year, the inventory is further reduced by introducing additional control measures for total and reactive VOC emissions and possibly for emissions of NO_x. Only after the target-year modeling demonstrates attainment with the NAAQS and after a period of public comment will EPA approve the SIP.

In the demonstration phase, considerable reliance is placed on the air-quality model and its required inputs to determine the emission reductions necessary to achieve the NAAQS. There are always uncertainties in the base-year inventory and in the projected control measure estimates. These uncertainties are further compounded by the projection of modeling inputs to a future year for attainment demonstration.

The tracking mechanism used in the implementation phase is also unsatisfactory. In past SIPs, states were required only to estimate emissions reductions periodically and track them in relation to the RFP line shown in Figure 3-1. Projected air-quality improvements were not checked with models. It is anticipated that at least one mid-course modeling demonstration will be required for the upcoming SIPs. Limited federal guidance was incorporated in the tracking process, as the federal government's major emphasis shifted from oversight to rule development and implementation strategies. Limited use of audits was employed by EPA, and those that were conducted were compromised by their lack of independence: the agency conducting the audit was the same one held responsible for the success of the process. The committee concludes that the lack of an adequate mechanism for tracking progress and for taking corrective action, if needed, significantly limits the ability of areas to effectively monitor and maintain progress toward achievement of the NAAQS.

Other processes—less demanding of resources than the process based on air quality and emissions—could be used. For example, the accommodative SIP would rely on emission reductions from federally mandated control measures and existing controls and thus allow for expected emissions growth in some areas. This approach would be applied only in the most marginal non-attainment areas. These areas would not need to develop more detailed demonstrations by using models and instead could implement control strategies only as needed; emissions tracking along with ambient air-quality monitoring might be all that is required. This approach might not be appropriate, however, for marginal nonattainment areas that are near more severe nonattainment areas and are expected to sustain large growths in emissions because of population growth and the attraction of new sources of pollution.

The use of a technology- or regulatory-based approach also has been suggested. In this approach, whenever a new technology is demonstrated, it is required immediately for all new sources and for existing sources undergoing modification or renewal of operating permits. Section 173 of the 1990 Clean Air Act amendments allows for offsets or a market-based approach to be used to reduce emissions from existing sources. These approaches are similar to the acid rain provisions of the Clean Air Act, which allow sources to comply with emission requirements by obtaining offsetting reductions from other sources. However, without a specific set of technology-forcing requirements (for example, demonstration programs and such incentives as excess emission fees) there is no certainty that advanced control technologies will be demonstrated, and therefore compliance with a target date is uncertain.

Ambient Monitoring

Ambient air-quality measurement is the basis for determining attainment of the NAAQS. EPA has developed strict guidelines for site location, instrumentation, and quality assurance. State and local agencies are required to maintain standard operating procedures for air-quality monitoring in accordance with National Air Monitoring Systems/State and Local Air Monitoring Systems, known as the NAMS/SLAMS network. Currently, the network consists of 231 NAMS and 420 SLAMS sites (EPA, 1990c).

The number and spatial distribution of ozone-monitoring sites in an area is governed by population. Each air-quality control region is required to have at least two monitoring sites: one that is generally upwind of the urban population center during episodes of high ozone concentrations, and one that is generally downwind. A monitoring subset of 15 cities in the NAMS/SLAMS network is shown in Table 2-2 of Chapter 2. Although each city complies with EPA criteria, the limited number of sites or the placement of monitors calls into question the validity of some city trends. For many rural areas there are no state or local ozone-monitoring requirements. It therefore is likely that there is insufficient monitoring to characterize rural areas and areas at the upwind boundary of many urban locations. Furthermore, past EPA criteria for ozone monitoring have been based on inadequate modeling. Information from grid-based urban and regional models should be used in designing an enhanced monitoring network. With the expanded domains of regional models, a broader, more coordinated ozone network with an appropriate number and redistribution of monitors will be needed to verify modeling demonstrations and compliance with SIPs.

Because of the need to determine VOC/NO_x ratios for different cities (see Chapter 6), EPA will publish enhanced monitoring guidelines. At least one monitoring site for VOC and NO_x is now required in every major metropolitan nonattainment area. By the summer of 1989, there were 25 such sites in 21 cities. Ambient data on precursor emissions have thus been collected in only a small portion of the nonattainment areas. However, the number of monitors is likely to increase significantly. The need for enhanced monitoring of NO_x and VOCs is discussed in later chapters.

The current EPA method for monitoring VOCs is a canister sampling system, which captures a 3-hour (6-9 a.m.) averaged sample for subsequent laboratory analysis. This method measures the quantities of VOCs that are nonmethane hydrocarbons. It is now standardized for the multiple-city program; however, identification of specific compounds remains a problem.

In addition, given the use of hourly averaging in air-quality models for consistency with the NAAQS averaging period, sampling averaged over three hours may pose problems in model verification. Altshuller (1989) pointed out the highly variable spatial and temporal components of the VOC/NO_x ratio at ground level and aloft and the resulting implications for modeling.

Continuous VOC monitoring methods are being developed to coincide with, and eventually to replace, periodic sampling methods. The continuous methods should be specific for the classes of reactive VOCs that ultimately control ozone production. In some cases special monitoring studies might be helpful to determine the transport of pollutants into urban areas. Particular monitoring methods and recommendations are described in Chapter 7.

Emissions Inventories

The Clean Air Act and ozone SIPs place specific emphasis on the development of reliable estimates of VOC and NO_x emissions. The emissions inventory is used to determine source types by area, the quantity and rate of pollutants emitted, and the kinds of processes and controls used at each source. EPA has published *Procedures for Emissions Inventory Preparation*, Volumes I-V (EPA, 1981), which recommends procedures for estimating emissions from point, area, and mobile sources. Emissions are aggregated by county, by municipality, or, when used for modeling, by model grid.

For base-year modeling, the best estimate of emissions during the time of the chosen episode of high ozone concentrations is used. Actual emissions are estimated from the operational data on emission sources (e.g., a factory) or from emissions monitoring data. This inventory is date-specific, with temperature adjustments that depict the meteorological episode. Emissions are temporally allocated and pollutants are identified within the appropriate grid. Biogenic emissions are not typically included but should be (see Chapter 9).

Target-year emissions estimates are determined from the amount of emissions legally allowed by operating permits. Growth factors for the target year are developed for each category and are derived from U.S. Department of Labor statistics. For point sources, the difference between actual and allowed emissions usually is significant, and each is determined independently. The base-year inventory should reflect emission conditions during the worst ozone pollution episode, and the target-year emissions estimate should focus on control measures for worst-case emissions. SIPs put in place after 1987 are likely to retain the existing concept of using allowed emissions. However, EPA is developing adjustments of the base-year inventory that will be used to determine requirements, modeling, and projection estimates.

Cross comparisons of emissions trend estimates by nonattainment area or city are limited and subject to considerable uncertainty. EPA's annual emission report, *National Air Pollution Emission Estimates: 1940-1988* (EPA, 1990d), cautions:

> The principal objective of compiling these data is to identify probable overall changes in emissions on a national scale. It should be recognized that these estimated national trends in emissions are not meant to be representative of local trends in emissions or air quality.

The lack of cross comparison data is disturbing given the importance of emissions data as the basis for SIP demonstration and implementation. The Office of Technology Assessment (OTA), on behalf of the Congressional subcommittees considering the reauthorization of the Clean Air Act, undertook a comprehensive evaluation of the costs and uncertainties in attaining the ozone NAAQS (OTA, 1989). With regard to estimating emissions, OTA reported:

> Our estimates of emissions throughout the analysis are subject to potentially significant uncertainty. We estimate that VOC emissions in nonattainment cities could be as low as 8 million or as high as 14 million tons per year in 1985 depending on several important mobile and stationary source assumptions.

Approaching absolute certainty is impossible, but selective bias in estimating emissions can be reduced or estimated through a reliable, statistically valid survey. Such a survey would complement but not replace the standard methods. The major emphasis would be directed at improving accuracy and establishing confidence levels. The survey would include all possible conditions found in the sample and be used to adjust the state emissions estimates. Fortunately, EPA has recognized this problem and is directing resources to improve the quality assurance of post-1987 SIPs (EPA, 1990e). However, a definition of a specific range of uncertainty or "reasonableness" has not yet been established. Emissions inventories for stationary, area, and mobile sources are discussed in Chapter 9.

Relationships Between Emissions and Air Quality

A major element of SIPs is a means to relate VOC and NO_x emissions to ozone concentrations. This relationship is elucidated through an air-quality model—a mathematical simulation of atmospheric transport, mixing, chemical reactions, and removal processes. Ozone air-quality models are discussed in Chapter 10. Past EPA guidance allowed the use of a one-dimensional model, the empirical kinetic modeling approach (EKMA), which requires as input only the VOC/ NO_x ratio measured at an upwind location and the measured peak ozone concentration to fix the percentage precursor reductions needed to reach the NAAQS. (Nonmethane hydrocarbons are used to represent VOCs.) The Clean Air Act Amendments of 1990 require ozone nonattainment areas designated as extreme, severe, serious, or multistate moderate to demonstrate attainment of the ozone NAAQS through photochemical grid-based modeling or any other analytical method determined by EPA to be at least as effective. The act does not specify the method for demonstrating attainment in marginal and within-state moderate areas. EPA has determined that the use of the EKMA may be sufficient for these areas, but prefers the use of grid-based models (EPA, 1991b). The more severe nonattainment areas are now moving toward the use of three-dimensional, grid-based air-quality models, which require sizable data bases on meteorology and emissions. Expensive field studies are generally needed to obtain useful input data for these models; at issue is the extent to which routinely collected data can substitute for data obtained from expensive studies without causing unacceptable declines in model performance. Because the air-quality model is the only means to estimate the effect of future emission reductions, a great deal of care has been taken to assess the accuracy and deficiencies of such models. These issues are addressed in Chapter 10.

Control Strategies

The primary purpose of the SIP is to set forth a control program of NAAQS attainment strategies that are legally enforceable at both the state and the federal level. As case examples for this report, a generic process for developing controls is illustrated by an OTA report (OTA, 1989; Rapoport, 1990), and an extreme example is demonstrated in the 1988 air-quality management plan (AMP) for California's South Coast Air Quality Management District (SCAQMD, 1989).

The OTA study characterized the emissions reduction potential of various control strategies and then applied these reductions to many of the nonattainment areas listed in Table 3-2. Each area was ranked by ozone design value, in ppb (130-140, 150-170, 180-260, >260) , and estimates were made of each area's potential to achieve the NAAQS by 1994 and by 2004. Although major assumptions were made with the emissions data, and the EKMA model was used in the comparison, the OTA study provides a common data base to compare control strategies nationally and examine their ability to attain the NAAQS.

Near-term VOC strategies were analyzed first. These included application of all reasonable available control technology (RACT) controls now required by any state to all large (>25 tons/yr) sources in nonattainment areas. OTA also estimated the cumulative benefits of additional strategies including emission controls on hazardous waste treatment, storage, and disposal facilities (TSDFs); federally regulated controls on architectural surface coatings; on-board motor vehicle controls; modification of gasoline station pumps to trap escaping vapors; enhanced inspection and maintenance (I/M) programs; more stringent motor vehicle exhaust standards and gasoline volatility standards; and the use of methanol in centrally owned motor vehicle fleets. The estimated results of these strategies were projected to target years 1994 and 2004, taking into account increased vehicle mileage and economic activity. Figure 3-3 shows the estimated reductions for these years.

The OTA report concluded that given the most optimistic estimates of these potential VOC emission reductions, nonattainment areas with design values of less than 160 ppb have some prospect of achieving the NAAQS. It was concluded that additional reductions are needed in areas with design values of 160 ppb or greater, even in the most optimistically modeled cases.

OTA then expanded the control options using categories that might not yet be available. These included NO_x controls on major existing stationary sources; I/M for NO_x; more stringent NO_x standards for motor vehicle exhaust, control of organic solvent evaporation; alternative fuels for passenger vehicles; and transportation control measures (TCMs). (TCMs include modified work schedules, highway lanes for carpools, bicycle lanes, road use tolls, and improved public transit.) Emission reductions from these future controls , however, were not estimated or modeled.

OTA estimated that emissions from solvents, highway motor vehicles, and gasoline refueling will account for 70% of the remaining VOC inventory, suggesting the need for longer-term strategies that include lowering or restricting organic solvent emissions, implementing long-term TCMs, and using alternative motor vehicle fuels such as methanol and compressed natural gas.

OTA also concluded that in many nonattainment areas, exhaustive exploration of potential control measures will be required to identify reductions to attain "or come as close as possible to" the NAAQS. In areas where TCMs are needed, close cooperation with state and local transportation agencies and land-use planners is essential.

The South Coast Air Quality Management Plan (SCAQMP) is an example of a plan to identify potential control measures that could be available in the near term (Tier I) and longer term (Tiers II and III). Tier I controls are those measures that can be reasonably adopted in the next 5 years and implemented over the next 20 years with currently available technology. Tier II would require significant advancement of current technology and further regulatory controls through technology-forcing standards or emission fees. Tier III would require technologic advances in industrial, commercial, and residential solvent and coating applications and the use of essentially emission-free vehicles. Major transportation and land use planning efforts also would be needed. All tiers together were calculated to result in an 84% and an 80% reduction for VOCs and NO_x, respectively.

Recently, EPA proposed for the South Coast area a FIP that is designed to augment the SCAQMD's 20-year plan (CFR, July 21, 1990). It was issued in compliance with a court order that resulted from a suit filed by several environmental organizations for the SCAQMD's failure to attain the NAAQS. The plan is believed to represent EPA's view of what future SIPs should contain in severe and extreme areas.

The EPA plan is designed to achieve the NAAQS by 2010 by promoting market-based incentives and by conforming with the 1990 amendments to the Clean Air Act. Similar to the tier concept, the FIP calls for the promulgation of "core" measures and compliance with a schedule for reasonable further progress. The program will implement backstop measures, such as innovative technology standards and economic incentive, programs if the area does not achieve emission reductions on schedule.

The core measures, to be implemented during the first 5 years of the FIP, include further limitations on the seasonal changes in gasoline volatility, use of oxygenated fuels and reformulated gasoline (as described in Chapter 12),

VOC control method

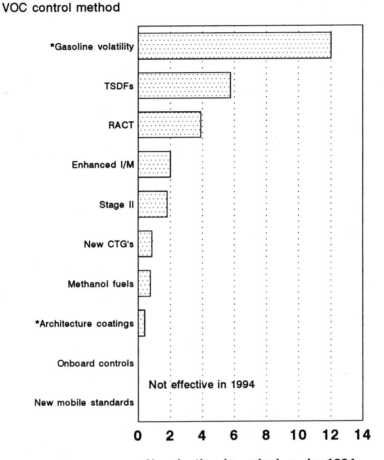

FIGURE 3-3   Predicted percent reductions of VOC emissions in 1994 and
2004 compared with 1985 emissions, by control method.

*Indicates that emissions reductions would also be achieved in areas in
attainment of the ozone NAAQS. Percent reductions from methanol fuels
would occur only in areas where use is required. TSDF = hazardous waste

VOC control method

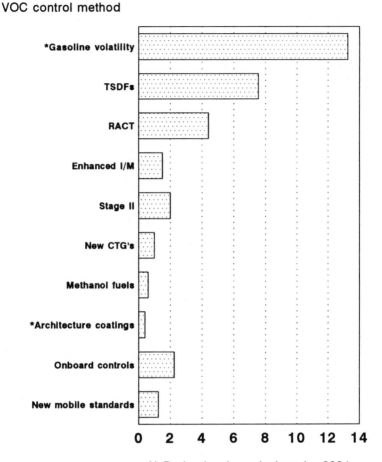

% Reduction in emissions by 2004

FIGURE 3-3   (continued).

treatment, storage, and disposal facilities. RACT = reasonable available control technology required for existing stationary sources. Enhanced I/M = inspection and maintenance of motor vehicles. Stage II = control devices on gas pumps to capture gasoline vapor during motor vehicle refueling. CTG = new control technique guidelines for RACT on existing stationary sources. Onboard controls = emission controls on motor vehicles.

Source: OTA, 1989.

controls on marine vessel tanks, and further control of evaporative emissions from gasoline-fueled motor vehicles. EPA has proposed to implement a regulatory-based FIP to obtain the remaining reductions needed after the first 5 years. Under this approach, highly restrictive performance standards would be developed, and as new technologies become available the regulations would be adjusted as needed.

The regulatory FIP also calls for an ultra-clean-vehicles program that will require new vehicle standards that can be met through any combination of fuel mixes and new vehicle design. Additionally, the program calls for a composite in-use standard instead of separate measures for exhaust, evaporative, and running loss and for loss during refueling. The motor vehicle program would allow marked-based practices such as banking, averaging, and trading of emissions.

The stationary-source program would encourage pollution prevention and market-based approaches. Public education would encourage the use of fewer consumer product solvents. The EPA plan also would allow sources to decide how to obtain VOC reductions, including the use of such measures as reformulation or substitution of compounds contributing to ozone formation, use of control equipment, or purchase of emission reduction credits from more effectively controlled facilities. In addition, industrial and commercial sources would be required to reduce emissions at a rate of 6% per year. Even with these measures, EPA estimates that because of growth, there will need to be further reductions. EPA will monitor growth projections at least every 3 years and propose such regulations as needed.

These future controls are likely to be far less cost effective than the controls now available (Wilson et al., 1990), and regional control flexibility could be a necessary ingredient to maximize cost effectiveness in attaining the ozone NAAQS. Likewise, compliance by an emission source with air-quality permits would not be fully effective, especially when the maximum emissions allowed by the permit are exceeded due to unusual operating conditions or high-temperature days. It is likely that with further controls in severe and extreme nonattainment areas becoming increasingly limited, these areas will follow the SCAQMP and EPA approach.

Rule Effectiveness

A major fallacy of existing SIPs is the presumption that control measures achieve 100% effectiveness. The number and diversity of VOC sources, the uncertain reliability of existing controls, and limitations in state and local resources make 100% compliance extremely unlikely. When determining the extent of emissions controls required to meet the NAAQS. States need to account for the fact that rules are less than fully effective.

Specific data or studies that compare the effectiveness of rules have been limited. The SCAQMD conducts a periodic audit that includes engineering assessments, field inspections, and source testing. In 1988 it found that more than 70% of the 180 facilities visited by the audit teams had underestimated emissions by an average of 15% (Guensler, 1990).

An American Petroleum Institute (API) evaluation of SIPs in seven cites reported the most common SIP deficiencies and inadequacies in implementation (API, 1989). The report concluded:

> Actual effectiveness of implemented control measures was less than that estimated in control strategies. This was observed for both motor vehicle I/M programs (which in some cases fell far short of targets) and controls on stationary and area VOC sources. Technological and enforcement weaknesses are both likely to contribute to this problem. The magnitude of shortfalls in effectiveness is uncertain, due to the general lack of data in all areas regarding in-use controls.

Moreover, the seven-city report found that the RFP reports were not a reliable means of identifying weaknesses during the SIP implementation phase. Even when deficiencies in the 1982 SIP estimates had been found and corrected, there was little feedback into the regulatory process, so there was not enough effort to upgrade the measures or implement further controls.

EPA has released a guidance document for new SIPs that provides the basis for preparing base-year inventories for stationary sources (EPA, 1990e). In addition, the Clean Air Act amendments of 1990 require EPA to review emission factors at least every 3 years. Emission factors must be established for sources that have not had them in the past. Also, the 1990 amendments require stationary sources to submit annual statements of VOC and NO_x emissions.

EPA's guidance procedure presumes that each rule is 80% effective. However, the real effectiveness is highly variable, and one option would allow tracking and establishment of an appropriate effectiveness level for each area. As an example, during the EPA study of regional ozone modeling for northeast transport (ROMNET), each of the 12 participating states took part in a survey to evaluate its own rule effectiveness for existing measures (EPA, 1988a). Three aspects of source categories were investigated: technology control efficiency, enforcement compliance, and projected losses due to rule cutoff levels, waivers, and exemptions. An example of technology control effectiveness is shown in Table 3-3. The states reported that outright prohibi-

TABLE 3-3   Maximum Technology Control Levels for VOC Area Sources

Description	Control level, % efficiency
Gasoline marketing	
Stage I	95
Stage II	86
Prescribed forest burning[a]	100
Agriculturala	100
Degreasing	83
Drycleaning	70
Graphic arts (printing)	85
Rubber and plastics manufacturing	83
Architectural coating	52
Auto body repair	88
Motor vehicle manufacturing	88
Paper coating	90
Fabricated metals coating	57
Machinery manufacturing	90
Furniture manufacturing	90
Flat wood products coating	90
Other transportation equipment	88
Electrical equipment	90
Ship building and repairing[a]	47
Miscellaneous industrial manufacturing[a]	85
Miscellaneous industrial solvent use[a]	85
Miscellaneous nonindustrial solvents	20
Publicly owned treatment works	90
Cutback asphalt paving	100
Fugitive emissions, SOCMI[b]	56
Bulk terminals/bulk plants	91
Refinery fugitive emissions	93
Process emissions, bakeries	90
Pharmaceutical emissions[a]	90
Synthetic fibers[a]	85
Crude oil- and gas-products fields	93
Hazardous-waste TSDFs[c]	90

[a]Values provided by states based on in-use experience; [b]synthetic organic chemical manufacturing industry; [c]transportation, storage, and disposal facility.

tions that require minimal enforcement provide the most effective control, and small-source controls, which require the most enforcement, provide the least effective control. The states reported that the product of the three aspects reduced the overall control effectiveness for many source categories to below 50%.

During the initial phase of the upcoming SIPs, an effort will be made to upgrade existing rules. EPA has indicated that efforts will be made to require consistency among the states and regions with regard to the effectiveness of similar rules. The more stringent operating permit and enforcement programs mandated in the 1990 amendments of the Clean Air Act also should improve effectiveness. In addition, the 1990 amendments contain an improved RFP tracking requirement that imposes a fixed annual percentage reduction averaged over a set number of years. Failure to meet this requirement would subject an area to the sanctions provided for in the act. Possible sanctions include more stringent offset requirements for new construction, withholding of federal highway funds, and, in extreme and severe areas, fines on major stationary sources. Major emphasis will be placed on effectiveness improvement during the implementation phase of the SIP.

SUMMARY

The heart of the Clean Air Act for ozone attainment is the State Implementation Plan (SIP), a plan for emission reductions designed to reduce ambient concentrations of ozone to concentrations that do not exceed the National Ambient Air Quality Standard (NAAQS). Despite considerable efforts over the past 2 decades in developing and implementing SIPs, NAAQS violations are still widespread.

The essential components of SIP development include

• monitoring of ambient pollutant concentrations to determine whether the NAAQS has been exceeded and, if so, by how much and where;
• collection and analysis of meteorological data and air-quality data needed to develop the appropriate emissions-air quality relationship;
• inventorying of emissions from point, area, and mobile sources to determine the emission reductions necessary to attain the NAAQS;
• projecting the emissions inventory to future years;
• identifying and selecting specific emissions control measures and demonstrating that the control strategy will be adequate to achieve the air-quality goal.

However, the SIP must be effectively implemented. The necessary steps in implementation include adoption and enforcement of regulations, tracking of progress in achieving emission reductions, and adjustment of the plan as necessary to achieve emission reduction targets. If the system works properly, implemented emission reductions should provide attainment.

Failure to attain the ozone NAAQS can result from ineffective implementation of the SIP or from defects in the SIP itself. The SIP will not succeed if it does not properly estimate emissions, if it does not adequately relate emissions to ambient concentrations, or if it fails to identify sufficient emission reductions.

Although the SIP appears to be a fundamentally correct approach to air-quality management, major weaknesses in both its methodology and its application are apparent. Base-year emissions inventories have underestimated actual emissions, in some cases by substantial amounts (see Chapter 9). As a result, future-year emissions inventories have been consistently underpredicted, and emission reductions brought about by controls often have a proportionately smaller effect on total emissions than the SIPs originally estimated. In addition, the reductions required to attain the ozone NAAQS were in most cases developed using EKMA (empirical kinetic modeling approach), an emissions-air quality model that did not account sufficiently for regional characteristics (see Chapter 6). Finally, individual control measures for motor vehicles and stationary sources were falsely assumed to meet their emission reduction targets, and provisions were not made for evaluating actual reductions. The magnitude of shortfalls in effectiveness is uncertain because of the general lack of data in all areas regarding in-use controls. Thus, SIPs have overstated the effectiveness of controls in reducing ozone concentrations and have understated the emission reductions needed to attain the NAAQS. Improvements in implementation must be directed toward the use, where possible, of realistic descriptions of the relationships between emissions and air quality and toward control programs that account for uncertainties and potential shortfalls in the effectiveness of controls. Procedures are needed for tracking actual emissions and reductions. Finally, and most important, feedback must be provided from the implementation phase to the SIP development phase.

The following actions are recommended to improve the effectiveness of State Implementation Plans:

· Establish air pollutant transport commissions in appropriate areas and determine their effectiveness.

· Concentrate SIP modeling resources in nonattainment areas requiring relatively large efforts to achieve compliance with the NAAQS, including multistate areas. Accommodative or technology-based SIPs should generally be used elsewhere.

- Conduct a review of the appropriate numbers and siting of ozone monitors for urban, suburban, and rural areas.
- Establish a reference method for a continuous total VOC (volatile organic compound) monitoring system that can speciate major VOC classes, and establish VOC monitoring in the most severe nonattainment areas (see Chapter 7).
- Develop independent validation techniques for the SIP components. For example, statistically based surveys could compare stationary source inventories, and roadside screening surveys should be compared with mobile source inventories.
- Establish an effective audit program to track SIP progress.
- Establish feedback between the SIP development and implementation phases. If control measures are not being implemented effectively for technical or other reasons, adjustments to the plan must be made.

4

The Effects of Meteorology on Tropospheric Ozone

INTRODUCTION

Meteorological processes directly determine whether ozone precursor species are contained locally or are transported downwind with the resulting ozone. Ozone can accumulate when there are high temperatures, which enhance the rate of ozone formation (as discussed in Chapter 2), and stagnant air. Some processes, such as those that lead to cloud formation, can disperse or transport ozone and its precursors. This chapter examines the effects of weather on tropospheric ozone formation, accumulation, and transport and discusses aspects of those processes that are important for predicting ozone concentrations through the use of mathematical models. The chapter also includes a discussion of rural ozone data for the United States, with a focus on the effects of meteorology.

OZONE ACCUMULATION

Major episodes of high concentrations of ozone are associated with slow-moving, high-pressure weather systems. These systems are associated with high concentrations of other chemical pollutants such as sulfur dioxide. There are several reasons that slow-moving, high-pressure systems promote high concentrations of ozone:

- These systems are characterized by widespread sinking of air through

93

most of the troposphere. The subsiding air is warmed adiabatically and thus tends to make the troposphere more stable and less conducive to convective mixing. Adiabatic warming of the air occurs as the air compresses while sinking; no heat is added to it.

• The subsidence of air associated with large high-pressure systems creates a pronounced inversion of the normal temperature profile (normally temperature decreases with height in the troposphere), which serves as a strong lid to contain pollutants in a shallow layer in the troposphere, as is common in the Los Angeles basin, for example. During an inversion, the temperature of the air in the lower troposphere increases with height, and the cooler air below does not mix with the warmer air above.

• Because winds associated with major high-pressure systems are generally light, there is a greater chance for pollutants to accumulate in the atmospheric boundary layer, the turbulent layer of air adjacent to the earth's surface.

• The often cloudless and warm conditions associated with large high-pressure systems also are favorable for the photochemical production of ozone (see Chapter 5).

In the eastern United States and Europe, the worst ozone pollution episodes occur when a slow-moving, high-pressure system develops in the summer, particularly around the summer solstice. This is the time with the greatest amount of daylight, when solar radiation is most direct (the sun is at a small zenith angle) and air temperatures become quite high (greater than 25°C) (RTI, 1975; Decker et al., 1976). As the slow-moving air in the shallow boundary layer passes over major metropolitan areas, pollutant concentrations rise, and as the air slowly flows around the high-pressure system, photochemical production of ozone occurs at peak rates. Major high-pressure systems at the earth's surface are associated with ridges of high-pressure surfaces in the middle and upper troposphere. Forecasting the onset of a major episode of ozone pollution in the eastern United States involves predicting the development of ridges of high pressure at 500 millibars (mb). These ridges are generally well predicted by global numerical prediction models for periods of 3-5 days (Chen, 1989; van den Dool and Saha, 1990). High ozone episodes are often terminated by the passage of a front that brings cooler, cleaner air to the region.

The accumulation of ozone in the Los Angeles basin illustrates the importance of meteorology. The weather in that area is dominated by a persistent Pacific high, which causes air subsidence and the formation of an inversion that traps the pollutants emitted into the air mass. The local physical geography exaggerates the problem, because the prevailing flow of air in the upper atmosphere is from the northeast, which enhances the sinking motion of air on the leeward, western side of the San Gabriel Mountains into the basin. The low-level flow of air is controlled by daytime sea-breeze and nighttime land-breeze circulations. During the day the sea-breeze is channeled by the coastal Puente Hills and the San Gabriel Mountains and the southern entrance to the San Fernando Valley (Glendening et al., 1986). Because of its low latitude (~34 degrees) and prevailing subsiding flow in the upper atmosphere, the basin experiences long hours of small-zenith-angle sunlight and relatively few clouds. These conditions are ideal for the photochemical production of ozone.

CLOUDS AND VENTING OF AIR POLLUTANTS

Clouds play an important role in mixing pollutants from the atmospheric boundary layer into the lower, middle and upper troposphere, a process known as "venting" (e.g., Gidel, 1983; Chatfield and Crutzen, 1984; Greenhut et al., 1984; Greenhut, 1986; Ching and Alkezweeny, 1986; Dickerson et al., 1987; Ching et al., 1988). They also influence chemical transformation rates and photolysis rates. The effect of clouds on vertical transport depends on their size and type. Although major high pressure systems may be cloud-free, weaker systems may permit the formation of a variety of cloud types. Consequently, regional models for ozone need to simulate cloud formation and vertical redistribution.

Ordinary cumulus clouds, such as fair weather cumulus, are relatively shallow, small-diameter clouds. They typically form from masses of warm air that develop in the boundary layer, and they can be modeled using approaches similar to those used for the dry boundary layer (Cotton and Anthes, 1989). Greenhut (1986) analyzed turbulence data from over 100 aircraft penetrations of fair-weather cumulus clouds, with the goal of developing parameterizations of cloud transport for EPA's Regional Oxidant Model (ROM) (discussed in Chapter 10). He found that the net ozone flux in the cloud layer was a linear function of the difference in ozone concentration between the boundary layer and the cloud layer, and that cloud turbulence contributed about 30% of the total cloud flux. Ozone fluxes in the regions between clouds were usually smaller than the cloud fluxes, but their contribution to the net transport of ozone was important because they occur over a larger area.

Cumulonimbus clouds are convective clouds of significant height (often the entire height of the troposphere). Precipitation is important in their life cycle, organization, and energy transformation. These clouds may function as wet chemical reactors and provide a source of NO_x from lightning. The fundamental unit of a cumulonimbus is a cell, shown on radar as a region of concentrated precipitation, and characterized as a region of coherent updraft and downdraft. Cumulonimbus clouds are classified by their cells, organization, and life cycles. Ordinary cumulonimbi contain a single cell which has a life cycle of 45 minutes to an hour. Many thunderstorms are composed of a number of cells, each having lifetimes of 45-60 minutes. These multicell storms can last for several hours and vertically redistribute large quantities of ozone and its precursors. Supercell storms, composed of a single steady cell, with strong updrafts and downdrafts, can last two to six hours and inject large quantities of pollutants into the upper troposphere.

Dickerson and coworkers demonstrated the role of cumulonibus clouds in transporting polluted boundary layer air to the upper troposphere using CO as a tracer (Dickerson et al., 1987; Pickering et al., 1989), but noted that not all cases of convection cause such transport (for example, convective clouds above a cold front [Pickering et al., 1988]). Pickering et al. (1990) argued that convective redistribution of ozone precursors may lead to an increase in the production rate of ozone averaged through the troposphere. Venting of NO_x from the boundary layer leads to lower concentrations, and the efficiency of ozone production per molecule of NO_x is higher for lower NO_x (Liu et al., 1987).

Occasionally, thunderstorms organize into systems several hundred kilometers across, called mesoscale convective systems (MCSs), that can last 6-12 hours or more. Lyons et al. (1986) provided a dramatic example of the effect of MCSs on the polluted boundary layer. They described a case when a massive complex of thunderstorms swept through the eastern U.S., which was under the influence of a stagnant high pressure system. The MCSs removed over half a million square kilometers of polluted boundary layer air and replaced it with cleaner middle tropospheric air, leading to significant decreases in ozone and sulfate concentrations and increases in visibility.

REGIONAL AND MESOSCALE PREDICTABILITY OF OZONE

Here we examine the predictability of ozone concentrations on the mesoscale (scales of a few tens of kilometers to a few hundred kilometers) and on the scale of major regions of the United States (i.e., the East Coast, or central U.S., or West Coast) from a meteorological perspective.

Major high-concentration episodes provide a good opportunity for predicting the transport and dispersion of ozone. That is because the greatest ozone concentrations occur when there are stagnant, high-pressure weather systems in which the larger-scale patterns of air flow vary slowly. Under these conditions, regional and mesoscale numerical prediction models can predict flow fields of air driven primarily by local circulations caused by features such as land-sea temperature differences, mountain topography, and differences in land use such as irrigated and nonirrigated land, or forested regions (Pielke, 1984). Likewise, during major episodes, the uncertainties of cloud transport are minimized because stagnant high-pressure regions are not favorable for deep convection; the main difficulty in prediction is in determining the times of the onset and termination of the stagnant high. The strength and position of a stagnant high-pressure system in the lower atmosphere are related to the presence of a ridge of high pressure in the middle and upper troposphere, which is reasonably well predicted by current global forecasting models for several days (van den Dool and Saha, 1990). Perhaps the greatest uncertainty in the prediction of ozone concentrations when there are strong stagnant high-pressure systems is whether a large mesoscale convective system will form on the periphery of the stagnant high and invade the interior of the system, sweeping large volumes of boundary layer ozone and other species into the middle and upper troposphere.

A much greater uncertainty in ozone predictability exists with weaker surface high- pressure systems. The strength, beginning, and end of ozone episodes associated with these systems are influenced by smaller-scale atmospheric disturbances, which are not as easy to predict with current global models. Moreover, because deep convection is more prevalent in weaker high-pressure systems, it is difficult to predict how much ozone will be removed from the atmospheric boundary layer by cumulonimbus transport. With the development of global and regional models that account more accurately for convection and surface processes and provide finer resolution, the ability to predict ozone concentrations during weaker high-pressure episodes will improve, but the ability to predict beyond about three days will probably remain marginal.

Another factor that limits the accuracy of model predictions is the transport and dispersion of pollutants from local plumes where high concentrations of pollutants can occur immediately downwind of emission sources. Because individual plumes are normally smaller than the grid size for regional and mesoscale models, these processes may have to be treated on the sub-grid scale to account for the different concentration regimes in the plumes (Sillman et al., 1990a).

GLOBAL AND LONG-TERM PREDICTABILITY OF OZONE

There is considerable interest in predicting the effects of global climate change and air quality legislation on concentrations of lower tropospheric ozone for the next several decades. What is our ability to predict the climate changes that will affect concentrations of ozone over one or more decades? As noted above, important to the ability to predict major episodes of high concentrations of ozone in a particular region is the ability to predict the global middle- and upper- atmospheric pressure ridge-trough pattern (Grotch, 1988). A persistent ridge in a particular region is favorable for establishing a major high-pressure system at the earth's surface, and if such a ridge occurs during early summer with low solar zenith angles and long days, a major high-ozone episode is likely to result. The problem is that the longwave ridge-trough pattern varies with the seasons and over periods of decades and longer in ways that cannot yet be predicted.

For example, general circulation models that simulate global greenhouse warming predict major shifts in the longwave ridge-trough pattern (Grotch, 1988). Despite the fact that higher average temperatures are associated with higher rates of ozone production, if the large-scale pressure pattern shifted such that the East Coast were preferentially under a trough in the early summer, that region would likely experience reduced concentrations of ozone. Unfortunately, although general circulation models predict a shift in the large-scale ridge-trough pattern associated with global warming, no two models predict the same shift in patterns, and none of the predictions of pattern shifts should be viewed with confidence.

Unfortunately, the preferred pressure ridge-trough pattern is unpredictable for periods beyond about 10 days. As a result, ozone concentrations cannot be predicted for longer periods.

OZONE IN THE EASTERN UNITED STATES

In the eastern United States, high concentrations of ozone in urban, suburban, and rural areas tend to occur concurrently on scales of over 1000 km. This blanket of ozone can persist for several days, and the concentrations can stay high (greater than 80 ppb) for several hours each day.

The major characteristics of episodes of high ozone concentration in the East were identified first during rural field studies sponsored by EPA from 1972 to 1975 (RTI, 1975; Decker et al., 1976). Ozone concentrations above 80 ppb were found for several consecutive days over areas larger than 100,000 km^2. The episodes generally were associated with slow-moving, high-pressure systems, when the weather was particularly favorable for photochemical formation of ozone; there were warm temperatures, clear skies, and light winds. The highest ozone concentrations often were found on the trailing side of the center of the high-pressure system. These early studies showed that average ozone concentrations at rural sites in the Midwest are higher than at urban sites, and they suggested a gradient in rural ozone values from west to east, with higher values in the eastern United States. Subsequent case studies documented the occurrence of high concentrations of ozone in the Midwest, Northeast, South, and on the Gulf Coast (Vukovich et al., 1977; Wolff et al., 1977; Spicer et al., 1979; Wolff et al., 1982; Altshuller, 1986). In one of the more dramatic cases, high ozone concentrations (>100 ppb) were found to extend from the Gulf Coast, throughout the Midwest, and up to New England (Wolff and Lioy, 1980). High concentrations are found throughout the atmospheric boundary layer when such conditions occur (e.g., Vukovich et al., 1985).

Some of the highest concentrations of ozone are found in plumes of pollutants downwind of urban and industrial areas. Studies that use surface and aircraft data have shown that the high concentrations are superimposed on elevated background concentrations during high-ozone episodes. The higher background concentrations are presumably due to enhanced photochemical production of natural and anthropogenic ozone in the warm, cloud-free conditions that characterize such episodes. The plumes may maintain their integrity for 12 hours, and they can cover an area larger than 150 by 50 km; the length of a plume is typically three times its width (White et al., 1976; Spicer et al., 1979, 1982; Sexton and Westberg, 1980; Clarke and Ching, 1983). Small cities (approximately 100,000 population) can generate 10-30 ppb ozone over background concentrations (Spicer et al., 1982; Sexton, 1983). Concentrations found in plumes from larger cities (St. Louis, Boston, Chicago, or Baltimore) are more typically elevated by 30-70 ppb over background (White et al., 1976; Sexton and Westberg, 1980; Spicer, 1982; Clark and Clarke, 1984; Altshuller, 1988). The most extreme cases, where concentrations are 60-150 ppb higher than background, are found over Connecticut, downwind of the New York-New Jersey industrial and metropolitan area (Rubino et al., 1976; Cleveland et al., 1977; Spicer et al., 1979). Plume studies were reviewed by EPA (1986a) and Altshuller (1986).

Analyses of data from rural sites in the United States have focused on the seasonal and diurnal behavior of ozone and the frequency distributions of its concentration (RTI, 1975; Decker et al., 1976; Singh et al., 1978; Evans et al., 1983; Pratt et al., 1983; Fehsenfeld et al., 1983; Evans, 1985; Logan, 1985, 1988, 1989; Lefohn and Mohnen, 1986; Lefohn and Pinkerton, 1988; Aneja et al., 1990). The behavior of ozone is illustrated here with data taken from two measurement programs, the Sulfate Regional Experiment (SURE) and its continuation as the Eastern Regional Air Quality Study (ERAQS), with nine sites (Mueller and Watson, 1982; Mueller and Hidy, 1983); from the National Air Pollution Background Network (NAPBN), with eight sites (Evans et al.,

1983; Evans, 1985), and from Whiteface Mountain, New York (Mohnen et al., 1977; Lefohn and Mohnen, 1986).

The annual cycle of monthly mean and monthly maximum values of ozone and the diurnal cycle in July are shown in Figure 4-1 for typical rural sites in the eastern and western United States. Cumulative probability distributions (providing percentile rank scores) for April 1 to Sept. 30 are shown in Figure 4-2 (Logan, 1988, 1989). Concentrations of ozone are highest in spring and summer, and average values are similar at all sites, about 30-50 ppb. Monthly maximum concentrations (the average of the daily maxima) are much higher at the SURE sites in the East, however, than at remote sites in the West, 60-85 ppb versus 45-60 ppb. The higher maxima are not reflected in the daily average values because the diurnal variation is much more pronounced at most of the eastern sites, with lower minima compensating for higher maxima. Values within a few ppb of the daily maximum persist for 7-10 h, from late morning until well into the evening at some sites. The cumulative probability distributions show that ozone almost never (probability <0.5%) exceeds 80 ppb at the three western sites, whereas concentrations above 80 ppb are quite common at the eastern sites. Ozone concentrations exceeded 80 ppb on 39% of days between May and August at the nine SURE sites and Whiteface Mountain in 1978, and on 26% of days in 1979. Concentrations occasionally exceed 120 ppb, the maximum allowed concentration set by the National Ambient Air Quality Standard (NAAQS) for ozone (Figure 4-3). Western sites affected by urban plumes also can show concentrations over 80 ppb (Fehsenfeld et al., 1983). The highest concentrations are observed at the central and eastern rural sites influenced by major urban and industrial sources of pollution (those in northern Indiana, Pennsylvania, Delaware, and Massachusetts in this case); high concentrations are less common at the more remote central and eastern sites (in Wisconsin, Louisiana, and Vermont).

Daily maximum concentrations of ozone for all of 1979 are shown in Figure 4-3 for four rural sites within 500 km of one another in the northeast. The high concentrations usually occur in periods a few days long, and high- (and low-) ozone days tend to occur concurrently. Ozone concentrations stay elevated for several hours each day during the high periods.

The data from the SURE/ERAQS program were used in an analysis of episodes of high concentrations for a region extending from Indiana east to Massachusetts, and south to Tennessee and North Carolina (Logan, 1989). Variations in ozone concentrations were highly correlated over distances of several hundred kilometers, and the highest concentrations tended to occur concurrently, or within 1-2 days of one another, at widely separated stations. There were 10 and 7 ozone pollution episodes of large spatial scale (>600,000 km^2) in 1978 and 1979, respectively, between the months of April and Septem-

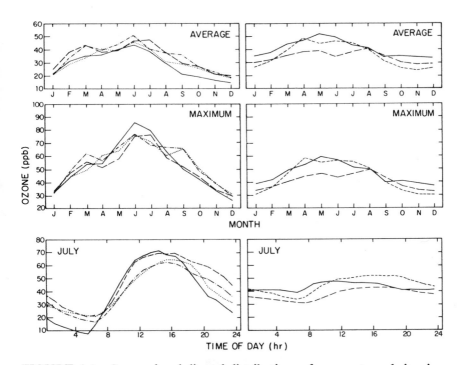

FIGURE 4-1   Seasonal and diurnal distributions of ozone at rural sites in
the United States. The upper panels show the seasonal distribution of
daily average values; the middle panels show monthly averages of the
daily maximum values. The lower panels show the diurnal behavior of
ozone in July. The left panels show results for four sites in the eastern
United States from Aug. 1, 1977 to Dec. 31, 1979: Montague, Massachusetts (solid); Scranton, Pennsylvania (dashed); Duncan Falls, Ohio
(dot dashed); and Rockport, Indiana (dot). The right panels show results for three sites in the western United States from four years of
measurements: Custer, Montana (1979-1982, site at 1250 meters, short
dashes); Ochoco, Oregon (1980-1983, 1350 meters, long dashes); and
Apache, Arizona (1980-1983, 2500 meters, solid).

Source: Logan, 1988.

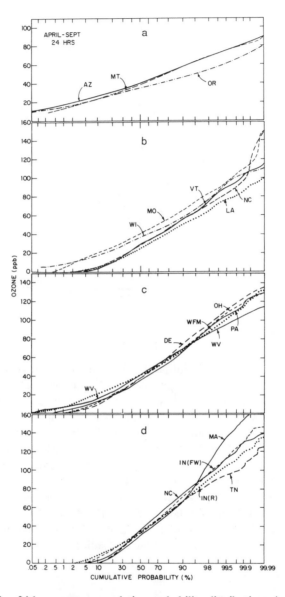

FIGURE 4-2   24-hour ozone cumulative probability distributions April 1-Sept. 30.
(a) Western NAPBN sites; (b) eastern NAPBN sites; (c) SURE sites; (d) Whiteface Mountain. Sites are identified by state; results are plotted on probability
paper; normally distributed data define a straight line. Source: Logan, 1989.

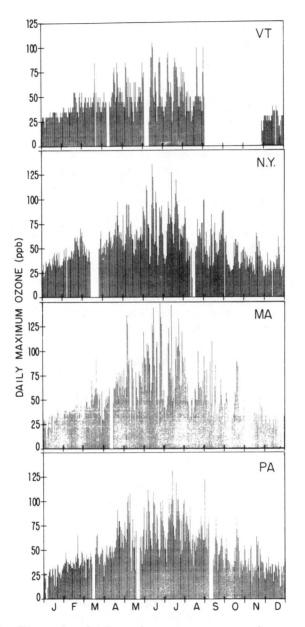

FIGURE 4-3   Time series of daily maximum ozone concentrations at rural sites in the northeastern United States in 1979. Source: Logan, 1989.

ber; they persisted for 3-4 days on average, with a range of 2-8 days, and were most common in June. Daily maximum ozone concentrations exceeded 90 ppb at more than half of the sites during these episodes and often were greater than 120 ppb at one or more sites. An analysis of the weather for each episode shows that high-ozone episodes were most likely in the presence of weak, slow-moving, persistent high-pressure systems as they migrated from west to east, or from northwest to southeast, across the eastern United States. Fast-moving and intense anticyclones (highs) were much less likely to promote the occurrence of ozone pollution episodes. The analysis of 2 complete years of data strengthens the conclusions of the case studies discussed earlier. There are no indications that either the weather or the ozone concentrations in 1978 and 1979 were particularly anomalous, although the concentrations could have been somewhat above average in 1978 (Logan, 1989).

The influence of the paths of anticyclones on the spatial pattern of ozone in the eastern two-thirds of the United States was examined by Vukovich and Fishman (1986). They showed maps of the mean diurnal maximum values of ozone for July and August of 1977-1981, using rural data where possible, and typical paths of anticyclones for each of these months. They concluded that if there is a persistent path for migratory high-pressure systems, the regions of high concentrations of ozone are associated with that pathway. An analysis of the climatology of anticyclones in July for 1950-1977 shows that the preferred track is across the northeast rather than across the southeast United States (Zishka and Smith, 1980). There was a downward trend in the number of anticyclones during this period.

The data discussed above show that there is a persistent blanket of high ozone in the eastern United States several times each summer, generally associated with stagnant high-pressure systems. Since rural ozone values commonly exceed 90 ppb on these occasions, an urban area need cause an ozone increment of only 30 ppb over the regional background to cause a violation of the NAAQS in a downwind area. Such increments have been demonstrated in the plume studies discussed earlier and in systematic studies of three urban areas (Kelly et al., 1986; Altshuller, 1988; Lindsay and Chameides, 1988).

Kelly et al. (1986) compared ozone concentration at a rural site outside Detroit with a site typically in the Detroit plume. For the upper quartile of ozone days in 1981, the ozone daily maximum was 104 ppb at the plume site, and the concentration at 1100 h was 47 ppb at the rural site. Kelly et al. argued that the plume generated 57 ppb ozone; the ozone maximum at the rural site was 73 ppb, 31 ppb below the plume's maximum value. On the day with the highest maximum ozone, 180 ppb, ozone concentrations were about 90 ppb at rural sites. Altshuller (1988) examined ozone formation in the St.

Louis plume using 2 years of surface data from 12 sites. He compared the maximum ozone concentration at the station nearest the plume center and the average of the maximum ozone at upwind stations to obtain the change in ozone concentration (ΔO_3). Monthly mean values of ΔO_3 were 26-53 ppb, with an overall average of 45 ppb; high concentrations were most common in July and August, and the 90th-percentile value of ΔO_3 was 80 ppb. In another study of the same data, Shreffler and Evans (1982) showed that upwind concentrations were 40-100 ppb and that ΔO_3 appeared to be independent of the upwind concentrations. Finally, Lindsay and Chameides (1988) compared maximum ozone concentrations from stations upwind and downwind of Atlanta and at a rural site 125 km away. On days when urban ozone concentrations exceeded 100 ppb, the ozone concentration was 80-85 ppb at the upwind station and 110-125 ppb at the downwind station, suggesting that the city contributed 30-40 ppb above the immediate background. On these days, the ozone concentration at the rural site was 65 ppb, 20 ppb higher than average.

A study of the meteorological conditions associated with high-ozone days (above 80 ppb) in 17 cities demonstrated the regional nature of the problem, at least in the Northeast. Samson and Shi (1988) examined the wind flow for all days in 1983-85 when ozone exceeded 80 ppb in these cities, using trajectory calculations integrated backwards to the source region. They found that days with concentrations above 120 ppb were generally associated with low wind speeds, with the exception of Portland, Maine, where high-ozone days were moderately windy, presumably due to long-range transport of ozone from the south and west. The median distance the air had traveled in the previous 24 hours was about 500 km for the northeastern cities, suggesting long-range transport, but only 250 km for the southern cities. High-ozone days tended to occur over a longer season for the southern cities than for the northeastern cities (Figure 4-4).

SUMMARY

Weather patterns play a major role in establishing conditions conducive to ozone formation and accumulation and in terminating episodes of high ozone concentrations. High ozone episodes are typically associated with weak, slow-moving high pressure systems traversing the central and eastern United States from west to east or from northwest to southeast. These episodes usually end with a frontal passage that brings cooler, cleaner air to the region. Clouds play an important role in the vertical redistribution of ozone and its precursors.

High ozone episodes last from 3-4 days on average, occur as many as 7-10

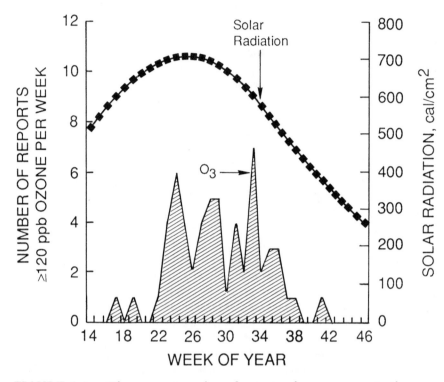

FIGURE 4-4a   The average number of reports of ozone concentrations
≥120 ppb at the combined cities of New York and Boston from 1983 to
1985 (1 April = week 14, 1 May = week 18, 1 June = week 22, 1 July =
week 27, 1 August = week 31, 1 September = week 35, 1 October =
week 40, 1 November = week 44). A representation of the annual variation in solar radiation reaching the earth's surface at 40°N latitude
(units, calories/cm^2) is shown. Average over 1983-1985.

times a year, and are of large spatial scale: >600,000 km^2. Maximum values
of non-urban ozone commonly exceed 90 ppb during these episodes, compared with average daily maximum values of 60 ppb in summer. An urban
area need contribute an increment of only 30 ppb over the regional background during a high ozone episode to cause a violation of the National Ambient Air Quality Standard (NAAQS) in a downwind area. Such increments
have been demonstrated in the studies described in this chapter. Given the
regional nature of the ozone problem in the eastern United States, a regional
model is needed to develop control strategies for individual urban areas. This

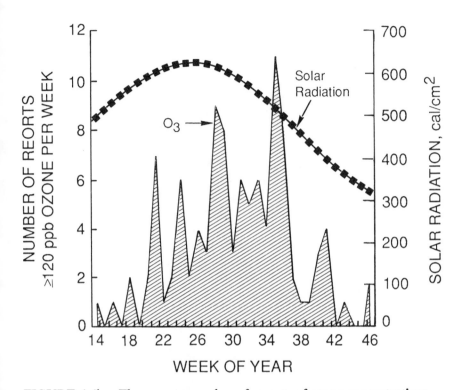

FIGURE 4-4b   The average number of reports of ozone concentrations ≥ 120 ppb at the combined cities of Dallas and Houston, from 1983 to 1985. (1 April = week 14, 1 May = week 18, 1 June = week 22, 1 July = week 27, 1 August = week 31, 1 September = week 35, 1 October = week 40, 1 November = week 44). A representation of the annual variation in solar radiation reaching the earth's surface at 30°N latitude (units, calories/cm^2) is shown. Source: Samson and Shi, 1988.

need was recognized by EPA and led to the development of the Regional Oxidant Model, discussed in Chapters 10 and 11.

Regional models for ozone require a meteorological component that realistically describes the atmospheric wind field and its turbulence and mixing characteristics. Such a description is generally provided by prognostic meteorological models, as discussed in Chapter 10.

5

Atmospheric Chemistry of Ozone and Its Precursors

INTRODUCTION

Large quantities of chemical compounds are emitted into the atmosphere as a result of anthropogenic and biogenic activities. These emissions lead to a complex spectrum of chemical and physical processes that result in such diverse effects as photochemical air pollution (including the formation of ozone in urban, suburban, and rural air masses), acid deposition, long-range transport of chemicals, stratospheric ozone depletion, and accumulation of greenhouse gases. Over the past 15 to 20 years, many laboratory and ambient atmospheric studies have investigated the physical and chemical processes of the atmosphere. Because these processes are complex, computer models are often used to elucidate and predict the effects of anthropogenic and biogenic emissions—and of changes in these emissions—on the chemistry of the atmosphere.

In this chapter, the gas-phase chemistry of the relatively unpolluted, methane-dominated troposphere is summarized, and the additional complexities of the chemistry of polluted atmospheres are discussed. The tropospheric chemistry of organic compounds of anthropogenic and biogenic origin, respectively, is discussed in detail, and the calculated tropospheric lifetimes of these compounds are presented. The formulation and testing of chemical mechanisms for use in urban and regional airshed computer models are discussed briefly, and the reactivities of organic compounds with respect to ozone formation, as calculated using these models, are discussed.

GENERAL SCHEMES OF TROPOSPHERIC CHEMISTRY

Ozone is present in the natural, unpolluted troposphere, and its tropospheric column density is approximately 10% of the total atmospheric (troposphere + stratosphere) ozone column density (Logan, 1985; Brühl and Crutzen, 1989; Fishman et al., 1990). The ozone present in the stratosphere absorbs short-wavelength radiation (≤ 290 nm (nanometers or 10^{-9} meters)) from the sun and allows only those wavelengths ≥ 290 nm to penetrate into the troposphere (Peterson, 1976; Demerjian et al., 1980). The sources of ozone in the natural troposphere are downward transport from the stratosphere and in situ photochemical production. Losses result from photochemical processes and from deposition and destruction at the earth's surface. The rates of downward transport, production, and losses are estimated to be of the same order of magnitude (Logan, 1985). The ozone present in the troposphere is important in the atmospheric chemistry because the OH radical is generated from the photolysis of ozone at wavelengths <319 nm (Levy, 1971; DeMore et al., 1990). The formation of OH radicals leads to cycles of reactions that result in the photochemical degradation of organic compounds of anthropogenic and biogenic origin, the enhanced formation of ozone, and the atmospheric formation of acidic compounds (see, for example, Heicklen et al., 1969; Stedman et al., 1970; Finlayson-Pitts and Pitts, 1986; WMO, 1986). The generation of the OH radical from ozone is shown in the following reactions:

$$O_3 + h\nu \rightarrow O_2 + O(^1D) \tag{5.1}$$

$$O(^1D) + M \rightarrow O(^3P) + M \tag{5.2}$$

$$O(^1D) + H_2O \rightarrow 2OH \tag{5.3}$$

Energy from solar radiation is represented by $h\nu$, the product of Planck's constant, h, and the frequency, ν, of the electromagnetic wave of solar radiation. $O(^1D)$ is an excited oxygen atom, and M is an inert compound, such as N_2 or O_2. $O(^3P)$ is a ground state oxygen atom. The chemistry of the clean, unpolluted troposphere is dominated by the chemistry of methane (CH_4) and its degradation products, formaldehyde (HCHO) and carbon monoxide (CO) (see, for example, Levy, 1972; Crutzen, 1973; Fishman and Crutzen, 1977; Logan et al., 1981).

Tropospheric Methane Oxidation Cycle

A sequence of reactions (Ravishankara, 1988; Atkinson et al., 1989a; Atkinson, 1990b) starts with the reaction of the OH radical with methane

$$OH + CH_4 \rightarrow H_2O + \overset{\bullet}{C}H_3 \tag{5.4}$$

The tropospheric lifetime of methane, τ_{CH_4}, is controlled by reaction with the OH radical,

$$\tau_{CH_4} = (k_{OH}^{CH_4}[OH])^{-1} \tag{5.5}$$

where $k_{OH}^{CH_4}$ is the rate constant for the reaction of the OH radical with methane and [OH] is the OH radical concentration. It should be noted that $k_{OH}^{CH_4}$ depends on temperature, and hence on altitude, and that the OH radical concentration is temporally and spatially dependent. The lifetime of methane in the troposphere is long enough that a diurnally and annually averaged concentration of global tropospheric OH radical can be used to calculate the lifetime of methane (and of other similarly long-lived trace species). Based on methylchloroform (CH_3CCl_3) emissions and atmospheric budgets and an equation analogous to Equation 5.5, Prinn et al. (1987) derived a tropospheric lifetime for methylchloroform of 6.3 years and a globally averaged tropospheric OH radical concentration of 7.7×10^5 molecule/cm^3. From this OH radical concentration, the methane lifetime is calculated to be approximately 12 years (Vaghjiani and Ravishankara, 1991). For organic compounds that react more rapidly with the OH radical and have much shorter lifetimes (≤ 1 year), the temporal and spatial variations of the OH radical concentrations need to be considered in the calculation of tropospheric lifetimes.

Under tropospheric conditions, the methyl radical rapidly, and solely, adds oxygen to form the methyl peroxy radical (CH_3O_2):

$$\overset{\cdot}{C}H_3 + O_2 \overset{M}{\rightarrow} CH_3O_2\cdot \qquad (5.6)$$

which can then react with nitric oxide (NO), nitrogen dioxide (NO_2), hydroperoxyl radical (HO_2), and organic peroxy radicals (RO_2)

$$CH_3O_2\cdot + NO \rightarrow CH_3O\cdot + NO_2 \qquad (5.7)$$

$$CH_3O_2\cdot + NO_2 \overset{M}{\rightleftharpoons} CH_3OONO_2 \qquad (5.8)$$

$$CH_3O_2\cdot + HO_2 \rightarrow CH_3OOH + O_2 \qquad (5.9)$$

$$CH_3O_2\cdot + RO_2\cdot \rightarrow \text{products} \qquad (5.10)$$
$$\text{(including} CH_3O\cdot, \text{ HCHO, and } CH_3OH)$$

Methyl peroxynitrate, CH_3OONO_2, thermally dissociates back to the reactants with a lifetime of methyl peroxynitrate with respect to thermal decomposition of ~ 1 sec at room temperature and atmospheric pressure, which increases to ~ 2 days for the temperature and pressure conditions in the upper troposphere (Atkinson et al., 1989a; Atkinson, 1990b). Because the reactions of the CH_3O_2 radical with NO and NO_2 have comparable rate constants for the temperatures and pressures encountered in the troposphere (Atkinson, 1990a), methyl peroxynitrate can act as a temporary reservoir of NO_2 and CH_3O_2 radicals in the upper troposphere.

The reaction of the methylperoxy radical with NO will dominate over reaction with the HO_2 radical for tropospheric NO mixing ratios equal to or greater than approximately 10-30 parts per trillion (ppt) (Logan et al., 1981). However, in the clean, unpolluted lower troposphere, NO mixing ratios are generally <30 ppt (see, for example, Kley et al., 1981; Logan, 1983; Ridley et al., 1987, 1989; Drummond et al., 1988; and Chapter 8), and under these conditions the HO_2 radical reaction to form methyl hydroperoxide, CH_3OOH, is important.

The subsequent reactions of CH_3OOH under tropospheric conditions are photolysis and reaction with the OH radical (Ravishankara, 1988; Atkinson, 1989, 1990a,b)

$$CH_3OOH + h\nu \rightarrow CH_3O\cdot + OH \qquad (5.11)$$

$$OH + CH_3OOH \begin{cases} \rightarrow H_2O + CH_3O_2\cdot & (67\% \text{ at } 298k) \\ \rightarrow H_2O + \underset{\downarrow}{CH_2OOH} & (33\% \text{ at } 298k) \\ \qquad\quad HCHO + OH \end{cases} \qquad (5.12)$$

These two processes are comparable in importance, and they reform the CH_3O and CH_3O_2 radicals. Wet deposition of methyl hydroperoxide and its incorporation into cloud, fog, and rain water also could be important (Hellpointner and Gäb, 1989). For a discussion of cloud chemistry, see, for example, Chameides (1984), Jacob (1986), Jacob et al. (1989), and Pandis and Seinfeld (1989).

The sole loss process for the methoxy radical in the clean troposphere is through reaction with oxygen to generate formaldehyde (Atkinson et al., 1989a; DeMore et al., 1990)

$$CH_3O\cdot + O_2 \rightarrow HCHO + HO_2 \qquad (5.13)$$

The HO_2 radical can lead to the regeneration of the chain-carrying OH radical by reaction with NO

or react with peroxy (RO_2) radicals (including HO_2) or ozone

$$HO_2 + NO \rightarrow OH + NO_2 \qquad (5.14)$$

$$HO_2 + RO_2\cdot \rightarrow ROOH + O_2 \qquad (5.15)$$

$$HO_2 + O_3 \rightarrow OH + 2O_2 \qquad (5.16)$$

The self-reaction of HO_2 radicals forms hydrogen peroxide, which, like methyl hydroperoxide, can undergo wet deposition and incorporation into cloud, fog, and rain water.

$$HO_2 + HO_2 \rightarrow H_2O_2 + O_2 \qquad (5.17)$$

When enough NO is present that the reactions of CH_3O_2 and HO_2 radicals with NO dominate over the reactions of these peroxy radicals with HO_2 (or other peroxy radicals) or of HO_2 radicals with ozone, then the overall methane photooxidation reaction is given by

$$(OH +) CH_4 + 2NO + 2O_2 = $$
$$H_2O + HCHO + 2NO_2 (+ OH) \qquad (5.18)$$

with methane being degraded to formaldehyde, two molecules of NO being converted to NO_2, and the OH radical being regenerated. (An equal sign is used instead of an arrow to indicate that the net overall process shown as Reaction 5.18 represents many individual reactions.) When the reaction of the CH_3O_2 radical with HO_2 dominates, then the oxidation of methane becomes a net sink for OH and HO_2 radicals, with approximately

$$CH_4 + 0.75 \ OH + 0.75 \ HO_2 + 0.25 \ O_2 = $$
$$HCHO + 1.75 \ H_2O \qquad (5.19)$$

The NO_x concentrations in the atmospheric boundary layer over continental areas in the northern hemisphere are generally high enough that the reactions of RO_2 and HO_2 peroxy radicals with NO dominate over the reactions of the RO_2 and HO_2 radicals with HO_2 and the reaction of the HO_2 radical with ozone. The result is net ozone formation. Only in remote locations such as the mid-Pacific Ocean and portions of the southern hemisphere are the NO_x concentrations low enough that the reactions of HO_2 radicals with ozone (Reaction 5.16) and other peroxy (RO_2) radicals dominate, leading to net ozone removal.

Formaldehyde also undergoes reaction in the troposphere by photolysis and reaction with the OH radical (Atkinson et al., 1989a; Atkinson, 1990a)

$$HCHO + h\nu \quad \Big[\begin{array}{l} \rightarrow HCO + H \\ \rightarrow H_2 + CO \end{array} \tag{5.20}$$

$$OH + HCHO \rightarrow H_2O + HCO \tag{5.21}$$

followed by

$$H + O_2 + M \rightarrow HO_2 + M \tag{5.22}$$

$$HCO + O_2 \rightarrow HO_2 + CO \tag{5.23}$$

and

$$OH + CO \rightarrow H + CO_2 \tag{5.24}$$

For HCHO, photolysis dominates over reaction with the OH radical (Atkinson, 1988), and the calculated lower tropospheric lifetime of HCHO due to photolysis and, to a lesser extent, reaction with the OH radical is ~4 hours at the sun's zenith angle of 0° (Rogers, 1990). The tropospheric removal of CO is by reaction with the OH radical, with a calculated lower tropospheric lifetime of ~2 months. The tropospheric lifetimes of HCHO and CO are thus both much shorter than that of methane.

Photochemical Formation of Ozone

In the troposphere, ozone formation occurs to any significant extent only from the photolysis of NO_2 at wavelengths <424 nm, when sufficient solar energy is absorbed by NO_2 to cause it to photodissociate

$$NO_2 + h\nu \rightarrow NO + O(^3P) \tag{5.25}$$

$$O(^3P) + O_2 + M \rightarrow O_3 + M \ (M - air) \tag{5.26}$$

$$NO + O_3 \rightarrow NO_2 + O_2 \tag{5.27}$$

In the absence of other processes that convert NO to NO_2, and assuming steady-state conditions, then

$$[O_3] - j_1[NO_2]/k_2[NO] \tag{5.28}$$

and the ozone concentration is linked to the NO_2/NO concentration ratio during daylight hours. (Here j_1 is the diurnally, seasonally, and latitudinally dependent rate of photolysis of NO_2, and k_2 is the rate constant for Reaction 5.27.) For an NO_2/NO concentration ratio of one, a reasonable mid-day value in the clean lower troposphere, and a temperature of 298 K, the resulting ozone concentration is $\sim 5 \times 10^{11}$ molecule/cm^3 (20 parts per billion (ppb) mixing ratio).

As discussed above for the methane oxidation cycle, the presence of volatile organic compounds (VOCs) causes enhanced NO-to-NO_2 conversion and hence the production of concentrations of ozone that exceed those encountered in the clean background troposphere (see, for example, Parrish et al., 1986). This is discussed further below. For example, for the OH radical-initiated reaction of methane in the presence of NO given above, the overall reaction is

$$(OH +) \ CH_4 + 2NO + 2O_2 - $$
$$H_2O + HCHO + 2NO_2(+ \ OH) \qquad (5.29)$$

This leads to a net reaction of

$$CH_4 + 4O_2 - H_2O + HCHO + 2O_3 \qquad (5.30)$$

Other Reactions in the Tropospheric Nitrogen Cycle

In addition to Reactions 5.25 and 5.27 and the reaction of the HO_2 radical with NO to regenerate the OH radical,

$$HO_2 + NO \rightarrow OH + NO_2 \qquad (5.31)$$

other tropospherically important reactions involve oxides of nitrogen (Finlayson-Pitts and Pitts, 1986; WMO, 1986; Atkinson et al., 1989a; DeMore et al., 1990). The recombination reactions

$$OH + NO \xrightarrow{M} HONO \qquad (5.32)$$

$$HO_2 + NO_2 \xrightarrow{M} HOONO_2 \qquad (5.33)$$

to form nitrous acid (HONO) and pernitric acid ($HOONO_2$) are of little importance because of the rapid photodissociation of HONO

$$HONO + h\nu \rightarrow OH + NO \qquad\qquad (5.34)$$

and the thermal decomposition of $HOONO_2$ back to reactants. However, the combination reaction of the OH radical with NO_2

$$OH + NO_2 \xrightarrow{\text{M}} HONO_2 \qquad\qquad (5.35)$$

is the major gas-phase route to the formation of nitric acid (HNO_3), and it is the major homogeneous gas-phase sink for NO_x (oxides of nitrogen) in the troposphere. This reaction also serves as a sink for OH and HO_2 radicals (odd hydrogen) for NO_x mixing ratios ≥ 0.1 ppb, and under these conditions the removal of OH radicals by Reaction 5.35 balances the formation of HO_x (oxides of hydrogen) radicals from the photolysis of ozone and HCHO.

The major reactions involved in the oxidation of methane in the presence of NO_x are diagrammed in Figure 5-1, which emphasizes the chain-cycle nature of this overall reaction process.

Ozone also reacts with NO_2 to form the nitrate (NO_3) radical,

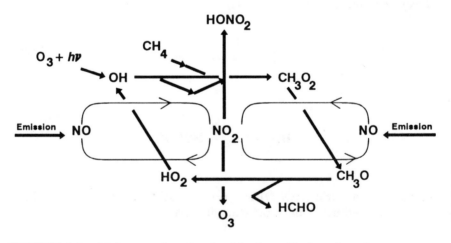

FIGURE 5-1   Major reactions involved in the oxidation of methane (CH_4) in the presence of NO_x.

$$NO_2 + O_3 \rightarrow NO_3 + O_2 \qquad (5.36)$$

and the NO_3 radical is interconverted with NO_2 and dinitrogen pentoxide (N_2O_5) through the reactions

$$NO_2 + NO_3 \overset{M}{\rightleftharpoons} N_2O_5 \qquad (5.37)$$

Because NO_3 radicals rapidly photolyze (with a photolysis lifetime of ~ 5 seconds at a solar zenith angle of $0°$)

$$NO_3 + h\nu \quad \begin{cases} \rightarrow NO + O_2 \\ \rightarrow NO_2 + O(^3P) \end{cases} \qquad (5.38)$$

and react rapidly with NO,

$$NO + NO_3 \rightarrow 2NO_2 \qquad (5.39)$$

concentrations of the NO_3 radical, and hence of N_2O_5, remain low during the daytime but can increase during evening and nighttime hours (Platt et al., 1981, 1984; Pitts et al., 1984a).

The homogeneous gas-phase reaction of N_2O_5 with water vapor to form nitric acid

$$N_2O_5 + H_2O \rightarrow 2HONO_2 \qquad (5.40)$$

is slow enough that only an upper limit can be placed on the rate constant (Atkinson et al., 1989a; Hatakeyama and Leu, 1989), but the wet and dry deposition of N_2O_5 or of NO_3 radicals provides a potentially important nighttime route to the removal of gas-phase NO_x and the formation of acid deposition (see, for example, Heikes and Thompson, 1983; Chameides, 1986; Mozur-kewich and Calvert, 1988).

$$N_2O_5 \text{ or } NO_3 \xrightarrow{\text{deposition}} HONO_2 \qquad (5.41)$$

The formation of aqueous-phase nitric acid subsequent to wet deposition of the NO_3 radical is expected to proceed via the intermediate formation of the nitrate (NO_3^-) ion (Chameides, 1986). In addition to this nighttime heterogeneous (involvement of at least two physical phases) removal process for NO_x through the intermediary of NO_3 radicals and N_2O_5, heterogeneous chemistry, including cloud chemistry, could be important in the chemical processes that occur in the troposphere (see, for example, Chameides, 1984; Jacob, 1986; Jacob et al., 1989; Pandis et al., 1989; Lelieveld and Crutzen, 1990). For example, Lelieveld and Crutzen (1990) have postulated that in the presence of clouds the formation of ozone in the troposphere is significantly diminished by the scavenging of HO_2 radicals and HCHO from the gas phase into cloud water. Clearly, further work is necessary to elucidate the role of heterogeneous reactions and aqueous-phase reactions in the chemistry of the troposphere and in the formation and destruction of ozone.

As noted above, nitrous acid (HONO) photolyzes to generate the OH radical

$$HONO + h\nu \rightarrow OH + NO \qquad (5.42)$$

and this photolysis reaction is rapid ($\sim 10^{-3}$ s^{-1} at a $0°$ zenith angle of the sun). In urban areas, HONO is formed at night, probably by the heterogeneous hydrolysis of NO_2 (Sakamaki et al., 1983; Pitts et al., 1984b; Akimoto et al., 1987; Svensson et al., 1987; Jenkin et al., 1988; Lammel and Perner, 1988).

$$2NO_2 + H_2O \rightarrow HONO (+ HONO_2) \qquad (5.43)$$

Under laboratory conditions, this heterogeneous formation of HONO is first-order in the NO_2 concentration. Because comparable amounts of nitric acid are not seen in the gas phase, nitric acid is thought to remain on the reaction vessel surfaces. Direct emission of HONO from combustion sources (Pitts et al., 1984c, 1989) also could contribute to the presence of HONO in a polluted atmosphere. The build-up of HONO at night can lead to substantial predawn concentrations of HONO—up to ~10 ppb (Harris et al., 1982; Winer et al., 1987; Rodgers and Davis, 1989). The rapid photolysis of HONO in the early morning can then lead to a pulse of OH radicals and to rapid initiation of photochemical activity (Harris et al., 1982; Lurmann et al., 1986a).

Chemistry of the Polluted Troposphere

In the lower troposphere, and especially in polluted urban areas, the chemical reactions of biogenic and anthropogenic VOC and anthropogenic NO_x emissions dominate over those of methane and its degradation products (Logan et al., 1981; Brewer et al., 1983; Finlayson-Pitts and Pitts, 1986; Seinfeld, 1989). Although in principle an extension of the chemistry of the clean, methane-dominated troposphere, the chemistry of the polluted troposphere, including urban and rural air masses, is significantly more complicated because of the presence of many VOCs of various classes (alkanes, alkenes, and aromatic hydrocarbons) and the added complexities in the chemistry of these organic species (see, for example, Atkinson, 1990a).

In the troposphere, VOCs undergo photolysis and reaction with OH and NO_3 radicals and ozone (and, for some aldehydes, also with HO_2 radicals) (Finlayson-Pitts and Pitts, 1986; Atkinson, 1988, 1990a). As with methane (Equation 5.5), the lifetime, τ, of a chemical with respect to reaction with a species X is given by

$$\tau - (k_x[X])^{-1} \qquad (5.44)$$

and depends on the rate constant k_x for reaction with X and the ambient tropospheric concentration of X ([X]). The OH and NO_3 radical and ozone concentrations vary temporally and spatially, and hence the "instantaneous" lifetime τ and the loss rate, τ^{-1}, of a chemical also vary with space and time. The variations in the OH radical, NO_3 radical, or ozone concentrations at any given time and place translate directly into variations in the instantaneous loss rate and lifetime of a chemical.

Direct ambient measurements of the OH radical in the lower troposphere (Hübler et al., 1984; Perner et al., 1987; Platt et al., 1988) give concentrations that range from $<5 \times 10^5$ to 9×10^6 molecule/cm^3, and these data are reasonably consistent with indirect measurements of OH radical concentrations (Roberts et al., 1984; Ayers and Gillett, 1988; Arey et al., 1989a) and with the diurnally and annually averaged concentration of global tropospheric OH radical (Prinn et al., 1987). Although at a given time and place, it might be possible to specify the concentrations of OH radical and ozone reasonably well, this is not the case for the NO$_3$ radical. Nighttime maximum NO$_3$ radical mixing ratios measured in the lower troposphere over continental areas range from <2 ppt to 430 ppt; the mixing ratio in marine air masses has been measured to be <0.5 ppt (Atkinson et al., 1986, and references therein). Nighttime concentrations of NO$_3$ in the troposphere are uncertain to at least an order of magnitude. Furthermore, as discussed by Winer et al. (1984), reaction with the NO$_3$ radical can be a removal process for the reacting organic compound or NO$_x$, depending on the relative strengths of the emission rates or the formation rates of the VOCs and NO$_3$ radicals.

In the remainder of this chapter, lifetimes are calculated assuming specified ambient concentrations of OH and NO$_3$ radicals and ozone. Table 5-1 gives the calculated tropospheric lifetimes of selected organic compounds from anthropogenic and biogenic sources with respect to the reactions that degrade them.

TABLE 5-1   Calculated Tropospheric Lifetimes of Selected VOCs Due to Photolysis and Reaction with OH and NO$_3$ Radicals and Ozone

| VOC | Lifetime due to reaction with | | | |
	OH	NO$_3$	O$_3$	hν
Methane	~12 years$_b$	>120 years	>4,500 years	
Ethane	60 days	>12 years	>4,500 years	
Propane	13 days	>2.5 years	>4,500 years	
n-Butane	6.1 days	~2.5 years	>4,500 years	
n-Octane	1.8 days	260 days	>4,500 years	
Ethene	1.8 days	225 days	9.7 days	

VOC	Lifetime due to reaction with			
	OH	NO_3	O_3	hν
Propene	7.0 hours	4.9 days	1.5 days	
Isoprene	1.8 hours	50 min	1.2 days	
α-Pinene	3.4 hours	5 min	1.0 days	
Acetylene	19 days	$\geq$2.5 years	5.8 years	
Formaldehyde	1.6 days	77 days	>4.5 years	
Acetaldehyde	1.0 days	17 days	>4.5 years	4 hours
Acetone	68 days	c	>4.5 years	15 days
Methyl ethyl ketone	13.4 days	c	>4.5 years	
Methylglyoxal	10.8 hours	c	>4.5 years	2 hours
Methanol	17 days	>77 days	c	
Ethanol	4.7 days	>51 days	c	
Methyl t-butyl ether	5.5 days	c	c	
Benzene	12.5 days	>6 years	>4.5 years	
Toulene	2.6 days	1.9 years	>4.5 years	
m-Xylene	7.8 hours	200 days	>4.5 years	

[a]OH, 12-hour average concentration of 1.5×10^6 molecule/cm^3 (0.06 ppt) (Prinn et al., 1987); NO_3 12-hour average concentration of 5×10^8 molecule/cm^3 (20 ppt) (Atkinson, 1991); O_3 24-hour average concentration of 7×10^{11} molecule/cm^3 (28 ppb) (Logan, 1985). Calculated from room temperature rate data, except for methane, of Atlkinson (1988, 1990a, 1991), Plum et al. (1983), and Rogers (1990).

[b]From Vaghjiani and Ravishankara (1991).

[c]Expected to be of negligible importance.

All of the tropospheric processes represented in Table 5-1 lead to the formation of organic peroxy radicals (RO_2). For example, for the reactions of OH and NO_3 radicals with alkanes (Reactions 5.45 and 5.46) where RH represents an alkane, and with alkenes (Reactions 5.47 and 5.48) where $>C=C<$ represents an alkene

$$\left.\begin{array}{c} OH \\ NO_3 \end{array}\right\} + RH \rightarrow R\cdot + \left\{\begin{array}{l} H_2O \\ HONO_2 \end{array}\right. \tag{5.45}$$

$$R\cdot + O_2 \xrightarrow{O_2} RO_2\cdot \tag{5.46}$$

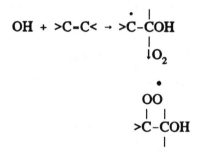

$$\tag{5.47}$$

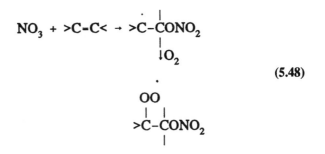

$$\text{(5.48)}$$

As with the CH_3O_2 radical formed from methane, these more complex RO_2 radicals react with NO, NO_2, and HO_2 radicals. The difference is that for the RO_2 radicals with more than two carbon atoms, the reaction with NO also can lead to the formation of organic nitrates,

$$RO_2{\cdot} + NO \longrightarrow \begin{cases} \rightarrow RO{\cdot} + NO_2 \\ \rightarrow RONO_2 \end{cases} \qquad \text{(5.49)}$$

with this organic nitrate formation increasing with increasing pressure, decreasing temperature, and (for the n-alkane series) the carbon number of the alkane (see, for example, Harris and Kerr, 1989; Carter and Atkinson, 1989a). At 298 K and atmospheric pressure the alkyl nitrate yields from the OH radical-initiated reactions of the n-alkanes increase from ~4% for propane to ~33% for n-octane (Carter and Atkinson, 1989a).

The alkoxy or substituted alkoxy (RO_2) radicals can react with O_2 (as for the CH_3O radical formed from methane); they can undergo unimolecular decomposition; or, for the alkoxy radicals with four or more carbon atoms, they can isomerize (Atkinson, 1990a). For example, neglecting the combination reactions with NO and NO_2, which are generally of negligible importance under tropospheric conditions (Atkinson, 1990a), the following reactions are possible for the 2-pentoxy radical formed from n-pentane,

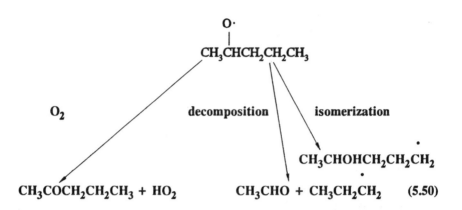

$$\text{CH}_3\text{COCH}_2\text{CH}_2\text{CH}_3 + \text{HO}_2 \qquad \text{CH}_3\text{CHO} + \text{CH}_3\text{CH}_2\text{CH}_2 \qquad (5.50)$$

The alkyl radicals formed (C_3H_7 and $CH_3CHOHCH_2CH_2CH_2$, in this case) then react further.

The reaction mechanisms of the aromatic hydrocarbons are not well understood (Atkinson et al., 1989b; Atkinson, 1990a). In the troposphere, benzene and the alkyl-substituted benzenes react only with the OH radical (Atkinson 1988, 1990a), and the kinetics and initial reaction mechanisms of these OH radical reactions are well understood (Atkinson, 1989). The major pathway of the OH radical reaction involves initial OH radical addition to the aromatic ring to yield a hydroxycyclohexadienyl-type radical (Atkinson, 1989). The subsequent reactions of these hydroxycyclohexadienyl radicals under tropospheric conditions are not well understood (see, for example, Atkinson et al., 1989b). Laboratory studies show that the hydroxycyclohexadienyl radicals react rapidly with NO_2 (Zellner et al., 1985; Knispel et al., 1990; Zetzsch et al., 1990) and that the reactions of these radicals with oxygen are slow (Knispel et al., 1990). At present, the relative importance of the reactions of the hydroxycyclohexadienyl radicals with oxygen and NO_2 under ambient tropospheric conditions is not totally clear.

The degradation reactions for all classes of VOCs, in addition to the conversion of NO to NO_2 and the formation of ozone, lead to the formation of carbonyl compounds (aldehydes, ketones, hydroxycarbonyls, and dicarbonyls), organic acids, organic nitrates (including peroxyacyl nitrates), and the inorganic acids, $HONO_2$ and (in the presence of SO_2) H_2SO_4. In most cases these first-generation products undergo further tropospheric degradation reactions leading to a further spectrum of organic products, NO-to-NO_2 conversion, and ozone formation. Because the carbonyl compounds are the major first-generation products, their subsequent reactions are important.

The simplest aldehyde, formaldehyde (HCHO, the tropospheric reactions of which have been presented above regarding the tropospheric methane oxidation cycle), has chemistry that is somewhat different from the higher aldehydes, such as acetaldehyde (Atkinson, 1990b). In the troposphere the photolysis of HCHO is calculated to be more important than reaction with the OH radical, in contrast to the higher aldehydes for which the OH radical reactions are more important than photolysis (Atkinson, 1990a). Furthermore, the HCO radical formed from the photolysis and OH radical reaction of HCHO reacts with oxygen to form the HO_2 radical and CO,

$$HCO + O_2 \rightarrow HO_2 + CO \qquad (5.51)$$

whereas the acyl (RCO) radicals formed from the higher aldehydes

$$OH + RCHO \rightarrow H_2O + R\overset{\bullet}{C}O \qquad (5.52)$$

react with O_2 by addition to form the corresponding acylperoxy (RC(O)OO) radicals

$$R\overset{\bullet}{C}O + O_2 \overset{M}{\rightarrow} RC(O)OO\cdot \qquad (5.53)$$

These acylperoxy radicals react with NO, NO_2, or HO_2 radicals

$$RC(O)OO + NO \rightarrow RC(O)O\cdot + NO_2$$
$$\qquad\qquad\qquad \rightarrow R\cdot + CO_2 \qquad (5.54)$$

$$RC(O)OO \cdot + NO_2 \overset{M}{\rightleftharpoons} RC(O)OONO_2 \qquad (5.55)$$

$$RC(O)OO \cdot + HO_2 \longrightarrow \begin{cases} RC(O)OOH + O_2 & (a) \\ RCOOH + O_3 & (b) \end{cases} \qquad (5.56)$$

Reaction 5.55 leads to the formation of peroxyacyl nitrates, the simplest member of which is peroxyacetyl nitrate (PAN, $CH_3C(O)OONO_2$), which thermally decomposes back to the reactants with a lifetime of ~ 30 min at 298 K and atmospheric pressure (Atkinson et al., 1989a). PAN and certain of its homologues, such as peroxypropionyl nitrate (PPN) and peroxybenzoyl nitrate (PBzN), have been observed in ambient air (Roberts, 1990, and references therein).

Until recently, the temperature dependence of the ratio of the rate constants for Reactions 5.54 and 5.55 was not well known (Atkinson et al., 1989a). That uncertainty led to different temperature dependencies assumed for those reactions for $R = CH_3$ in the chemical mechanisms developed for use in airshed computer models (Carter et al., 1986a; Gery et al., 1988a, 1989). The output of such models led to widely differing predictions for ozone (and PAN) formation at temperatures below 298 K (Dodge, 1989). Experimental data of Kirchner et al. (1990) and Tuazon et al. (1991) show that the ratio of rate constants for Reactions 5.54 and 5.55 ($R = CH_3$) is 2.2, independent of temperature over the range ~ 280-320 K. For the acetylperoxy radical, the ratio of the rate constant for Reaction 5.56a divided by the sum of rate constants for Reactions 5.56a and 5.56b is 0.67, independent of temperature (Moortgat et al., 1989).

The small ($\leq C_4$) alkanes (RH) have fairly simple reaction schemes after their initial reactions with the OH radical (their only significant tropospheric removal process). For example,

$$OH + RH \rightarrow H_2O + R \cdot \qquad (5.57)$$

$$R\cdot + O_2 \xrightarrow{M} RO_2\cdot \qquad (5.58)$$

$$RO_2\cdot + NO \rightarrow RO\cdot + NO_2 \qquad (5.59)$$

$$RO\cdot + O_2 \rightarrow \text{carbonyl} + HO_2 \qquad (5.60)$$

$$RO\cdot \rightarrow \text{carbonyl} + R'\cdot \qquad (5.61)$$

$$HO_2 + NO \rightarrow OH + NO_2 \qquad (5.62)$$

where R', an alkyl radical with fewer carbon atoms than the parent RH alkane, then undergoes an analogous series of reactions that lead to the formation of carbonyl compounds (which react further in the atmosphere by photolysis and reaction with the OH radical), the conversion of NO to NO_2, and the regeneration of OH radicals.

It should be noted that, apart from the losses of certain product species onto surfaces through wet and dry deposition (for example, N_2O_5, $HONO_2$, aldehydes, H_2O_2, hydroperoxides, and SO_2) (Heikes and Thompson, 1983; Leuenberger et al., 1985; Betterton and Hoffmann, 1988; Mozurkewich and Calvert, 1988), heterogeneous reactions of intermediate radical species have generally not been considered important in the chemistry of the troposphere. However, there is a growing appreciation of the importance of heterogeneous scavenging reactions that involve radical species in the global budgets of ozone and of the various NO_x species (Chameides, 1986; Lelieveld and Crutzen, 1990).

The general reaction scheme for the degradation of a VOC in the troposphere can be written in a very approximate way as

$$VOC (+ h\nu, HO, NO_3, O_3) \rightarrow \alpha RO_2\cdot \qquad (5.63)$$

where RO_2 can also be HO_2, followed by

$$RO_2\cdot + \beta NO \rightarrow \gamma NO_2 + \delta OH \qquad (5.64)$$

Reaction (5.63) includes all loss processes of the VOC under atmospheric conditions, and α, β, γ, and δ are coefficients (which can be greater than or less than one, including zero) that generally depend on the relative importance of the various loss processes and on the VOC/NO_x concentration ratio. Reaction process 5.63 determines the lifetime of an organic compound in the troposphere (refer to Table 5.1). Subsequent reactions (Reaction 5.64) lead to conversion of NO to NO_2, to the generation or regeneration of OH radicals, and to the formation of ozone.

ATMOSPHERIC CHEMISTRY OF ANTHROPOGENIC VOCS

The general features of the atmospheric chemistry of alkanes, alkenes, and aromatic hydrocarbons emitted from anthropogenic sources are understood, although there are still some significant uncertainties (Atkinson, 1990a). The kinetics of the initial reactions of the majority of anthropogenic VOCs with OH and NO_3 radicals and ozone, and their photolysis rates, have either been determined experimentally or can be calculated reliably (Atkinson, 1989, 1990a, 1991). Table 5-1 lists the calculated lifetimes of a series of anthropogenic VOCs with respect to reaction in the troposphere with the important reactive species.

In the sections below, the salient features of the atmospheric chemistry of the alkanes, alkenes, aromatic VOCs, and oxygenates are briefly discussed, including the chemistry of the potential alternative fuels. This discussion is largely based on the recent review and evaluation of Atkinson (1990a), which should be consulted for more detail.

Alkanes

In the troposphere, the alkanes react essentially only with the OH radical; the nighttime NO_3 radical reaction is of minor significance in terms of the overall removal of the alkanes (Atkinson, 1990a). The OH (and NO_3) radical reactions proceed by H-atom abstraction,

$$OH + RH \rightarrow H_2O + R\cdot \qquad (5.65)$$

followed for the simple alkanes (those with fewer than four carbon atoms) by the sequence of reactions (for example, for a secondary alkyl radical R_1R_2CH in the presence of NO_x)

$$R_1\overset{\bullet}{R_2}CH + O_2 \rightarrow R_1CH(O_2\cdot)R_2 \qquad (5.66)$$

$$R_1CH(O_2\cdot)R_2 + NO \rightarrow R_1CH(O\cdot)R_2 + NO_2 \qquad (5.67)$$

$$R_1CH(O\cdot)R_2 + O_2 \rightarrow R_1C(O)R_2\cdot + HO_2 \qquad (5.68)$$

$$R_1CH(O\cdot)R_2 \rightarrow R_1CHO + R_2\cdot \qquad (5.69)$$

The carbonyl compounds R_1CHO and $R_1C(O)R_2$ and the fragment alkyl radical R_2 undergo further reactions. For the alkanes composed of more than three carbon atoms, alkyl nitrate formation from the reactions of the alkyl peroxy radicals with NO,

$$RO_2\cdot + NO \overset{M}{\rightarrow} RONO_2 \qquad (5.70)$$

in competition with the formation of NO_2 and the corresponding alkoxy radical, becomes increasingly important, and the alkyl nitrate formation yields at 298 K and 760 Torr total pressure increase from ~4% from propane to ~33% from n-octane (Carter and Atkinson, 1989a).

The isomerization of alkoxy radicals involving a six-membered transition state (thus requiring a carbon chain of four or more), also is expected to become important for the $\geq C_4$ alkanes; for example,

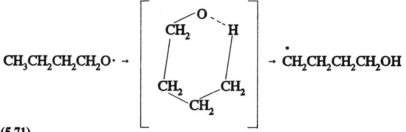

(5.71)

followed by a sequence of reactions (Atkinson, 1990a) that leads to the formation of δ-hydroxycarbonyl compounds, for example, $HOCH_2CH_2CH_2CHO$ from the above n-butoxy radical. This isomerization reaction is in competition with unimolecular decomposition of the alkoxy radical or reaction of the alkoxy radical with O_2 (Atkinson, 1990a).

The major uncertainties in the atmospheric chemistry of the alkanes concern the alkyl nitrate formation yields from the reactions of the various alkyl peroxy and substituted alkyl peroxy radicals with NO, and the importance of, and reactions subsequent to, alkoxy radical isomerization. A further important area of uncertainty concerns the atmospheric chemistry of the carbonyl compounds formed as first-generation products from the alkanes.

Alkenes

In the troposphere, the chemical removal of the alkenes proceeds by reaction with OH and NO_3 radicals and ozone, and all removal pathways must be considered. The rate constants for the initial reactions of these species are reasonably well defined and the initial steps of the reaction mechanisms are known (Atkinson, 1990a). The major uncertainties in the alkene chemistry (apart from the chemistry of isoprene and the monoterpenes discussed below) involve

• The reaction mechanism and the products formed from the long-chain alkenes, such as the 1-alkenes composed of more than four carbon atoms. For example, it is not known whether isomerization of the β-hydroxyalkoxy radicals occurs (Atkinson and Lloyd, 1984).

• The reaction mechanisms of the ozone reactions and the radical formation yields in these reactions. The only alkene for which the reaction mechanism appears to be reasonably well understood is ethene; the experimental data are much less definitive for the higher alkenes (Atkinson, 1990a). Experimental data on the reaction of propene with ozone in air lead to a yield of radical species that is significantly higher than the yield estimated from computer modeling of environmental chamber data (Carter et al., 1986a; Carter, 1990a). For the alkenes composed of four or more carbon atoms, few experimental data on reaction mechanisms or products are available, and further studies are needed.

• The reaction mechanisms and products formed from the reactions of the NO_3 radical with the alkenes under tropospheric conditions (Atkinson, 1991), although these reactions generally are important only for the internal alkenes, such as the 2-butenes.

Aromatic VOCs

The greatest uncertainties in the atmospheric chemistry of anthropogenic VOCs concern the aromatic compounds. The aromatic hydrocarbons react only with the OH radical under tropospheric conditions, by two pathways, one involving H-atom abstraction from the substituent groups (or, for benzene, from the aromatic ring C-H bonds)

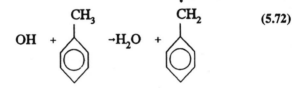

$$(5.72)$$

and the other involving initial OH radical addition to the aromatic ring to form a hydroxycyclohexadienyl radical

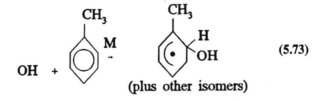

$$(5.73)$$

(plus other isomers)

The rate constants for Reactions 5.72 and 5.73 and the ratios of the two rate constants are known (Atkinson, 1989), and the reaction sequence that follows

the H-atom abstraction pathway (Reaction 5.72) is reasonably well understood. (It leads to the formation of aromatic aldehydes, benzyl nitrates, and
peroxybenzoyl nitrates (Atkinson, 1990a).) It also is known that the OH radical addition pathway (Reaction 5.73) leads to the formation of ring-retaining
products, such as phenols and nitroaromatics (the latter in low yield), and to
the formation of ring-cleavage products, including α- and γ-dicarbonyls. The
formation yields of many of these products have been measured (Atkinson,
1990a).

However, the reactions of the hydroxycyclohexadienyl and alkyl-substituted
hydroxycyclohexadienyl radicals formed from the initial addition of the OH
radical to the ring under tropospheric conditions are not understood. Recent
kinetic data (Zellner et al., 1985; Knispel et al., 1990; Zetzsch et al., 1990)
show that the hydroxycyclohexadienyl radicals react rapidly with NO_2, but that
their reactions with NO and O_2 are slow. The study of the products formed
from the OH radical-initiated reactions of benzene and toluene by Atkinson
et al. (1989b) is consistent with these kinetic data, and it leads to the conclusion that in the presence of NO_2 concentrations $\geq 1.5 \times 10^{13}$ molecule/cm^3
(≥ 600 ppb) the hydroxycyclohexadienyl radicals react with NO_2 and not with
O_2. It is possible that this is also the situation under conditions that are representative of less-polluted areas. This finding, that the hydroxycyclohexadienyl-type radicals react rapidly with NO_2

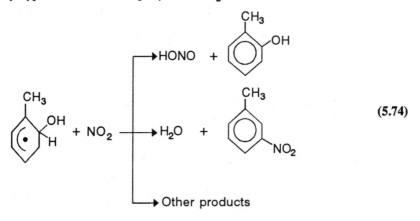

(5.74)

and only very slowly with O_2, differs from the reaction sequences in the current chemical mechanisms of Gery et al. (1988a, 1989) and Carter et al.
(1986a). Clearly, further experimental and mechanism development work on
the tropospheric chemistry of the aromatic hydrocarbons is necessary.

Carbonyl Compounds

It is evident from the discussions of the atmospheric chemistry of VOCs of anthropogenic and biogenic origin that carbonyl compounds are formed during the atmospheric degradation of all VOCs. The chemistry of these various carbonyl compounds needs to be known. Unfortunately, there are several areas of uncertainty concerning the atmospheric chemistry of all of the carbonyl compounds other than formaldehyde and acetaldehyde. In particular, there is a need for data concerning the absorption cross-sections and photolysis products and the photodissociation quantum yields (as a function of wavelength) for these carbonyl compounds. These data are necessary to assess the importance of photolysis as a tropospheric degradation route for these carbonyl compounds.

Reactions of Organic Peroxy (RO_2) Radicals

Under conditions where the mixing ratio of NO is less than approximately 30 ppt, the reactions of organic peroxy radicals with HO_2 radicals and other peroxy radicals dominate over reaction with NO (Logan et al., 1981). To date, however, there are few data concerning the kinetics and products of the reactions of the HO_2 radical with organic peroxy radicals or of the various combination reactions of organic peroxy radicals (Atkinson, 1990a).

Oxygenates Proposed as Alternative Fuels

Oxygenated organic compounds are being investigated as alternative fuels, either as single compounds or as blends with present gasolines. Methanol (CH_3OH), ethanol (CH_3CH_2OH) and methyl t-butyl ether [$CH_3OC(CH_3)_3$] are now used as additives to gasoline, and the alcohols could also be used alone. The chemistry of these compounds is briefly discussed below. Alternative fuels are discussed further in Chapter 12.

Methanol

The only important gas-phase reaction of methanol is with the OH radical, with a rate constant at 298 K of 9.3×10^{-13} cm^3/molecule-sec (Atkinson, 1989). This reaction proceeds by H-atom abstraction (the percentages are for room temperature),

$$OH + CH_3OH \begin{cases} -H_2O + CH_3O\bullet & (15 \pm 8\%) \\ -H_2O + C\bullet H_2OH & (85 \pm 8\%) \end{cases} \quad (5.75)$$

followed by the reactions

$$CH_3O\bullet + O_2 \rightarrow HCHO + HO_2 \qquad (5.76)$$

$$\overset{\bullet}{C}H_2OH + O_2 \rightarrow HCHO + HO_2 \qquad (5.77)$$

Hence the overall reaction of the OH radical with methanol under atmospheric conditions leads to the formation of HCHO and the HO_2 radical.

$$OH + CH_3OH + O_2 - H_2O + HCHO + HO_2 \qquad (5.78)$$

Ethanol

As with methanol, the only important reaction for ethanol under tropospheric conditions is with the OH radical. This reaction has a rate constant at 298 K of 3.3 x 10^{-12} cm^3/molecule-s (Atkinson, 1989). The OH radical reaction can proceed by three channels (the percentages are for room temperature):

$$OH + CH_3CH_2OH \begin{cases} - H_2O + \overset{\bullet}{C}H_2CH_2OH & (\sim 5\%) \\ - H_2O + CH_3\overset{\bullet}{C}HOH & (\sim 90\%) \\ - H_2O + CH_3CH_2O\bullet & (\sim 5\%) \end{cases} \quad (5.79)$$

Under tropospheric conditions, the major reactions of these initially formed radicals are:

for CH_2CH_2OH

$$HOCH_2\overset{\bullet}{C}H_2 + O_2 \overset{M}{\rightarrow} HOCH_2CH_2OO\bullet \qquad (5.80)$$

$$HOCH_2CH_2OO\bullet + NO \rightarrow HOCH_2CH_2O\bullet + NO_2 \qquad (5.81)$$

$$HOCH_2CH_2O\bullet + O_2 \rightarrow HOCH_2CHO + HO_2 \qquad (5.82)$$

$$HOCH_2CH_2O\bullet \rightarrow HCHO + C\bullet H_2OH$$
$$\downarrow O_2 \qquad (5.83)$$
$$HCHO + HO_2$$

for CH_3CHOH

$$CH_3\overset{\bullet}{C}HOH + O_2 \rightarrow CH_3CHO + HO_2 \qquad (5.84)$$

for CH_3CH_2O

$$CH_3CH_2O\bullet + O_2 \rightarrow CH_3CHO + HO_2 \qquad (5.85)$$

At room temperature and 760 Torr total pressure of air in the presence of NO, the overall OH radical reaction is

$$OH + CH_3CH_2OH + 0.05\ NO + 1.05\ O_2 \rightarrow$$
$$H_2O + 0.95\ CH_3CHO + 0.078\ HCHO + \qquad (5.86)$$
$$0.011\ HOCH_2CHO + 0.05\ NO_2$$

Methyl *t*-butyl ether

The only significant reaction under tropospheric conditions is with the OH radical, with a rate constant at room temperature of $2.8 \times 10^{-12}\ cm^3/mole$-cule-s (Atkinson, 1989). This reaction proceeds by H-atom abstraction

$$OH + CH_3OC(CH_3)_3 \left\{ \begin{array}{l} \rightarrow H_2O + \overset{\bullet}{C}H_2OC(CH_3)_3 \quad (\sim 80\%) \\[12pt] \rightarrow H_2O + CH_3OC(CH_3)_2\overset{\bullet}{C}H_2 \quad (\sim 20\%) \end{array} \right. \qquad (5.87)$$

The subsequent reactions will involve addition of O_2 to form the peroxy radicals, followed by, in the presence of NO, the conversion of NO to NO_2 to yield the alkoxy radicals $OCH_2OC(CH_3)_3$ and $CH_3OC(CH_3)_2CH_2O$. The $OCH_2OC(CH_3)_3$ radical then reacts with O_2 to generate mainly *t*-butyl formate $[(CH_3)_3COCHO]$ (Japar et al., 1990; Tuazon et al., 1991).

$$(CH_3)_3COCH_2O\bullet + O_2 \rightarrow (CH_3)_3COCHO + HO_2 \qquad (5.88)$$

BIOGENIC VOCs

Measurements of the ambient concentrations of isoprene and other VOCs that are known to be emitted by vegetation, as well as estimates of the total inventory of biogenic VOC sources, suggest that these compounds help foster episodes of high concentrations of ozone in areas affected by anthropogenic NO_x (Lamb et al., 1987; Trainer et al., 1987; Chameides et al., 1988; Sillman et al., 1990b). This section focuses on the atmospheric chemistry of biogenic VOCs; they are discussed again in Chapters 8 and 9. The atmospheric chemistry of isoprene and most of the monoterpenes observed as vegetative emissions has been investigated over the past 10 years. In general, isoprene and the monoterpenes can be regarded as alkenes or cycloalkenes, and their gas-phase atmospheric reactions are generally analogous to those for the alkenes such as propene and *trans*-2-butene. Rate constants have been determined at room temperature for the gas-phase reactions of isoprene, a series of monoterpenes, and related compounds with OH and NO_3 radicals and ozone; these data are given in Table 5-2.

TABLE 5-2   Room-Temperature Rate Constants for the Gas-Phase Reactions of a Series of Organic Compounds of Biogenic Origin with OH and NO_3 Radicals and Ozone

VOC	Structure	Rate constant, cm^3/molecule-s, for reaction with		
		OH^a	$NO_3{}^b$	$O_3{}^c$
Isoprene		1.0×10^{-10}	5.9×10^{-13}	1.4×10^{-17}
Camphene		5.3×10^{-11}	6.5×10^{-13}	9.0×10^{-19}

| VOC | Structure | Rate constant, cm^3/molecule-s, for reaction with | | |
		OH^a	NO_3^b	O_3^c
2-Carene		8.0×10^{-11}	1.9×10^{-11}	2.4×10^{-16}
Δ^3-Carene		8.8×10^{-11}	1.0×10^{-11}	3.8×10^{-17}
d-Limonene		1.7×10^{-10}	1.3×10^{-11}	2.1×10^{-16}
Myrcene		2.2×10^{-10}	1.1×10^{-11}	4.9×10^{-16}
Ocimene		2.5×10^{-10}	2.2×10^{-11}	5.6×10^{-16}
α-Phellan-drene		3.1×10^{-10}	8.5×10^{-11}	1.9×10^{-15}
α-Pinene		5.4×10^{-11}	5.8×10^{-12}	8.7×10^{-17}
β-Pinene		7.9×10^{-11}	2.4×10^{-12}	1.5×10^{-17}

VOC	Structure	Rate constant, cm^3/molecule-s, for reaction with		
		OH^a	NO_3^b	O_3^c
Sabinene		1.2×10^{-10}	1.0×10^{-11}	8.8×10^{-17}
α-Terpinene		3.6×10^{-10}	1.8×10^{-10}	8.7×10^{-15}
γ-Terpinene		1.8×10^{-10}	2.9×10^{-11}	1.4×10^{-16}
Terpino-lene		2.3×10^{-10}	9.6×10^{-11}	1.4×10^{-15}
1,8-Cineole		1.1×10^{-11}	1.7×10^{-16}	$<1.5 \times 10^{-19}$
p-Cymene		1.5×10^{-11}	9.9×10^{-16}	$<5 \times 10^{-20}$

[a]From Atkinson, 1989; Atkinson et al., 1990a, and Corchnoy and Atkinson, 1990.

[b]From Atkinson et al., 1988; Atkinson et al., 1990a; and Corchnoy and Atkinson, 1990.

[c]From Atkinson and Carter, 1984 and Atkinson et al., 1990b.

Isoprene and the monoterpenes are highly reactive toward all three of these reactive intermediates. The tropospheric lifetimes due to reaction with OH and NO_3 radicals and ozone can be calculated by combining the rate constant data with estimated ambient tropospheric concentrations of OH and NO_3 radicals and ozone. The resulting tropospheric lifetimes with respect to these gas-phase reactive loss processes are given in Table 5-3. Obviously, the calculated lifetimes of isoprene and the monoterpenes are short. The OH radical and ozone reactions are of generally comparable importance during the daytime, and the NO_3 radical reaction is important at night if NO_3 radicals are present at concentrations of $> 10^7$ molecule/cm^3 (> 0.4 ppt). (Over continental areas, lower tropospheric nighttime NO_3-radical mixing ratios range from < 2 ppt to 430 ppt [Atkinson et al., 1986]). As noted above, the NO_3 radical reactions act as a removal process for either the biogenic VOCs or NO_x, depending on the relative magnitudes of the biogenic emission fluxes and the formation rate of the NO_3 radical from the reaction of ozone with NO_2 (Winer et al., 1984).

TABLE 5-3   Calculated Tropospheric Lifetimes of VOCs

	Lifetime due to reaction with		
VOC	OH^a	O_3^b	NO_3^c
Isoprene	1.8 hr	1.2 days	1.7 days
Camphene	3.5 hr	18 days	1.5 days
2-Carene	2.3 hr	1.7 hr	36 min
Δ^3-Carene	2.1 hr	10 hr	1.1 hr
d-Limonene	1.1 hr	1.9 hr	53 min
Myrecene	52 min	49 min	1.1 hr
Ocimene	44 min	43 min	31 min
α-Phellandrene	35 min	13 min	8 min
α-Pinene	3.4 hr	4.6 hr	2.0 hr
β-Pinene	2.3 hr	1.1 days	4.9 hr
Sabinene	1.6 hr	4.5 hr	1.1 hr

α-Terpinene	31 min	3 min	4 min
γ-Terpinene	1.0 hr	2.8 hr	24 min
Terpinolene	49 min	17 min	7 min
1,8-Cineole	1.4 days	>110 days	16 yr
ϱ-Cymene	1.0 days	>330 days	2.7 yr

[a]For a 12-hr daytime average OH radical concentration of 1.5×10^6 molecule/cm^3 (0.06 ppt) (Prinn et al., 1987).
[b]For a 24-hr average O_3 concentration of 7×10^{11} molecule/cm^3 (30 ppb) (Logan, 1985).
[c]For a 12-hr average NO_3 radical concentration of 2.4×10^7 molecule/cm^3 (1 ppt) (Atkinson et al., 1986)

Few definitive data are available concerning the products formed from the atmospheric reactions of isoprene and the monoterpenes. The most studied of the biogenic compounds have been isoprene and its major degradation products methacrolein and methyl vinyl ketone (Arnts and Gay, 1979; Kamens et al., 1982; Niki et al., 1983; Gu et al., 1985; Tuazon and Atkinson, 1989, 1990a,b; Paulson et al., 1992a,b); these two degradation products have recently been observed and measured in ambient air (Pierotti et al., 1990; Martin et al., 1991). However, the products and reaction mechanisms of the atmospherically important reactions of isoprene and the monoterpenes are not well understood; for the monoterpenes few products have been identified and even fewer have been quantified. Based on the aerosol formation observed in recent product studies from the OH radical-initiated and ozone reactions with α- and β-pinene (Hatakeyama et al., 1989, 1991; Pandis et al., 1991), it is calculated that the atmospheric degradation reactions of the biogenic monoterpene VOCs can account for a significant, and often dominant, fraction of the secondary aerosol observed in urban and rural areas (Pandis et al., 1991). In contrast, the atmospheric photooxidation of isoprene is expected to be a negligible pathway for the formation of secondary aerosol (Pandis et al., 1991). The product data reported in the literature are summarized below.

NO_3 Radical Reaction

Barnes et al. (1990) have used Fourier transform infrared (FT-IR) absorption spectroscopy to investigate the gas-phase reactions of isoprene, α- and β-pinene, Δ^3-carene, and d-limonene in the presence of one atmosphere of air. Formaldehyde (HCHO), CO, and methacrolein were identified from the NO_3 radical reaction with isoprene; the HCHO and CO yields were 11% and 4%, respectively (Barnes et al., 1990). The FT-IR spectra indicated the presence of $>C=O$ and $-ONO_2$ groups, and the intensities of these FT-IR bands allowed an estimated formation yield of ~80% of nitrate-containing products. The NO_3 radical reactions with the monoterpenes led to the formation of aerosols, although for α- and β-pinene, spectral features indicated the presence of $>C=O$ and $-ONO_2$ groups. It should be noted that the initial isoprene and monoterpene concentrations in these experiments were ~5 × 10^{14} molecule/cm^3 (20,000 ppb), to be compared with ambient concentrations of less than 20 ppb (see Petersson, 1988)).

Kotzias et al. (1989) also used FT-IR absorption spectroscopy and mass spectrometry (MS) to study the reaction of the NO_3 radical with β-pinene. Their results are similar to those of Barnes et al. (1990) in that both the FT-IR and MS data indicate the presence of organic nitrates.

The initial reaction steps in the NO_3 reactions are expected to involve initial NO_3 radical addition to a $>C=C<$ bond

$$NO_3 + >C\text{-}C< \rightarrow >C\cdot\text{-}\overset{|}{\underset{|}{C}}\text{-}ONO_2 \qquad (5.89)$$

followed by

$$>C\cdot\text{-}\overset{|}{\underset{|}{C}}\text{-}ONO_2 + O_2 \xrightarrow{M} OO\text{-}\overset{|}{\underset{|}{C}}\text{-}\overset{|}{\underset{|}{C}}\text{-}ONO_2 \qquad (5.90)$$

$$\overset{\cdot}{OO}\overset{|}{\underset{|}{C}}\text{-}\overset{|}{\underset{|}{C}}ONO_2 + NO_2 \rightarrow O_2NOO\overset{|}{\underset{|}{C}}\text{-}\overset{|}{\underset{|}{C}}ONO_2 \qquad (5.91)$$

$$\overset{\bullet}{O}O\overset{|}{C}-\overset{|}{C}ONO_2\!\!\rightarrow\; \rightarrow\; \overset{\bullet}{O}\overset{|}{C}-\overset{|}{C}ONO_2 \qquad (5.92)$$

$$\overset{\bullet}{O}\overset{|}{C}-\overset{|}{C}ONO_2 \;\rightarrow\; {>}C\text{-}O \;+\; {>}\overset{\bullet}{C}ONO_2 \qquad (5.93)$$
$$\downarrow$$
$${>}C\text{-}O \;+\; NO_2$$

$$\underset{H}{\overset{\bullet}{O}\overset{|}{C}}-\overset{|}{C}ONO_2 \;+\; O_2 \;\rightarrow\; O\text{-}\overset{|}{C}-\overset{|}{C}ONO_2 \;+\; HO_2 \qquad (5.94)$$

The nitratoperoxynitrate ($O_2NOROONO_2$) formed in Reaction 5.91 is thermally unstable, and the organic peroxy radicals undergo radical-radical reactions with other peroxy (RO_2) and HO_2 radicals (Atkinson, 1991).

Ozone Reactions

There have been few quantitative product studies of the reactions of ozone with isoprene and the monoterpenes. For isoprene, Kamens et al. (1982) and Niki et al. (1983) observed the formation of HCHO, methacrolein, and methyl vinyl ketone. Both groups reported HCHO, methyl vinyl ketone, and methacrolein yields (in molar units) of 85-96%, 13-18%, and 33-42%, respectively. The use of isotope labeling allowed Niki et al. (1983) to conclude that the majority of the HCHO formed arose from secondary reactions. The recent study of Paulson et al. (1992b) has provided evidence that the O_3 reaction with isoprene leads to the formation of OH radicals and $O(^3P)$ atoms in large amounts, with molar yields of 65% and 45%, respectively. The formation of OH radicals and $O(^3P)$ atoms leads to secondary reactions which complicate the analysis of the O_3-isoprene reaction. Based on computer modeling of product data, Paulson et al. (1992b) concluded that the products formed from the O_3 reaction with isoprene are methacrolein, methyl vinyl ketone, and propene, with yields of 68%, 25%, and 7% respectively.

In general, the initial reaction sequence is expected to be (Atkinson and Lloyd, 1984; Atkinson and Carter, 1984; Atkinson, 1990a),

$$O_3 + >C_1=C_2< \rightarrow \left[\begin{array}{c} \cdot O \\ O \quad O \\ | \quad | \\ >C_1-C_2< \end{array} \right]^{\neq}$$

$$[>C_1OO]^{\neq} + >C_2=O \qquad >C_1=O + [>C_2OO]^{\neq}$$

followed by decomposition or stabilization of the initially energy-rich biradicals [>COO]$^{\neq}$, and this accounts for the HCHO, methyl vinyl ketone and methacrolein observed. The reported product distributions account for only ~60% of the overall products formed.

Several studies have investigated the products of the ozone reactions with monoterpenes (see, for example, Wilson et al., 1972; Schwartz, 1974; Schuetzle and Rasmussen, 1978; Hull, 1981; Yokouchi and Ambe, 1985; Hatekayama et al., 1989); the two most recent studies (Yokouchi and Ambe, 1985; Hatakeyama et al., 1989) are the most definitive. Yokouchi and Ambe (1985) used high concentrations, ~(3-15) × 10^{15} molecule/cm^3 (~120,000-600,000 ppb) , of ozone and the monoterpenes (α- and β-pinene and d-limonene), and observed ready formation of aerosols, as expected from the high concentrations of reactants (Finlayson-Pitts and Pitts, 1986; Izumi et al., 1988). Using gas chromatography (GC) and GC/MS techniques, they identified pinonaldehyde (2',2'-dimethyl-3'-acetylcyclobutyl ethanol) and, to a lesser extent, pinonic acid (2',2'-dimethyl-3'-acetylcyclobutyl acetic acid) from α-pinene and 6,6-dimethylbicyclo[3.1.1]heptan-2-one from β-pinene. No products were identified from the d-limonene reaction.

The most recent product study of Hatakeyama et al. (1989) was carried out at much lower reactant concentrations (typically ~3 × 10^{13} molecule/cm^3 [~1,000 ppb]), using FT-IR absorption spectroscopy and GC/MS for analysis. From the α-pinene reaction, CO, CO$_2$, HCHO, pinonaldehyde and nor-pinonaldehyde, were identified, with molar formation yields of 9%, 30%, and 22%, respectively, for CO, CO$_2$, and HCHO; the "total aldehydes" yield was ~51% (mainly pinonaldehyde and nor-pinonaldehyde). The products identified from the β-pinene reaction were CO$_2$, HCHO, and 6,6-dimethylbicyclo-[3.1.1]heptan-2-one with molar yields of 27%, 76%, and 40%, respectively. Aerosol formation accounted for 14-18% of the overall reaction. The products observed in these two recent studies can be explained with the general reaction scheme outlined above, although only a fraction of the overall product distribution has been accounted for.

OH Radical Reactions

Few quantitative studies have dealt with the OH-radical-initiated reactions of isoprene or the monoterpenes. Prior to recent studies on the reactions of isoprene (Tuazon and Atkinson, 1990a; Paulson et al., 1992b), methacrolein (Tuazon and Atkinson, 1990b), methyl vinyl ketone (Tuazon and Atkinson, 1989), and a series of monoterpenes (Arey et al., 1990), the only quantitative product studies were those of Arnts and Gay (1979) for isoprene and monoterpenes and of Gu et al. (1985) for isoprene. Gu et al. (1985) reported the formation of methacrolein, methyl vinyl ketone, and 3-methylfuran from isoprene with formation yields of 23%, 17%, and 6%, respectively. This is in agreement with the yields of 29%, 21%, and 4.4%, respectively, obtained by Tuazon and Atkinson (1990a) and Atkinson et al. (1989c). (These yields require upward revision by ~10-15% [Atkinson, pers. comm., 1991; Paulson et al., 1992a], because of the neglect of the $O(^3P)$ atom reaction in the Tuazon and Atkinson [1990a] study.) Arnts and Gay (1979) used irradiated mixtures of NO_x, VOC, and air to generate OH radicals, and secondary reactions involving ozone and possibly NO_3 radicals were undoubtedly important. In general, the carbon balances they determined by FT-IR absorption spectroscopy were low; <17% for the monoterpenes and 44% for isoprene. Only for isoprene did they identify organic products other than HCHO, CH_3CHO, $CH_3C(O)OONO_2$ [PAN], HCOOH, and CH_3COCH_3, these being methacrolein and methyl vinyl ketone (Arnts and Gay, 1979).

As shown in Figure 5-2, the major products observed by Tuazon and Atkinson (1989, 1990a,b) and by Atkinson et al. (1989c) from the OH-radical-initiated reactions of isoprene, methyl vinyl ketone, and methacrolein are as follows: from isoprene, methyl vinyl ketone, methacrolein, formaldehyde (HCHO), and 3-methylfuran; from methyl vinyl ketone, glycolaldehyde ($HOCH_2CHO$), methylglyoxal (CH_3COCHO), HCHO, and peroxyacetyl nitrate ($CH_3C(O)OONO_2$, PAN); and from methacrolein, hydroxyacetone ($HOCH_2COCH_3$), CH_3COCHO, a peroxyacyl nitrate identified as $CH_2=C(CH_3)C(O)OONO_2$, CO_2, and HCHO. Other unidentified products were observed from isoprene, and these were characterized from their IR spectra as organic nitrates (yield ~10-15%) and as carbonyls or hydroxycarbonyls (yield ~25%)

These product yield data show (Tuazon and Atkinson, 1990a; Paulson et al., 1992b; and Figure 5-2) that the OH radical reaction with isoprene in the presence of NO_x leads to the formation of methacrolein + HCHO (~24%), methyl vinyl ketone + HCHO (~34%), and 3-methylfuran (~5%); the HCHO yield is equal to the sum of the methacrolein and methyl vinyl ketone yields. For the OH radical-initiated reaction of methyl vinyl ketone, the reaction pathways are essentially totally accounted for; the OH radical addition to

148

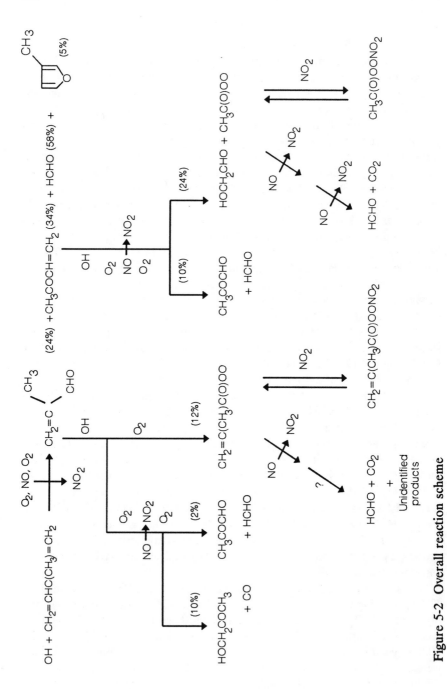

Figure 5-2 Overall reaction scheme

the terminal carbon atom of the $>C=C<$ bond (which leads to the formation of $HOCH_2CHO$) accounts for $72 \pm 21\%$ of the overall reaction. The product yield data for the OH radical reaction with methacrolein show that OH radical addition to the $>C=C<$ bond accounts for approximately 50% of the overall reaction; the remaining $\sim 50\%$ proceeds by H-atom abstraction for the -CHO group. In particular, the OH radical addition pathway leads to the formation of hydroxyacetone in high yield, in contrast to expectations provided in the literature (Lloyd et al., 1983) that assume $CH_3COCHO + HCHO + HO_2$ would be the major product.

A reaction scheme for the OH radical reaction of isoprene in the presence of NO_x, including the subsequent reactions of methyl vinyl ketone and methacrolein, is shown in Figure 5-2.

Arey et al. (1990) have investigated the products of the OH radical-initiated reactions of a series of monoterpenes. 6,6-Dimethylbicyclo[3.1.1]heptan-2-one was observed as a product from the gas-phase reaction of β-pinene with the OH radical in the presence of NO_x, with a formation yield of $30.0 \pm 4.5\%$, and 4-acetyl-1-methylcyclohexene was observed as a product from the OH radical reaction of d-limonene with a formation yield of $17.4 \pm 2.8\%$. On the basis of mass spectral data, a number of ketone and keto-aldehyde products were tentatively identified in irradiated CH_3ONO-NO-air-monoterpene mixtures of d-limonene, α-pinene, Δ^3-carene, sabinene, and terpinolene. The estimated formation yields of the observed products ranged from a low of $\sim 29\%$ for α-pinene to a high of $\sim 45\%$ for d-limonene. No significant products were observed from the OH radical-initiated reaction of myrcene. Other unidentified products are obviously formed from these reactions in large overall yield.

The major features of the atmospheric degradations of anthropogenic and biogenic VOCs are shown in Figure 5-3.

DEVELOPMENT AND TESTING OF CHEMICAL MECHANISMS

Computer models that incorporate emissions of VOCs and NO_x, meteorology, and the chemistry of VOC/NO_x mixtures simulate the complex physical and chemical processes of the atmosphere and predict the effects of changes of emissions of anthropogenic VOCs, biogenic VOCs, or NO_x on photochemical air pollution. An essential component is the chemical mechanism that describes the series of reactions in the troposphere subsequent to emissions of VOCs and NO_x.

In general, chemical mechanisms are assembled using the available literature on kinetic, mechanistic, and product data for the atmospherically impor-

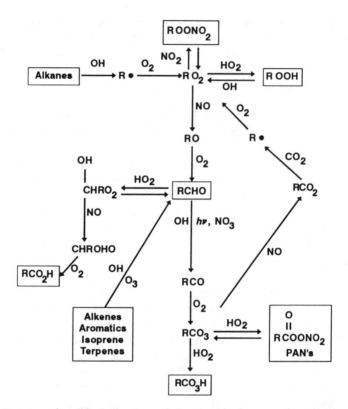

FIGURE 5-3   Simplified diagram of the chemical processing that occurs among VOCs.

tant inorganic and organic reactions, often using relevant reviews and evaluations (for example, the series of the National Aeronautics and Space Administration [NASA] and International Union of Pure and Applied Chemistry [IUPAC] data evaluation panels, with the most recent evaluations from these groups being those of DeMore et al. [1990] and Atkinson et al. [1989a], respectively). Apart from the initial reactions of VOCs with OH and NO_3 radicals and ozone, the vast majority of the tropospheric degradation reactions of VOCs are not well understood with regard to rate constant or products, so there are large areas of uncertainty, and reaction sequences are formulated by analogy.

The initial formulation(s) of the chemical mechanisms are then tested against environmental chamber data concerning the irradiations of single VOCs or of mixtures of VOCs in air in the presence of NO_x, and the mechanism predictions are compared with the experimental data. However, there are significant problems associated with the direct comparison of chemical mechanism predictions against environmental chamber data, because of the effects of the chamber itself (see, for example, Bufalini et al., [1977]; Carter et al., [1982a]; Joshi et al., [1982]; and Killus and Whitten, [1990]). It is necessary to take into account several chamber effects, such as light intensity and spectral distribution, sorption of chemicals to walls, degassing of chemicals from the chamber walls, and heterogeneous reactions that lead to the formation (upon irradiation) of radical species (Carter et al., 1982a; Killus and Whitten, 1990; Carter and Lurmann, 1991). (The radicals appear to be formed, at least in part, through the heterogeneous generation of nitrous acid.) Some of these effects are not completely understood at a physical or chemical level, and the necessity of including the array of chamber effects introduces additional uncertainties into the direct comparison of mechanism predictions with experimental data.

Through comparison of experimental and predicted data, the initial chemical mechanism is refined and adjusted to provide a good fit between the experimental data and mechanism predictions. In some areas of the chemical mechanism, such as the degradation reaction schemes of the aromatic VOCs, the chemical mechanism has been derived by providing the best fit of an assumed mechanism to the experimental data.

Two chemical mechanisms have been developed recently for use in urban airshed simulation models, and predictions of these two mechanisms (Carter et al., 1986a; Gery et al., 1988a, 1989) are in reasonably good agreement at room temperature (Dodge, 1989). Although there are significant areas of uncertainty in the chemical mechanisms for the tropospheric oxidation of VOCs, this general agreement arises because the chemical mechanisms have been tested against, and revised to be in agreement with, a common data base of environmental chamber experiments conducted at the University of North Carolina and the University of California, Riverside (Carter et al., 1986a; Gery et al., 1988a, 1989; Carter, 1990a).

Hence, although the unknown aspects of the atmospheric chemistry of alkanes, alkenes, and aromatics have in some cases been treated differently, the resulting chemical mechanisms are constrained to be in close agreement with one another and with the environmental chamber data (which are mostly available at room temperature). This agreement of the two most recent chemical mechanisms at room temperature does not guarantee their correctness; indeed, one or both mechanisms could be incorrect in their treatment of various aspects of the chemistry of the alkanes, alkenes, and aromatic hydrocarbons, and this could well be the case for the aromatic hydrocarbon chemistry. However, because the chemical mechanisms are tested against environmental chamber data, it is not clear to what extent even a totally incorrect treatment of the chemistry of a specific class of VOCs will result in incorrect predictions of the airshed model under actual atmospheric conditions.

The chemical mechanisms of Gery et al. (1988a) and Carter et al. (1986a) do, however, lead to significantly different predictions of ozone formation at temperatures below ~298 K for VOC/NO$_x$ ratios <10 (Dodge, 1989); the CB-IV mechanism of Gery et al. (1988a) predicts the formation of less ozone at lower temperatures than does the mechanism of Carter et al. (1986a). This discrepancy arises because of different assumptions for the temperature dependence of the ratio of the rate constants for Reactions 5.96 and 5.97, where the acetyl peroxy radical reacts with NO and NO$_2$, respectively.

$$CH_3C(O)OO\bullet + NO \rightarrow CH_3C(O)O\bullet + NO_2$$
$$\downarrow$$
$$\overset{\bullet}{C}H_3 + CO_2$$
(5.96)

$$CH_3C(O)OO\bullet + NO_2 \overset{M}{\rightleftharpoons} CH_3C(O)OONO_2$$
$$(PAN)$$
(5.97)

Gery et al. (1988a) assume a temperature dependence in the ratio of the rate constants for Reactions 5.96 and 5.97 of $e^{-5250/T}$, and Carter et al. (1986a) use a value of the rate constant ratio that is independent of temperature. At room temperature the rate constant ratio for Reactions 5.96 and 5.97 used in both mechanisms is similar. The more sensitive temperature dependence used by Gery et al. (1988a) leads to NO$_x$ being sequestered as PAN at lower temperatures and not being available to participate in the formation of ozone.

The data of Kirchner et al. (1990) and Tuazon et al. (1991) show that the rate constant ratio for Reactions 5.96 and 5.97 is equal to 2.2, independent of temperature over the range 283-313 K at 740 Torr total pressure of air. This will require revision of the Gery et al. (1988a) CB-IV mechanism and of the Carter et al. (1986a) mechanism.

OZONE FORMATION POTENTIAL OF VARIOUS VOCS

VOCs emitted from anthropogenic and biogenic sources react in the troposphere in the presence of NO_x and sunlight to lead to the photochemical formation of ozone. For two decades it has been known that VOCs vary widely in the speed with which they react in the troposphere and in the extent to which they promote or inhibit ozone formation (see, for example, Altshuller and Bufalini, 1971, and references therein; Dimitriades, 1974; Pitts et al., 1977). Several reactivity scales have been proposed to define the potential of VOCs to form ozone in the atmosphere, and these have included the maximum amount of ozone generated in irradiated mixtures of a single VOC, NO_x, and air for several hours (Wilson and Doyle, 1970; Laity et al., 1973; Dimitriades and Joshi, 1977; Joshi et al., 1982); the rate of NO photooxidation (Heuss and Glasson, 1968; Glasson and Tuesday, 1970a,b, 1971), the rate of ozone formation (Heuss and Glasson, 1968; Winer et al., 1979), the rate of VOC consumption (Heuss and Glasson, 1968), all in irradiated mixtures of VOCs and NO_x; and the rate constants for the reaction of the VOC with the OH radical (Darnall et al., 1976; Pitts et al., 1977). Although there are differences among these reactivity scales, the reactivities of VOCs as obtained from these scales are in a general order of increasing reactivity: alkanes and monoalkylbenzenes < 1-alkenes and dialkylbenzenes < trialkylbenzenes and internal alkenes (Heuss and Glasson, 1968; Darnall et al., 1976).

There are problems associated with basing reactivity scales on smog chamber experiments because of chamber effects (see, for example, Bufalini et al., 1977; Joshi et al., 1982; Carter et al., 1982a), and experimental chamber data are not directly applicable to ambient atmospheric conditions because they generally do not take into account the dilution and continuous input of VOCs and NO_x in ambient air (Carter and Atkinson, 1989b). The use of the OH radical reaction rate constant scale does not suffer from these effects (Darnall et al., 1976), but this scale does not take into account the reactions subsequent to the initial OH radical reaction and it ignores other tropospheric loss processes, such as photolysis and reaction with NO_3 radicals and ozone. Thus, for example, the formation of photoreactive products, such as formaldehyde, leads to increased overall reactivity with respect to ozone formation, whereas the generation of products such as organic nitrates, which act as sinks for NO_x and radical species, leads to a decreased ozone-forming potential (Carter and Atkinson, 1987, 1989b).

A useful definition of reactivity is that of incremental reactivity, defined as the amount of ozone formed per unit amount (as carbon) of VOC added to a VOC mixture representative of conditions in urban and rural areas in a given air mass (Dodge, 1984; Carter and Atkinson, 1987, 1989b; Carter, 1991),

$$Incremental\ reactivity - \{\Delta[ozone]/\Delta[VOC]\} \qquad (5.98)$$

where Δ[ozone] is the change in the amount of ozone formed as a result of the change in the amount of organic present, Δ[VOC] (note that Carter and Atkinson [1989b] used the quantity Δ([ozone] - [NO]) rather than Δ[ozone] under conditions where the maximum ozone was not attained and NO was not fully consumed). This concept of incremental reactivity corresponds closely to control strategy conditions, in that the effects of reducing the emission of a VOC or group of VOCs, or of replacing a VOC or group of VOCs with other VOCs, on the ozone-forming potential of complex mixture of VOC emissions are simulated.

The theoretical and experimental studies of Bufalini and Dodge (1983), Dodge (1984), and Carter and Atkinson (1987, 1989b) show that, in agreement with previous scales, VOCs exhibit wide variations in reactivity with respect to ozone formation. Furthermore, the absolute and relative calculated incremental reactivities of VOCs depend on VOC/NO_x ratios (Bufalini and Dodge, 1983; Dodge, 1984; Carter and Atkinson, 1989b) and on the "scenario" used (i.e., the VOC mix to which incremental changes are made and associated physical factors, such as the amount of dilution [Carter and Atkinson, 1989b]). Table 5-4 shows, as an example, the incremental reactivities calculated when selected VOCs are added to an eight-component urban VOC mixture at various VOC/NO_x ratios (Carter and Atkinson, 1989b). These data are in general agreement with the earlier modeling study of Dodge (1984) and show that the incremental reactivities of VOCs depend on the particular VOC and vary with the VOC/NO_x ratio. In particular, the incremental reactivities are generally independent of, or increase with, the VOC/NO_x ratio up to a VOC-/NO_x ratio of ~6-8. At higher ratios, the absolute magnitude of the incremental reactivities decreases with increasing VOC/NO_x ratio, becoming close to zero (or more negative) at high VOC/NO_x ratio, where the formation of ozone is NO_x-limited and VOC control becomes irrelevant (Carter and Atkinson, 1989b; Sillman et al., 1990b). The observed negative incremental reactivities are due to the presence of NO_x or radical sinks in the chemistry of the VOC, with the NO_x sinks being most important at high VOC/NO_x ratios and the radical sinks most important at low VOC/NO_x ratios (Carter and Atkinson, 1989b).

TABLE 5-4   Calculated Incremental Reactivities of CO and Selected VOCs
as a Function of the VOC/NO$_x$ Ratio for an Eight-Component VOC Mix
and Low-Dilution Conditions

Compound	VOC/NO$_x$, ppbC/ppb			
	4	8	16	40
Carbon monoxide	0.011	0.022	0.012	0.005
Ethane	0.024	0.041	0.018	0.007
n-Butane	0.10	0.16	0.069	0.019
n-Octane	0.068	0.12	0.027	-0.031
Ethene	0.85	0.90	0.33	0.14
Propene	1.28	1.03	0.39	0.14
trans-2-Butene	1.42	0.97	0.31	0.054
Benzene	0.038	0.033	-0.002	-0.002
Toluene	0.26	0.16	-0.036	-0.051
m-Xylene	0.98	0.63	0.091	-0.025
Formaldehyde	2.42	1.20	0.32	0.051
Acetaldehyde	1.34	0.83	0.29	0.098
Benzaldehyde	-0.11	-0.27	-0.40	-0.40
Methanol	0.12	0.17	0.066	0.029
Ethanol	0.18	0.22	0.065	0.006
Urban mix[a]	0.41	0.32	0.088	0.011

[a]Eight-component VOC mix used to simulate VOC emissions in an urban
area in the calculations. Surrogate composition, in units of ppb compound per
ppbC surrogate, was ethene, 0.025; propene, 0.0167; *n*-butane, 0.0375; *n*-pen-
tane, 0.0400; isooctane, 0.0188; toluene, 0.0179; *m*-xylene, 0.0156; formaldehyde, 0.0375; and inert constituents, 0.113.
Source: Adapted from Carter and Atkinson (1989b).

Theoretical studies of the ozone-forming potential of VOCs have investigated the factors that influence these reactivities. There are several approaches to dealing with this topic conceptually, but the computer modeling studies of Atkinson and Carter (1989b) and of Carter (1991) provide a useful framework for discussing the various factors involved. A generalized scheme for the degradation of VOCs in the troposphere in the presence of NO_x is

$$VOC\ (+\ h\upsilon,\ OH,\ NO_3,\ O_3) \rightarrow \alpha RO_2\bullet \qquad (5.99)$$

$$RO_2\cdot + \beta NO \rightarrow \gamma NO_2 + \delta OH \qquad (5.100)$$

The rate of formation of ozone and oxidation of NO are then determined by the rate of formation of RO_2 radicals and the number of molecules of NO converted to NO_2 per RO_2 radical generated. To a first approximation these processes can be dealt with independently, and the ozone-forming potential of a VOC then depends on

• The rate at which the organic compound reacts in the troposphere. This reaction rate is equal to the inverse of its lifetime (i.e., its decay rate). The quantity of interest is the fraction of the emitted organic that has reacted (by whatever route, photolysis, reaction with OH radicals, NO_3 radicals, ozone, etc.) during the time being considered.
• The reaction mechanism subsequent to the initial reaction(s) of the organic compound. Different aspects of the reaction mechanisms, for example, the VOC/NO_x concentration ratio, become important under different conditions.

The following features of a reaction mechanism affect the formation of ozone:

• The existence of NO_x sinks (low values of γ in Reaction 5.100) in the reaction mechanism lead to a lowering of reactivity with respect to ozone formation. Examples include the generation of alkyl nitrates from the reaction of alkyl peroxy radicals with NO (which competes with the pathway to form NO_2 and the corresponding alkoxy radical),

$$RO_2\bullet + NO \xrightarrow{M} RONO_2 \qquad (5.101)$$

the formation of peroxyacetyl nitrate (PAN) or its analogues,

$$RC(O)OO\bullet + NO_2 \xrightarrow{M} RC(O)OONO_2 \qquad (5.102)$$

and the formation of other organic nitrogen-containing compounds

$$C_6H_5O\cdot + NO_2 \rightarrow \text{nitrophenols} \qquad (5.103)$$
$$\text{(phenoxy)}$$

This aspect of the reaction mechanisms becomes important at high VOC/NO_x ratios, where the availability of NO_x becomes limiting, and the formation of organic nitro compounds competes with the formation of NO_2 (with subsequent photolysis to generate ozone).

• The generation or loss of radical species can lead to a net formation or loss of OH radicals ($\delta > 1$ or < 1, respectively, in Reaction 5.100), which in turn leads to an enhancement or suppression of OH radicals in the entire air mass and hence to an enhancement or suppression of overall reactivity of all chemicals through the effect on the formation rate of RO_2 radicals (Reaction 5.99). The effects of radical formation or loss are most important at low VOC/NO_x ratios, where the formation of ozone is determined by the rate at which RO_2 radicals are formed.

The alkenes, including isoprene and the monoterpenes of biogenic origin, react with ozone in addition to reacting with OH and NO_3 radicals (Atkinson and Carter, 1984; Atkinson, 1991; Atkinson et al., 1990b; Tables 5-1 and 5-2), and this reaction process can act as a sink for ozone, especially under the low NO_x conditions encountered in rural and clean atmospheres. However, because the reactions of ozone with alkenes lead to the generation of radicals from biogenic sources will act as ozone sinks unless NO_x mixing ratios are $<$ 0.1 ppb.

To a first approximation the overall reactivity of VOC towards ozone formation is

$$\frac{\Delta[O_3]}{\Delta[VOC]} - \text{incremental reactivity} - \begin{array}{l}[(\text{fraction of VOC reacted}) \\ \times (\text{mechanistic reactivity})]\end{array} \quad (5.104)$$

where

$$\text{mechanistic reactivity} - \Delta[O_3]/\Delta[VOC]_{\text{reacted}} \quad (5.105)$$

The mechanistic reactivity then reflects the presence of radical and NO_x sources or sinks in the VOC's reaction mechanism subsequent to the initial loss process. An extreme example is benzaldehyde, whose reaction mechanism subsequent to the initial OH radical reaction results in the loss of radical and NO_x sinks, with the overall OH radical reaction having $\delta = 0$ (no net OH radical formation in Reaction 5.100) and $(\beta-\gamma) = 1$ (consumption of one molecule of NO_x per molecule of benzaldehyde reacted in Reaction 5.100).

The approach in Chapter 8 to assess the importance of isoprene and other biogenic organic emissions in the formation of ozone in urban-suburban , rural, and remote air masses uses the kinetic reactivities of the VOCs measured in ambient air, and hence is based on the instantaneous rate of formation of RO_2 radicals (Reaction 5.99). While the differences in the mechanistic reactivities of the various VOCs are neglected, this simpification is not expected to significantly alter the conclusions drawn in Chapter 8, especially since isoprene has a high positive mechanistic reactivity compared with the reactivities of the alkanes and aromatic VOCs (Table 5-5) that comprise the bulk of anthropogenic emissions observed in ambient air.

TABLE 5-5   Calculated Incremental Reactivities and Kinetic and Mechanistic Reactivities for CO and Selected VOCs for Maximum Ozone Formation Conditions, Based on Scenarios for 12 Urban Areas in the U.S.

Compound	Incremental reactivity, mole O_3/mole C	Kinetic reactivity, fraction reacted	Mechanistic reactivity, mole O_3/mole C
Carbon monoxide	0.019	0.043	0.45
Methane	0.0025	0.0016	1.6
Ethane	0.030	0.049	0.61
Propane	0.069	0.21	0.34
n-Butane	0.124	0.37	0.34
n-Octane	0.081	0.75	0.107
Ethene	0.77	0.81	0.95
Propene	0.82	0.97	0.85
trans-2-Butene	0.81	0.99	0.82
Benzene	0.023	0.21	0.111
Toluene	0.106	0.64	0.17
m-Xylene	0.50	0.96	0.52
Formaldehyde	1.26	0.97	1.30
Acetaldehyde	0.70	0.92	0.77
Benzaldehyde	-0.29	0.95	-0.31
Acetone	0.055	0.058	0.95
Methanol	0.147	0.16	0.93
Ethanol	0.19	0.44	0.42
Isoprene	0.70	1.00	0.70
α-Pinene	0.21	0.99	0.21
Urban mix[a]	0.28		

[a]All-city average urban VOC mix.
Source: Adapted from Carter (1991)

Table 5-5 shows the calculated incremental reactivities and kinetic and mechanistic reactivities for selected VOCs for conditions that simulate those in 12 U.S. cities, with an all-city average VOC mix and the NO_x concentrations varied to yield the maximum amount of ozone (Carter, 1991).

SUMMARY

The general features of the atmospheric chemistry of ozone and its precursors are well understood. The chemistry of the polluted troposphere is considerably more complex than that of a less polluted, methane-dominated troposphere because of the presence of many VOCs of various classes. VOCs in the troposphere are photolyzed and react with OH and NO_3 radicals and ozone to form organic peroxy radicals (RO_2). Subsequent reactions lead to the conversion of NO to NO_2, the generation of OH radicals, and the formation of ozone.

The atmospheric chemistry of anthropogenic VOCs, including alkanes, alkenes, and aromatic hydrocarbons, is generally understood. The kinetics of the initial reactions of the majority of anthropogenic VOCs, and the photolysis rates of these VOCs, have been determined experimentally or can be reliably calculated. However, there are many uncertainties concerning the chemistry of aromatic hydrocarbons, carbonyl compounds, and long-chain alkanes and alkenes.

Biogenic VOCs, including isoprene and the monoterpenes, are believed to foster episodes of high ozone concentrations in the presence of anthropogenic NO_x. These VOCs are highly reactive toward OH and NO_3 radicals and ozone. The mechanisms and products of the important reactions of these compounds are not well understood.

An essential component of air quality models is the chemical mechanism that describes the series of reactions in the troposphere subsequent to emissions of VOCs and NO_x. Chemical mechanisms are constructed using kinetic, mechanistic, and product data and refined by comparison with environmental chamber data. Two mechanisms recently developed for use in urban airshed models (those of Gery et al. [1988a, 1989] and Carter et al. [1986a]) have been tested against a common base of environmental chamber data; hence their close agreement does not guarantee their correctness.

VOCs vary widely in the speed with which they react in the troposphere and the extent to which they promote or inhibit ozone formation. Several VOC reactivity scales have been proposed. One useful measure is incremental reactivity, defined as the amount of ozone formed per unit amount of VOC added to a VOC mixture representative of conditions in urban and rural areas. The ozone-forming potential of a VOC depends on the rate at which it reacts and on the reaction mechanism subsequent to the initial reaction. VOC reactivities generally decrease with increasing VOC/NO_x ratios; at high ratios, ozone formation is NO_x-limited and VOC control is irrelevant.

The following areas of uncertainty need to be addressed before the role of VOCs in the formation of ozone can be assessed in detail:

· The detailed tropospheric degradation reaction mechanisms of the aromatic VOCs are not well understood. In particular, the reactions of the hydroxycyclohexadienyl-type radicals under ambient tropospheric conditions require study.

· The tropospheric reaction mechanisms of biogenic VOCs (for example, isoprene and the monoterpenes) must be investigated, and the gas- and aerosol-phase products determined under realistic atmospheric conditions.

· Reactions under low-NO_x conditions, such that reactions of organic peroxy radicals with HO_2 and other RO_2 radicals and of HO_2 with ozone dominate over the reactions of HO_2 and RO_2 radicals with NO, require study. In addition to the need for further kinetic and mechanism data, environmental chamber studies carried out at low VOC/NO_x ratios, close to rural ambient ratios, would be extremely useful.

· The chemistry of long-chain alkanes and long-chain alkenes (for example, the $\geq C_5$ 1-alkenes) needs to be elucidated.

· The formation of carbonyl compounds during the atmospheric degradation reactions of VOCs and the atmospheric chemistry of these carbonyl compounds require further study. In particular, there is a need for absorption cross-sections, photodissociation quantum yields, and photodissociation product data (as a function of wavelength) for the carbonyl compounds.

· The role of heterogeneous reactions in the chemistry of tropospheric ozone needs to be clarified, expanding on issues raised by Lelieveld and Crutzen (1990).

Long-term research (over a period of 5 years or more) is needed to obtain the basic kinetic and product data required for formulation of detailed chemical mechanisms for the atmospheric degradation of anthropogenic and biogenic VOCs. Short- to medium-term programs (2-5 years) are needed to determine the reactivities of VOCs, either singly or as mixtures, with respect to ozone formation. Short-term studies (1-2 years) will be particularly useful for assessing the effects on ozone formation of conversion from gasoline to alternative fuels and for quantifying the effects of biogenic VOCs on urban, suburban, and rural ozone.

6

VOCs and NO$_x$:
Relationship to Ozone
and Associated Pollutants

INTRODUCTION

Ozone (O$_3$) is produced in the troposphere as a result of a complex set of reactions that involve volatile organic compounds (VOCs) and oxides of nitrogen (NO$_x$). These reactions are discussed in detail in Chapter 5. Because the initial atmospheric concentrations (and corresponding emissions) of VOCs and NO$_x$ are not directly proportional to the maximum ozone concentration ultimately formed, a principal question associated with the VOC-NO$_x$-O$_3$ system is "What is the maximum amount of ozone that can form from a given initial mixture of VOCs and NO$_x$?"

A.J. Haagen-Smit (see, for example, Haagen-Smit and Fox, 1954) first plotted maximum ozone concentrations that result from initial mixtures of VOCs and NO$_x$ on a graph, the axes of which are the initial VOC and NO$_x$ concentrations. Isopleths (lines of constant value) of the maximum ozone concentrations can be constructed by connecting points that correspond to various initial conditions. Each point on a particular isopleth represents the same ozone concentration. Because ozone isopleth diagrams are a concise way to depict the effect of reducing initial VOC and NO$_x$ concentrations on the peak ozone concentrations, they have been used quantitatively to develop control strategies for ozone reduction by the U.S. Environmental Protection Agency's EKMA (empirical kinetic modeling approach) (Dodge, 1977). Figure 6-1 shows a typical set of EKMA ozone isopleths.

In principle, isopleths can be generated directly from smog chamber experiments in which the initial VOC and NO$_x$ concentrations are systematically varied. However, for application to the atmosphere, such data need to be corrected, for example, for chamber-wall effects on the chemical reactions, the relatively high concentrations used, and the level of dilution. Although the ozone isopleth diagram was put forward by Haagen-Smit as an empirical representation of the $VOC-NO_x-O_3$ relationship, the chemistry that gives rise to the characteristic isopleth shape is now well understood. Isopleths are now generated by models that use photochemical reaction mechanisms, and they are tested against smog chamber data.

EKMA, which is largely being supplanted for use in ozone NAAQS attainment demonstration by grid-based models, simulates urban ozone formation in a hypothetical box of air that is transported from the region of most intense source emissions (a center city, for example) to the downwind point of maximum ozone accumulation. Emissions of VOCs and NO_x are assumed to be well mixed in the box, which varies in height, to account for dilution caused by changes in the height of the mixed layer of air; ozone formation is simulated using a photochemical mechanism. By simulating an air mass as a box of air over its trajectory for a large number of predetermined combinations of initial VOC and NO_x concentrations, EKMA generates ozone isopleths that are, to varying degrees, specific to particular cities. Once the maximum measured ozone concentration in a city has been identified, the VOC and NO_x reductions needed to achieve the National Ambient Air Quality Standard (NAAQS) are determined in EKMA from the distances along the VOC and NO_x axes to the isopleth that represents the 120 ppb (parts per billion) peak ozone concentration mandated by the NAAQS.

The location of a particular point on the ozone isopleth is defined by the ratio of the VOC and NO_x coordinates of the point, referred to as the VOC/NO_x ratio. Figure 6-1 shows that the shape of the ozone isopleths depends on the VOC/NO_x ratio. (The lines in Figure 6-1 correspond to VOC/NO_x ratios of 15, 8, and 4 ppb carbon (C)/ppb.) The 0.32 parts per million (ppm) (320 ppb) ozone isopleth, for example, spans a wide range of VOC/NO_x ratios. As a result, the degrees of VOC and NO_x reductions required to move from the 320 ppb isopleth to the 120 ppb isopleth vary considerably depending on the VOC/NO_x ratio at the starting point on the 320 ppb isopleth.

The VOC/NO_x ratio is important in the behavior of the $VOC-NO_x-O_3$ system. Moreover, it has a major effect on how reductions in VOC and NO_x affect ozone concentrations. This chapter examines how ozone isopleths depend on the VOC/NO_x ratio. Chapter 8 discusses the data on ambient VOC/NO_x ratios in urban, suburban, and rural areas of the United States, and Chapter 11 addresses the implications of these ratios for the effectiveness of VOC and NO_x control in reducing ozone.

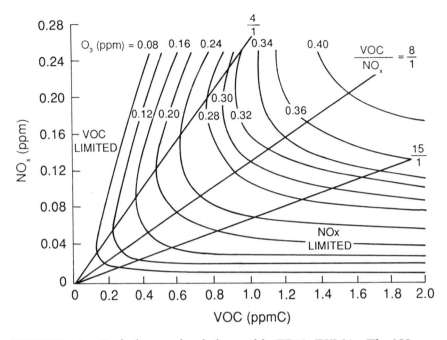

FIGURE 6-1 Typical ozone isopleths used in EPA's EKMA. The NO$_x$-limited region is typical of locations downwind of urban and suburban areas, whereas the VOC-limited region is typical of highly polluted urban areas. Source: Adapted from Dodge, 1977.

CHARACTERISTICS OF OZONE ISOPLETHS

An ozone isopleth diagram characteristically exhibits a diagonal ridge from the lower left to the upper right corner of the graph. The corresponding VOC/NO$_x$ ratio is typically about 8:1, although the shape of the isopleths, and hence the ratio, is sensitive to a number of factors.

It is useful to consider two areas on the graph: those to the right of the ridge line and those to the left. The variation of peak ozone concentration with the VOC/NO$_x$ ratio can be explained on the basis of the atmospheric chemistry discussed in Chapter 5 and summarized as follows:

$$NO_2 + h\nu(\lambda < 420 \text{ nm}) \rightarrow NO + O(^3P) \qquad (6.1)$$

$$O(^3P) + O_2 \xrightarrow{M} O_3 \qquad (6.2)$$

$$O_3 + NO \rightarrow NO_2 + O_2 \qquad (6.3)$$

$$O_3 + h\nu \ (\lambda < 320 \text{ nm}) \rightarrow O(^1D) + O_2 \qquad (6.4)$$

$$O(^1D) \xrightarrow{M} O(^3P) \qquad (6.5)$$

$$O(^1D) + H_2O \rightarrow 2OH \qquad (6.6)$$

$$RH + OH \rightarrow R + H_2O \qquad (6.7)$$

$$R + O_2 \xrightarrow{M} RO_2 \qquad (6.8)$$

$$RO_2(HO_2) + NO \rightarrow RO \ (OH) + NO_2 \qquad (6.9)$$

$$RO + O_2 \rightarrow RCHO + HO_2 \qquad\qquad (6.10)$$

$$RCHO + h\nu \rightarrow R + CHO \qquad\qquad (6.11)$$

$$CHO + O_2 \rightarrow HO_2 + CO \qquad\qquad (6.12)$$

$$OH + NO_2 \overset{M}{\rightarrow} HONO_2 \qquad\qquad (6.13)$$

For VOC/NO$_x$ ratios to the right of the ridge line (characteristic of rural areas and of suburbs downwind of center cities), lowering NO$_x$ concentrations either at constant VOC concentration or in conjunction with lowering VOCs results in lower peak concentrations of ozone. (Chapter 8 investigates VOC/NO$_x$ ratios of different geographic areas.) At these high VOC/NO$_x$ ratios, the system is said to be NO$_x$-limited. In this region of an isopleth, there is an ample supply of organic peroxy radicals (RO_2) and peroxy radicals (HO$_2$) to convert nitric oxide (NO) to nitrogen dioxide (NO$_2$). The only important tropospheric source of ozone is the photolysis of NO$_2$ (Reactions 6.1 and 6.2), so that decreasing the available NO$_x$ leads directly to a decrease in ozone. When the system is NO$_x$-limited, ozone concentrations are sensitive neither to reductions of VOC at constant NO$_x$ nor to the VOC composition.

At VOC/NO$_x$ ratios to the left of the ridge line (characteristic of some highly polluted urban areas) lowering VOC at constant NO$_x$ results in lower peak ozone concentrations; this is also true if VOCs and NO$_x$ are decreased proportionately and at the same time. However, the isopleths in Figure 6-1 indicate that lowering NO$_x$ at constant VOC will result in increased peak ozone concentrations until the ridge line is reached, at which point the ozone concentration begins to decrease. This seemingly contradictory prediction, that lowering NO$_x$ can, under some conditions, lead to increased ozone, results from the complex chemistry involved in ozone formation in VOC-NO$_x$ mixtures (see Chapter 5; Finlayson-Pitts and Pitts, 1986; Seinfeld, 1986). In this region of low VOC/NO$_x$ ratio, the radicals that propagate VOC oxidation and NO-to-NO$_2$ conversion are scavenged (Reaction 6.13) by the relatively high concentrations of NO$_x$. The NO$_2$ effectively competes with the VOCs for the OH radical, slowing RO_2 and HO_2 radical production (Reactions 6.7 and 6.8) relative to that at lower NO_x concentrations. As a result, as NO_x is decreased, more of the OH radical pool is available to react with the VOCs, leading to greater formation of ozone. Ozone is also removed by its rapid reaction with NO, although during the day, the subsequent photolysis of NO_2 (Reactions 6.1 and 6.2) regenerates ozone. The regeneration rate depends on the rate of NO_2 photolysis and hence on solar zenith angle and other factors.

An additional source of free radicals for NO-to-NO_2 conversion is aldehyde (RCHO) production from VOC oxidation (Reactions 6.7-6.10), followed by photolysis of the aldehyde (Reaction 6.11) and secondary Reactions 6.8, 6.9, and 6.12. Reducing the VOCs reduces aldehyde production, resulting in smaller RO_2 and HO_2 radical concentrations, which lowers the rate of NO-to-NO_2 conversion by Reaction 6.9.

The increase in peak ozone concentration at relatively low VOC/NO_x ratios that occurs when NO_x is reduced has been a major issue in the development of ozone control strategies. It is one reason that historically the major emphasis has been on reductions of VOC. One issue that this report addresses is the effectiveness of the "VOC only" approach to ozone control. The isopleth graph shows that NO_x reductions will have significantly different effects depending on the particular VOC/NO_x ratio, which varies significantly within an air basin. Because NO_x generally reacts more rapidly than VOCs in air masses, NO_x is removed preferentially, and areas downwind of major VOC and NO_x sources, such as rural areas, often have relatively high VOC/NO_x ratios. These are the places to the right of the ridge line in Figure 6-1. VOC/NO_x ratios smaller than those at the ridge line occur in some highly polluted urban areas.

UNCERTAINTIES AND SENSITIVITIES OF ISOPLETHS

Because ozone isopleths have been used to develop control strategies, it is important to understand the sensitivity of the shape and separation of the isopleths to uncertainties in the input variables and data used to generate them. These uncertainties include the details of the chemical mechanism used in the model and the initial VOC composition, the effects of which are discussed explicitly in the following sections. In addition, isopleths are sensitive to the ambient concentrations of VOCs, NO_x, and ozone that are available for entrainment into the volume of air being studied, and the VOC and NO_x composition of emissions into the air volume, which change the VOC/NO_x ratio.

Sensitivity to Chemical Mechanism

To predict ozone concentrations that will result from specific concentrations of VOCs and NO$_x$ requires a chemical reaction mechanism. To represent accurately the complete chemistry of the VOC-NO$_x$-O$_3$ system, one would need, in principle, to know which VOCs were present and all of their atmospheric reaction mechanisms and kinetics. Even if such detailed data were available, it is not now practical to include such detailed chemistry for each VOC in air quality models. Consequently, the approach to developing photochemical reaction mechanisms has involved the lumping of VOCs into groups. All reactions of a certain class of VOCs may be represented by those of a single species, or VOCs may be segmented according to the kinds of carbon bonds in the molecules. Because different mechanisms use somewhat different approximations in lumping the VOC chemistry, the ozone concentrations predicted for a given set of initial conditions by different chemical mechanisms will not agree exactly, and the resulting isopleths can differ.

Of the many chemical mechanisms developed in recent years (see, for example, a comparison of 20 models by Hough [1988]), three have gained wide acceptance in the modeling community: the carbon bond mechanism (CBM) (Gery et al., 1989), the CAL mechanism (Carter et al., 1986; Lurmann et al., 1987; Carter, 1990a), and the regional acid deposition model (RADM) (Stockwell 1986, 1988; Stockwell et al., 1990). These mechanisms include periodically updated descriptions of gas-phase reactions that closely reflect current understanding of atmospheric chemistry. The mechanisms have been evaluated against a common, large set of experimental smog chamber data. Statistical analysis of experimental data and model calculations with the CAL mechanism, for example, indicate agreement to $\pm 30\%$ for ozone concentrations (Carter and Atkinson, 1988). Because all of the current generation of models generally rely on the same kinetic and mechanistic data base and evaluations, however, it has been pointed out that agreement among them could be coincidental and cannot guarantee that the mechanisms are correct (Dodge, 1989).

The sensitivity of ozone isopleths to the chemical mechanism used was especially significant as the mechanisms were being developed, when they diverged significantly in their treatment of VOC chemistry (Dunker et al., 1984; Shafer and Seinfeld, 1986; Hough, 1988; Hough and Reeves, 1988; Dodge, 1989). Agreement of the predictions for ozone formation among various chemical mechanisms is much better now, although there are significant uncertainties in the mechanisms, and the model predictions for secondary pollutants other than ozone (e.g., hydrogen peroxide, H$_2$O$_2$) disagree more than do the predictions for ozone (Hough, 1988; Hough and Reeves, 1988; Dodge, 1989, 1990).

The most significant uncertainties are associated with aromatic oxidations and the treatment of carbonyl photolysis. (Until recently, the temperature dependence of peroxyacetyl nitrate (PAN) formation also was controversial, and this had an effect on ozone isopleths especially at low temperatures. However, recent measurements of the temperature dependence of the ratio of the rate constants for the reaction of the acylperoxy radical $[CH_3CO_3]$ with NO vs. NO_2 (see Chapter 5) have confirmed that there is no significant temperature dependence of this rate constant ratio.) The sensitivity to the aromatic oxidation mechanism of model predictions for peak ozone concentrations (as well as sensitivity to the earlier PAN temperature dependence discrepancy) has been examined in some detail by Dodge (1989, 1990), and the effect of VOC composition on the ozone isopleths predicted using each mechanism has been studied by F.W. Lurmann (pers. comm., Sonoma Technology, Santa Rosa, Calif., April 1990).

Sensitivity to VOC Composition

Because different VOCs show widely varying reactivities in terms of ozone formation (see Chapter 5), the peak ozone generated in a given VOC-NO$_x$ mixture and hence the shapes of the ozone isopleths, particularly when the VOC/NO$_x$ ratio is low, are sensitive to the initial VOC composition.

Figure 6-2, for example, shows the ozone isopleths predicted by F.W. Lurmann (pers. comm., Sonoma Technology, Santa Rosa, Calif., April 1990) using the LCC (Lurmann, Carter, and Coyner) mechanism for the base case, in which 5% of the initial VOCs in the boundary layer and 10.7% aloft are aldehydes and also for the case in which only 2% of the initial VOCs in the boundary layer and 4.3% of VOCs aloft are aldehydes. Lurmann generated these isopleths (and those in Figure 6-3) using baseline conditions for a trajectory leading to Glendora, California, on August 24, 1984, when a 390 ppb peak ozone concentration was observed. The VOC composition for the base case was the "all-city average" (see Table 6-1) reported by Jeffries et al. (1989). The mixing height was allowed to increase from 250 meters at 8:00 a.m. to 700 meters at 1:00 p.m., and the concentrations of ozone and VOCs aloft were taken to be 0.10 ppm (100 ppb) and 0.10 ppmC (100 ppbC), respectively. The ozone does not drop to 0 in the low-concentration portions of the graphs in Figures 6-2 and 6-3 because of the assumption of ozone and VOC entrainment into the air mass from aloft.

At a VOC concentration of 1,000 ppbC and a NO$_x$ concentration of 100 ppb, the peak ozone concentration changes from 300 ppb in the base case to 280 ppb for the lower aldehydes case; at 500 ppbC VOCs and 100 ppb NO$_x$,

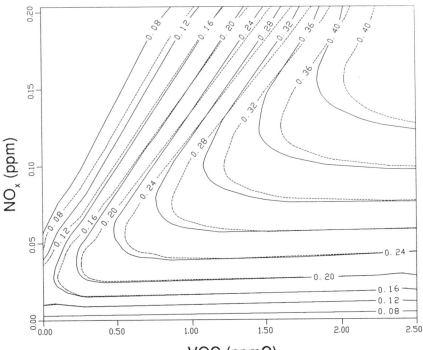

FIGURE 6-2 Ozone (ppm) isopleths generated using the Lurman, Carter, and Coyner (LCC) mechanism and assuming that of the total VOCs (excluding methane), the following percentages are aldeydes: for solid lines 5% in the atmospheric boundary layer (ABL), 10.7% aloft (base case) and for broken lines 2% in the ABL, 4.3% aloft. Source: F.W. Lurmann, Sonoma Technology, Santa Rosa, Calif., pers. comm., April 1990.

the change is from 160 to 120 ppb ozone. These reductions in the peak ozone concentration are expected because of the photochemical reactivity of aldehydes and their efficiency at generating free radicals. A similar sensitivity of the predicted peak ozone concentration to the initial aldehyde concentration and speciation has been noted by Dodge (1990).

Differences in the speciation of VOCs have been shown, using mixtures with compositions very different from those found in the atmosphere, to give significantly different isopleths, although the reactivity of the mixture with

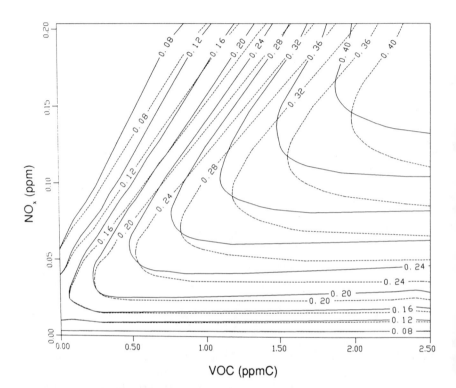

FIGURE 6-3   Ozone (PPM)isopleths generated using the Lurman, Carter, and
Coyner (LCC) mechanism and VOC compositions (including methane) typical
(Jeffries et al., 1989) of Washington D.C. (higher aromatic, lower alkane and
alkene content), solid lines, and Beaumont, Texas (higher alkane and ethene,
lower aromatic content), broken lines. Source: F.W. Lurmann, Sonoma Technology, Santa Rosa, Calif., pers. comm., April 1990.

respect to OH is kept constant (Carter et al., 1982). For changes in VOCs more typical of those observed in urban areas, there also is some sensitivity of the isopleths to changes in VOC composition. For example, Figure 6-3 shows the isopleths predicted using the LCC mechanism for two VOC profiles that differ significantly from an all-city average VOC composition (Jeffries et al., 1989). (The base case is shown in Figure 6-2). These two profiles are typical of two extremes: Washington, D.C. (solid link in Figure 6-3a), which has a higher aromatic but lower alkane and alkene content compared with the all-city average, and Beaumont, Texas, which has a higher than average alkane and ethene but lower aromatic content (see composition data in Table 6-1). The isopleths are particularly sensitive to VOC composition at low VOC/NO$_x$ ratios. For example, at concentrations of 1.00 ppmC (1,000 ppbC) for VOCs and 0.15 ppm (150 ppb) NO$_x$, peak ozone concentrations of ~220 ppb and 160 ppb were predicted for Washington, D.C., and Beaumont, Texas, mixtures, respectively.

OTHER LIMITATIONS OF ISOPLETHS
FOR EVALUATION OF CONTROL STRATEGIES

The ozone isopleth diagram, as generated in EPA's EKMA, has been used for determining the percentage reductions in VOCs and NO$_x$ needed to attain the NAAQS for urban areas. This approach has several well-recognized limitations. One is that such isopleth diagrams reflect a 1-day simulation in one location and do not apply to multiday episodes of high concentrations of ozone and downwind areas. Furthermore the predicted ozone concentrations from 1-day simulations can be quite sensitive to the initial conditions used. Another limitation in the application of these isopleth diagrams lies in the difficulty of selecting the appropriate VOC/NO$_x$ ratio in defining the base-year point on the peak ozone isopleth. VOC/NO$_x$ ratios are discussed in detail in Chapters 8 and 11. Ambient VOC/NO$_x$ ratios vary not only from one location to another in a particular area, but also with time of day. For example, VOC/NO$_x$ ratios from 2 to 40 have been measured in different locations in various urban areas (Baugues, 1986). In addition, the ratio can vary significantly from ground level to aloft as a result of emissions from tall stacks and pollutant carry-over from previous days (Altshuller, 1989).

One extremely important problem in evaluating control strategies with any air-quality model is that measured ambient VOC/NO$_x$ ratios are significantly larger than those calculated on the basis of current emissions inventories. This phenomenon, which could be a result of underestimation of anthropogenic VOC emissions, failure to sufficiently include biogenic VOCs, the trapping of NO$_x$ emissions aloft, differences in averaging times between measure-

TABLE 6-1   Speciation of VOCs for Washington, D.C., Beaumont, Texas, and an All-City Average[a] Used to Generate Figures 6-2 and 6-3

Species in LLC Mechanism	Washington	Beaumont	All-city average[a]
ALK4 (C$_4$ and C$_5$ alkanes)	0.219	0.305	0.214
ALK7 ($\geq$ C$_6$ alkanes)	0.257	0.285	0.280
ETH (ethene)	0.033	0.047	0.037
PRPE (terminal alkenes)	0.032	0.040	0.048
TBUT (internal alkenes)	0.017	0.024	0.029
TOL (mono-alkylbenzenes)	0.108	0.042	0.080
XYL (di-akylbenzenes)	0.102	0.048	0.079
TMB (tri-alkylbenzenes)	0.055	0.027	0.042
FORM (HCHO)	0.020	0.020	0.020
ALD (CH$_3$CHO)	0.030	0.030	0.030
NRHC (Nonreactive HC)	0.127	0.132	0.141

[a]Represents 47 city sites. (Five cities had two sites.)
Source: Jeffries et al., 1989

ments and emissions profiles, and VOC and NO$_x$ measurement errors (Seinfeld, 1988; Altshuller, 1989), is addressed in Chapter 9.

It must be concluded that the concept of a single VOC/NO$_x$ ratio for an urban area has severely limited utility in quantitative assessment of control strategy options. The major limitation in quantitatively applying the conventionally generated isopleths, such as those in Figure 6-1, to the development of control strategies is the difficulty in adequately representing an entire air basin or region, where the VOC/NO$_x$ ratio can vary significantly from areas close to emissions sources to those downwind. This limitation has been overcome through the use of three-dimensional airshed models that are used to generate isopleths as a function of location in the region. This second- generation approach to isopleth development for an air basin has major advantages in being able to take into account transport, meteorology, and local emissions far more realistically than is possible with a model like EKMA, and hence to allow control strategies to be examined from the perspective of an entire air basin rather than separate locations. It allows the effect of precursor controls on peak ozone concentrations to be examined regardless of where the maximum occurs in an air basin.

For example, Milford et al. (1989) applied a grid-based airshed model to the Los Angeles basin for Aug. 30-31, 1982. The effects of reducing precursor VOC and NO$_x$ emissions were examined for five locations within the air basin, from downtown Los Angeles in the western part of the air basin to San Bernardino in the east (see detailed discussion in Chapter 11.) Figure 6-4 shows the effect of VOC and NO$_x$ controls on peak ozone formation in the Los Angeles air basin as a whole, irrespective of the location of the peak. The isopleths in Figure 6-4 show that when the air basin as a whole is considered, as much as 80% control of VOCs alone will not result in attainment of the NAAQS; if biogenic VOC emissions were included, VOC control would be even less effective. In addition, in contrast to what the EKMA isopleths show, the greater number of "L" shaped isopleths in Figure 6-4 show that basin-average peak concentrations of ozone will not necessarily increase as NO$_x$ is decreased.

The second-generation methods also have limitations. For example, because by their nature they are episode-specific, carrying out such calculations for all regions in the United States is not now feasible. In addition, many of the uncertainties associated with the more conventionally generated isopleths, such as those associated with the chemical mechanisms, also apply to this approach.

EFFECTS OF VOC AND NO$_X$ CONTROL ON OTHER SPECIES

Although this report focuses on ozone control, reducing concentrations of VOCs, NO$_x$, or both can affect the concentrations of many secondary pollutants formed in VOC-NO$_x$ mixtures, some of which can also be hazardous to human health, plants, or materials. It is thus important to recognize that

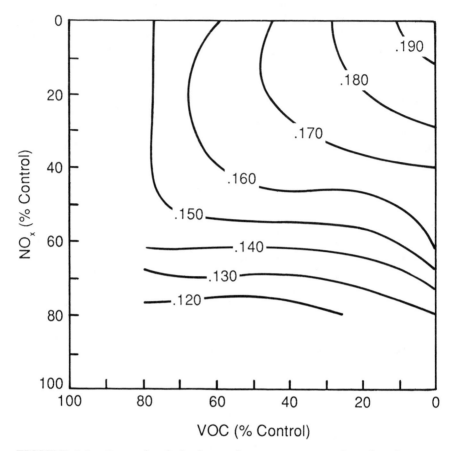

FIGURE 6-4 Ozone isopleths for peak ozone concentrations (ppm) regardless of lcoation in the Los Angeles air basin. A decrease in percent control along an axis corresponds to a higher concentration of a precursor in the atmosphere. Source: Milford et al., 1989.

concurrent with ozone control, changes in other secondary pollutants will occur as well. The next sections discuss some of the most important secondary pollutants. The intent is not to review all secondary pollutants in detail or to describe how each will respond to lower VOC and NO_x concentrations, but rather to illustrate other significant changes that may accompany reductions in ozone.

NO$_2$, HNO$_3$ and Particulate (Inorganic) Nitrate

Decreasing NO$_x$ emissions will lower concentrations of nitrogen dioxide, NO$_2$, which is also a criteria pollutant. Russell et al. (1988b) showed in simulations for Los Angeles that reducing NO$_x$ emissions should decrease both peak NO$_2$ and the associated nitric acid and nitrate aerosols that form from NO$_2$.

Nitrates from nitric acid, HNO$_3$, are a significant component of acid aerosols, which are under consideration for inclusion on the list of criteria air pollutants (Lipfert et al., 1989). These particles are typically in the respirable 0.1-1.0 micrometer (μm) size range, the size that also contributes the most to degradation of visibility. Nitric acid also is a significant component of acid rain and fog formed in ambient air from NO$_2$ through a variety of reactions (see Chapter 5):

$$OH + NO_2 \xrightarrow{M} HONO_2, \qquad (6.14)$$

$$O_3 + NO_2 \rightarrow NO_3 + O_2 \qquad (6.15)$$

$$NO_3 + NO_2 \leftrightarrow N_2O_5 \qquad (6.16)$$

$$N_2O_5 + H_2O \xrightarrow{surface} 2HNO_3 \qquad (6.17)$$

$$NO_3 + HCHO \rightarrow HNO_3 + CHO \qquad (6.18)$$

$$NO_3 + RH \rightarrow HNO_3 + R \qquad (6.19)$$

Reaction of HNO_3 with ammonia, NH_3, or absorption into aerosol and fog and cloud droplets incorporates nitrate into the condensed phase.

Most light extinction in urban atmospheres is the result of light scattering and absorption by particles, especially those in the 0.1-1.0 μm range. Several studies (see Finlayson-Pitts and Pitts, 1986) treat the relationship between the light-scattering coefficient (b_{sp}) associated with these fine particles measured in a number of urban areas and their chemical composition in the following form:

$$b_{sp} = a_o + \sum a_i M_i \qquad (6.20)$$

where M_i is the mass concentration of the ith chemical species (g/m^3), a_i is the mass scattering coefficient (m^2/g) associated with this particular species, and a_o is a constant for the data set. Light scattering depends strongly on the concentrations of sulfate, nitrate, and carbon in the air.

Table 6-2 lists some mass-scattering coefficients found for particles collected in a variety of locations. The increase in the a_i for nitrate with year could result from the collection of gaseous nitric acid as a filter artifact in earlier studies (Appel et al., 1985). The most recent studies, in which this artifact should be minimized, show that, on a mass basis, nitrate is of comparable importance to sulfate with respect to light scattering. Thus reductions in NO_x associated with ozone control are expected to affect the nitrate component of light scattering.

Nitrous Acid

Photolysis of nitrous acid, HONO, is believed to be an important early-morning source of OH, and it acts as a major initiator of VOC oxidation at dawn. Figure 6-5 shows the contribution to OH generation from three major sources under typical conditions in a polluted urban atmosphere as a function of time of day (Winer, 1986). HONO is predicted to be the greatest source of OH at dawn. Similar conclusions were reached by Rodgers (1986) for Atlanta, based on HONO measurements in that area (Rodgers and Davis, 1989).

The effects of HONO on ozone formation have not been studied extensively because there is no reliable large data base on ambient HONO concentrations or their sources. One study (Lurmann et al., 1986b) suggests that chang-

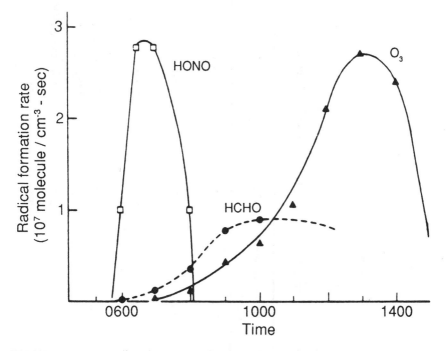

FIGURE 6-5   Predicted sources of HO radicals as a function of time of day for a typical polluted urban atmosphere. Source: Winer, 1986.

es in HONO will be reflected primarily in the timing of the ozone peak, rather than in its absolute value. Such a shift in the location of the ozone peak could change the population exposed to this peak.

The sources of HONO in ambient air are not clear, although laboratory studies (see Finlayson-Pitts and Pitts (1986) for a review to 1985; Akimoto et al., 1987; Svensson et al., 1987; Jenkin et al., 1988; Lammel and Perner, 1988) suggest that heterogeneous reactions of NO_2 are major contributors:

$$2NO_2 + H_2O \xrightarrow{\text{surface}} HONO + HNO_3 \qquad (6.21)$$

Table 6-2 Reported Mass Scattering Coefficients (a_i) in Units of m^2/g for Fine Particles Containing Sulfate, nitrate, and Carbon in Various Locations

Location	Sulfate as $(NH_4)_2SO_4^c$	Nitrate as $NH_4NO_3^c$	Carbonaceous particles[a]	ΔMass[i]	Reference
Los Angeles	5.4	$2.0 + 4.6\ \mu^2$	N.S.[b]	2.0	White and Roberts, 1977
Denver	5.9	2.5	3.2	1.5	Groblicki et al., 1981
Detroit	6.2	N.S.	3.1	1.7	Wolff et al., 1982
China Lake, California	4.3	d	1.5	1.0[e]	Ouimette and Flagan, 1982
Portland, Oregon	4.9	4.4	5.0	2.1	Shah et al., 1984
Average of San Jose, Louisiana, and Riverside	$3.6 + 5.9\ \mu$	$4.9 + 4.8\ \mu$	4.5[a]	0.4[f]	Appel et al., 1985[g]

| Western Netherlands | 4.8 | 8.6 | N.D.[h] | 4.7 | Diederen et al., 1985 |

[a]Combination of organic and elemental carbon except in studies of Appel et al. (1985) where the coefficient is for elemental carbon only.

[b]Not significant.

[c]μ = relative humidity/100.

[d]Present only at very low concentrations in these samples.

[e]In addition to sulfate, carbonaceous aerosol, and the remainder, b_{sp} was found to be correlated to the crustal species Fe, Ca, and Si with $a_i = 2.4$ m^2/g.

[f]Not significantly different from zero at $p = 0.90$.

[g]Data adjusted to reflect $(NH_4)_2SO_4$ and NH_4NO_3 stoichiometry. Correlation of b_{sp} with coarse sulfate also was found with $a_i = 13.4$ m^2/g; this could be due to a correlation of coarse sulfate with some other efficient light scatterer such as sea salt particles.

[h]Not determined.

[i]ΔMass = (Total mass - sulfate - nitrate - carbonaceous particles)

Source: Finlayson-Pitts and Pitts, 1986. Adapted from Shah et al., 1984.

181

(The reaction of OH with NO produces only a small steady-state concentration of HONO during the day because it is photolyzed rapidly). In addition, direct emissions of HONO from combustion sources have been observed (Pitts et al., 1984c, 1989).

Because NO_2 is the most likely precursor to HONO, reductions in NO_2 also should reduce ambient concentrations of this photochemically labile species. However, given the uncertainty in the kinetics and mechanisms of HONO production, quantitatively predicting the relationship between control of NO_2 and HONO is not now possible.

Peroxyacetyl Nitrate

PAN is formed from the reaction of acetylperoxy radicals with NO_2:

$$\underset{CH_3\overset{\overset{\textstyle O}{\|}}{C}OO}{} + NO_2 \leftrightarrow \underset{CH_3\overset{\overset{\textstyle O}{\|}}{C}OONO_2}{} \tag{6.22}$$

PAN thermally decomposes over the range of temperatures found in the atmosphere, reforming acetylperoxy radicals and NO_2. If NO is present, the acetylperoxy radical reacts with NO, so that PAN is permanently removed.

PAN is important as a plant toxicant and, through Reaction 6.22, transports NO_x over relatively large distances through the atmosphere. Its rate of decomposition significantly increases with temperature, so that it can be formed in colder regions, transported, and then decomposed to deliver NO_2 to warmer regions (Singh and Hanst, 1981; Hov, 1984a).

NO_3 and N_2O_5

The nitrate radical (NO_3) and dinitrogen pentoxide (N_2O_5) are both formed from the reactions of NO_2 with O_3:

$$NO_2 + O_3 \rightarrow NO_3 + O_2 \tag{6.23}$$

$$NO_3 + NO_2 \leftrightarrow N_2O_5 \qquad (6.24)$$

Both NO$_3$ and N$_2$O$_5$ are believed to be significant in nighttime atmospheric chemistry (NO$_3$ photolyzes rapidly at dawn, depleting both NO$_3$ and N$_2$O$_5$ by shifting the equilibrium between NO$_2$, NO$_3$, and N$_2$O$_5$ toward NO$_2$). As discussed above, hydrogen abstraction reactions of NO$_3$ form nitric acid, as does the hydrolysis of N$_2$O$_5$, which is a major source of nitric acid in the atmosphere (Russell et al., 1985).

NO$_3$ works in the nighttime oxidation of naturally produced organic compounds, such as isoprene and the pinenes, as well as dimethylsulfide and methyl mercaptan (Finlayson-Pitts and Pitts, 1986).

From Reactions 6.23 and 6.24 forming NO$_3$ and N$_2$O$_5$, if either ozone alone or ozone and NO$_2$ are reduced, lower concentrations of NO$_3$ and N$_2$O$_5$ would be expected, thus decreasing their contribution to acid deposition and the oxidation of organic compounds at night.

Other Nitrated Species

The formation of mutagenic nitrated polycyclic aromatic hydrocarbons (PAHs) has been observed in the atmospheric reactions of PAHs (see Finlayson-Pitts and Pitts, 1986; Pitts, 1987; Arey et al., 1989, for reviews). Some nitrated species are not found in significant amounts in particles directly emitted from combustion sources but are characteristic of atmospheric reactions of PAHs in air. For example, the OH-initiated oxidation of fluoranthene leads to formation of 2-nitrofluoranthene, which has been identified in organic extract of ambient particulate matter:

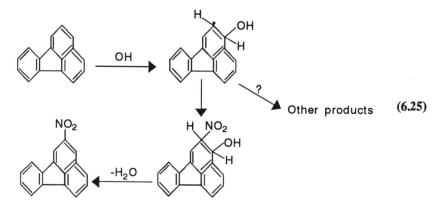

                                                                    Other products   (6.25)

This nitrated product is a direct mutagen, which in at least some samples of ambient PAHs, contributes significantly to the total observed mutagenicity (Pitts, 1987). Similarly, NO_3 reacts, at least in part, to form nitrogen derivatives of some PAHs (Atkinson et al., 1990):

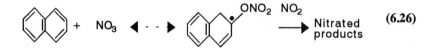

$$(6.26)$$

There also is evidence for the formation of one or more unidentified NO_y species (the sum of the oxides of nitrogen) in the atmosphere. For example, Singh et al. (1985), Fahey et al. (1986), Buhr et al. (1990), and Ridley (1989) have observed in ambient air measurements that the sum of individual measurements of the species ($NO + NO_2 + HNO_3$ + particulate NO_3^- + PAN) was less than the total NO_y measured by detecting NO (via its chemiluminescence with ozone) after reduction using a gold/CO converter. The discrepancy between the two measurements increases with the extent of photooxidation, suggesting that the unidentified species is a stable product of VOC-NO_x reactions, possibly an organic nitrate or nitrates.   Finally,   several   inorganic halogen-containing species, including $ClNO$, $BrNO$, $ClNO_2$, and $BrNO_2$, can be formed in coastal marine areas by the reactions of NO_2, N_2O_5, and possibly NO_3 with the components of sea salt (Finlayson-Pitts, 1983; Finlayson-Pitts et al., 1989a,b, 1990; Livingston and Finlayson-Pitts, 1991). These species are highly labile photochemically, and could play a role in ozone formation, in the case of the chlorine compounds, or ozone destruction, in the case of the bromine derivatives. For example, chlorine atoms will help initiate photochemical oxidation in much the same way that OH does, contributing to NO-to-NO_2 conversion and hence to ozone formation:

$$N_2O_5 + NaCl \rightarrow ClNO_2 + NaNO_3 \qquad (6.27)$$

$$ClNO_2 + h\nu \rightarrow Cl + NO_2 \qquad (6.28)$$

$$Cl + RH \rightarrow HCl + R \qquad (6.29)$$

Such oxidation of VOCs by chlorine atoms could account for 10% or more of the initiation of tropospheric VOC oxidation in marine urban areas such as Los Angeles (Livingston and Finlayson-Pitts, 1991). While the reaction of chlorine with ozone is also fast, the total concentration of VOCs in urban and rural areas is usually large enough compared with that of ozone that Reaction 6.29 is expected to dominate the tropospheric chlorine atom loss in those regions.

Atomic bromine, on the other hand, reacts only very slowly with organics but rapidly with ozone:

$$Br + O_3 \rightarrow BrO + O_2 \qquad\qquad (6.30)$$

The fate of the BrO radical depends in part on the concentration of NO$_2$; for example, in typical urban areas, NO$_2$ sequesters the bromine in the form of BrONO$_2$, so that significant chain destruction of O$_3$ does not occur. However, under NO$_x$-depleted conditions in remote areas, BrO could photolyze or react with itself or with HO$_2$, thus regenerating reactive bromine atoms and leading to a chain destruction of ozone. Reductions in NO$_x$ and ozone would be expected to decrease the formation of these photochemically labile species.

Secondary Organic Particles and Acids

Recent atmospheric measurements suggest that organic acids constitute a significant fraction of total acidity, even in areas such as Los Angeles, which has large NO$_x$ emissions (Keene et al., 1983; Grosjean, 1989, 1990; Grosjean et al., 1990a; Grosjean and Parmar, 1990). The sources are not well established, but both direct emissions from mobile sources and atmospheric oxidation of VOCs (reactions of Criegee biradicals from ozone-alkene reactions, for example) have been suggested. Reduction of VOCs and ozone will thus reduce the concentrations of these organic acids as well.

Organic particles are believed to be largely the result of atmospheric photo-oxidation of VOCs, particularly aromatic hydrocarbons and long-chain and cyclic alkenes (Grosjean and Seinfeld, 1989). Their ambient concentrations are expected to depend on the concentrations not only of precursor VOCs, but also of the oxidants ozone, OH, and perhaps NO$_3$. However, the relationship between organic particle concentrations and reductions in VOCs and NO$_x$ cannot be predicted quantitatively, given the current lack of understanding of the mechanisms of organic particle generation in the atmosphere.

SUMMARY

Isopleth diagrams are a convenient means of representing the complex relationship between initial concentrations of volatile organic compounds (VOCs) and oxides of nitrogen (NO_x) and peak concentrations of ozone subsequently formed via chemical reactions in the troposphere. Ozone isopleth diagrams generated for urban areas using laboratory experiments or models (either EKMA [empirical kinetic modeling approach] or three-dimensional, grid-based models) show that the extent of reduction of peak ozone concentrations resulting from reductions of precursor emissions depends on the initial VOC/NO_x ratio. At higher VOC/NO_x ratios (greater than ~8-10), ozone concentrations are relatively insensitive to VOC concentrations, and NO_x control is more effective in lowering ozone. Measurements of 6:00 a.m. to 9:00 a.m. average VOC and NO_x show that most urban areas appear to fall into this category, with measured VOC/NO_x ratios ≥8-10 (see chapter 8). At VOC/NO_x ratios less than ~8-10, found in some highly polluted urban areas, such as Los Angeles, lowering VOC reduces ozone, whereas NO_x control might actually increase ozone at some locations.

Isopleth diagrams generated in the traditional manner, using EKMA, have several shortcomings, especially their failure to treat the effects of VOC and NO_x controls throughout an airshed. Because the VOC/NO_x ratio generally increases as an air mass moves downwind from major NO_x sources, control strategies derived from the isopleths for upwind locations often are inappropriate for downwind areas within the same air basin.

This problem has recently been overcome by applying three-dimensional urban airshed models to generate ozone isopleth diagrams for some air basins where the requisite detailed model input is available. However, because of the limitations associated with developing information for urban airshed models (see Chapter 10), generating such isopleth diagrams for all regions in the United States is not now feasible. The relationship between VOC and NO_x control and ozone concentrations, as determined by three-dimensional models, is discussed further in Chapter 11.

Changes in VOCs and NO_x will, because of their complex chemical interactions, also lead to changes in a variety of other pollutants associated with ozone, such as nitric acid, peroxyacetyl nitrate, nitrogen dioxide, and aerosol particles. Some of these pollutants have known harmful effects on human health and welfare. Hence, it is important to recognize that control strategies implemented for ozone will simultaneously affect other species.

7

Techniques For Measuring Reactive Nitrogen Oxides, Volatile Organic Compounds, and Oxidants

INTRODUCTION

A key element in advancing the understanding of tropospheric production of ozone is the ability to make unequivocal measurements of the concentrations of the ozone precursors, the reactive nitrogen oxides (NO_y compounds) and volatile organic compounds (VOCs). Because tropospheric chemistry is shaped by oxidants, such as ozone, hydrogen peroxide (H_2O_2), and oxidizing radicals, the concentration of these compounds also must be measured to test present understanding of atmospheric oxidation mechanisms.

If measurements are to be meaningful, reliable instruments and techniques are necessary. Therefore, it is vital to have trustworthy estimates of the uncertainties in the observations, because these observations are the touchstones against which theoretical understanding is tested. With such estimates, observations and theory can be compared meaningfully; the results from separate studies can be merged reliably; gradients in observations over large distances can be characterized credibly from separate data sets; and time patterns from different monitoring networks can be used to establish longer trend records. In addition, the ability to make accurate measurements of the concentrations of these species, coupled with an adequate theoretical understanding, provides the ability to test whether control measures to regulate the emission of anthropogenic oxidant precursors are effectively curbing the concentrations of these compounds in the atmosphere.

The atmospheric chemistry community has devised a way to address measurement uncertainties that is arduous but effective. It uses formal, rigorous, and unbiased inter- and intramethod comparisons of techniques and instruments. The most instructive of those instrument intercomparisons have the following features:

- Several different techniques are used to measure the same species.
- Insofar as possible, measurements are made at the same place and time and under typical operating conditions.
- Accuracy and precision estimates are stated in advance of the study.
- Atmospheric samples are "spiked" with known amounts of species that are potential artifacts. Where possible, samples of ambient air as well as synthetic air are spiked.
- All investigators prepare their results independently and separately from the others.
- Investigators jointly (or through an independent party) compile separate results and assess the state of agreement or disagreement.
- Results and conclusions are published in a peer-reviewed journal.
- The process is repeated occasionally.

There have been several field studies devoted specifically to the assessment of instrument reliability, as opposed to obtaining data to answer a geophysical question (Hoell et al., 1987a,b; Hering et al., 1988; Fehsenfeld et al., 1987, 1990). Because these intercomparisons provide the only objective assessment of instrument capability, the scientific community's knowledge of the accuracy of current instrumentation depends heavily on these results. The instruments and techniques available for measurement of NO_y, VOCs, and oxidant species; the basic operating principles of these devices; and highlights of the tests done thus far to determine instrument reliability are summarized below.

MEASUREMENT TECHNIQUES
FOR OXIDES OF NITROGEN AND
THEIR OXIDATION PRODUCTS

The reactive oxides of nitrogen in the atmosphere are largely nitric oxide (NO) and nitrogen dioxide (NO_2), known together as NO_x. During the daytime, there is a rapid interconversion of NO and NO_2. One important by-product of this interconversion is the photochemical production of ozone in the troposphere. In addition, NO_x is converted to a variety of other organic and inorganic nitrogen species. These are the compounds that make up the reactive nitrogen family, NO_y (NO_x + organic nitrates + inorganic nitrates).

This chemistry is described in detail in Chapters 5 and 6. The techniques that have been developed to measure these compounds are discussed below.

Nitric Oxide

The reliability of techniques to measure NO has been established rigorously. Two fundamentally different methods have been compared: chemiluminescence (NO-O_3 chemical reaction and emission of radiation from the nitrogen dioxide product) and laser-induced fluorescence (absorption of radiation by nitric oxide and then reradiation at different wavelengths by the excited nitric oxide). During two separate tests of these techniques (Hoell et al., 1985, 1987a), chemiluminescence instruments and a laser-induced fluorescence instrument measured ambient concentrations simultaneously at a rural site and from an aircraft. The data agreed within 30% in all of these chemically different environments and over concentrations of NO spanning a range of 0.005 to 0.2 parts per billion (ppb). These results strongly indicate that NO can be measured reliably by either technique under most field conditions.

Nitrogen Dioxide

Many techniques have been developed to measure nitrogen dioxide, but few can measure NO_2 at concentrations below parts per billion, and few have been demonstrated to be free of interference from other atmospheric constituents. The standard way to measure NO_2 in almost all air quality studies has been to use surface-conversion techniques to convert NO_2 to NO and to subsequently detect the NO by chemiluminescence. The conversion techniques include the use of heated catalytic metal surfaces and surfaces coated with ferrous sulfate or other compounds. However, the development of the photolytic NO_2-to-NO converter several years ago (Kley and McFarland, 1980) offered a potentially more specific conversion technique, albeit less simple. A recent comparison (Fehsenfeld et al., 1987) made a detailed study of the performance of surface and photolytic methods. In this study, the ferrous sulfate and photolytic converters agreed well at NO_2 concentrations of 1 ppb and greater. However, the ferrous sulfate converter systematically reported higher values at lower concentrations, reaching a factor of 2 higher at 0.1 ppb. Spiking tests showed that the ferrous sulfate converter also was converting peroxyacetyl nitrate (PAN) to NO. Hence, whenever PAN is significant in comparison to the NO_2, the ferrous sulfate converter gives results that overestimate the concentration of NO_2. A heated molybdenum oxide surface converter was found to convert NO_2, PAN, and HNO_3 to NO, indicating that heated-surface converters also cannot be considered specific for NO_2 or for NO_x.

The photolytic converter/chemiluminescence and ferrous sulfate/chemiluminescence techniques were compared during aircraft flights over the eastern Pacific Ocean and the southwestern United States at altitudes of 0.6 to 7.3 kilometers (km) (Ridley et al., 1988). In agreement with the intercomparison discussed above, the ferrous-sulfate-equipped instrument was found to be much less specific for NO_2. It registered levels about three times larger than the photolytic converter, presumably because of the conversion of PAN and perhaps other organic nitrates to NO.

Newer technology is emerging to measure NO_2. Three techniques that show considerable promise are photofragmentation/2-photon laser induced fluorescence (LIF), tunable-diode laser absorption spectrometry (TDLAS), and luminol chemiluminescence. The LIF and TDLAS techniques provide specific spectroscopic methods to measure NO_2; the luminol technique provides a sensitive, portable method with low power requirements. Two recent studies have tested these techniques against the photolysis/chemiluminescence technique (Fehsenfeld et al., 1990; Gregory et al., 1990).

A ground-based comparison (Fehsenfeld et al., 1990) tested the photolysis/chemiluminescence technique against the TDLAS and luminol techniques. For NO_2 concentrations above 0.2 ppb, no interferences were found either for the photolytic converter/chemiluminescence technique or for TDLAS. However, interpretation of the results from TDLAS showed that correlation coefficients should not be used to select the data that are near the detection limit of NO_2 for the instrument (Fehsenfeld et al., 1990). At these levels the background noise is normally distributed about the reference NO_2 spectrum. Selection of the data with high correlation coefficients would lead to NO_2 concentrations that are too high (Fehsenfeld et al., 1990). This test indicated that interferences from PAN and ozone influence the NO_2 measurements made using the luminol technique. However, during the comparison those interferences were consistent enough that for NO_2 concentrations above 0.3 ppb, they could be corrected using simultaneously measured values of ozone and PAN (Fehsenfeld et al., 1990). Techniques are being developed to remove or separate interfering substances from the ambient air prior to analyses by the luminol detector. However, the effectiveness of these techniques has not been verified by field tests.

An airborne comparison of TDLAS with LIF and photolytic converter/chemiluminescence was conducted by Gregory et al. (1990). The intercomparison of these three instruments in ambient air for NO_2 >0.1 ppb indicated a general level of agreement among the instruments of 30-40%. For NO_2 < 0.05 ppb the results indicated that TDLAS overestimates the NO_2 mixing ratio, presumably because of the use of correlation coefficients as the data selection criterion. At these low concentrations, agreement between LIF and photolytic converter/chemiluminescence measurements was within 0.02 ppb with an equal tendency for one to be high or low compared to the other. This 0.02 ppb agreement is typically within the expected uncertainties of the two techniques at NO_2 mixing ratios < 0.05 ppb.

It is believed that, properly used, the LIF, TDLAS, and the photolytic converter/chemiluminescence techniques measure NO_2 concentrations well below 0.1 ppb, free of significant artifact or interference. These techniques should therefore be able to measure NO_2 concentrations throughout the troposphere above North America.

Peroxyacetyl Nitrate

Two instruments, both of which use cryogenically enriched sampling with electron-capture gas chromatography detection, have been intercompared in the remote maritime troposphere (Gregory et al., 1990). At mixing ratios of <0.1 ppb, the two instruments differ on average by 0.017 ppb with a 95% confidence interval of ± 0.009 ppb. At PAN mixing ratios of 0.1-0.3 ppb, the difference between the instruments was 25% ± 6%. A linear regression equation developed by comparing all data <0.3 ppb from the two techniques gave a line with a slope of 1.34 ± 0.12 and an intercept of 0.0004 ± 0.012 ppb. Although one instrument was consistently high relative to the other for ambient measurements, these levels of agreement were usually within the stated accuracy and precision of the two instruments. These results are re-assuring. Nevertheless, their significance is reduced by the similarity in the design and operation of the two instruments.

Nitric Acid

A test by Hering et al. (1988) focused on the capability to measure nitric acid (HNO_3). Over an 8-day period at a site with urban and suburban charac-teristics, six methods were used to make simultaneous measurements: filter pack, denuder difference, annular denuder, transition flow reactor, tunable-diode laser, and Fourier transform infrared spectrometer. The reported concentrations of HNO_3 varied by more than a factor of 2. These differences were substantially larger than the estimated precision of the instruments. The tests indicated that artifacts or interferences exist for some of the sampling methods associated with either the field sampling components (e.g., inlet lines), the operating procedures, detector specificity, or alteration during sampling in the physical or chemical make-up of the ambient air, such as shifts in the gas- and solid-aqueous-phase equilibrium of HNO_3, ammonia, and ammonium nitrates.

Several conclusions could be drawn from Hering's data set. The larger percentage differences in the techniques that were observed at higher HNO_3 concentrations and the dependence of the differences on day or night sampling suggest uncontrollable shifts of the equilibrium (ammonium nitrate evaporation) in samples obtained by some instruments. The annular denuder exhibited poor intramethod precision for HNO_3, and its average value was substantially below the means of the spectroscopic methods and those of all methods. The results from tungstic acid adsorption tubes and filter packs (>8-hr sample) deviated substantially from those two means. The filter packs exhibited a positive bias (systematically higher than average HNO_3 concentrations) that increased as the sampling time average increased, indicating an artifact due to ammonium nitrate particle evaporation to release HNO_3 (and ammonia). The denuder difference, transition-flow reactor, filter pack (<8-hr sample), and spectroscopic methods were in good agreement. This comparison provides a valuable start in assessing the problems of reliable measurement of HNO_3.

A more recent test involved three different measurement approaches: nylon filter collection, tungstic oxide denuder, and TDLAS (Gregory et al., 1990). In general, the filter measurements were high relative to those reported by the denuder. No correlation was observed between the filter and denuder techniques for HNO_3 < 0.15 ppb. Below 0.3 ppb, the difference between simultaneous measurements from the denuder and filter instruments was greater than the expected accuracy and precision stated for each instrument for more than 75% of the measurements. Comparing the denuder technique and TDLAS, TDLAS measurements were consistently higher; for HNO_3 > 0.3 ppb, TDLAS results were systematically higher by a factor of approximately two. There was only one instance of overlap among all three techniques at concentrations of HNO_3 well above detection limits. In that case, the measurements from the filter and TDLAS were in agreement, whereas those from the denuder (with only a 35% overlap) were about a factor of two lower. The paucity of simultaneous measurements from all three instruments prevented firm conclusions being drawn from the intercomparison. However, it was clear that there was substantial disagreement among the three techniques, even at mixing ratios well above their respective detection limits. These intercomparisons clearly indicate that current techniques do not allow the unequivocal determination of HNO_3 in the range of concentrations expected in the nonurban atmosphere.

One final comparison is worthy of note. A new technology is emerging to measure HNO_3 and other soluble gases. For HNO_3, this approach involves the absorption of HNO_3 contained in air into ultraclean water followed by analysis with ion chromatography. The approach has been obvious for decades, but only recently have water purification and handling techniques become sufficiently advanced to allow measurement of low levels of these compounds (Cofer et al., 1985). A recent application of this approach is the mist-chamber technique (Cofer et al., 1985; Talbot et al., 1990), which is used to measure nitric acid and other atmospheric acids (Talbot et al., 1988). The mist-chamber technique was recently compared with the nylon filter method (Talbot et al., 1990). Laboratory and field tests indicated that both techniques were capable of collecting and analyzing HNO_3 emitted from permeation tube sources. However, in field measurements made at a rural site, the nylon filter yielded HNO_3 mixing ratios 70% larger than those measured simultaneously by mist-chamber techniqes. Subsequent tests revealed a small positive interference for ozone on the nylon filter, but this interference could not account for the large discrepancy noted above. Talbot et al. (1990) suggested that the nylon filter may suffer interference from other species as well. In any event, this comparison shows the need for caution in interpreting the measurements of HNO_3 made with available techniques and underlines the need for further study to determine the reliability of the various methods.

Total Reactive Nitrogen Oxides

Understanding of reaction pathways can be advanced by the measurement of the total abundance of reactive nitrogen compounds, NO_y, as well as by the measurement of the individual NO_y species. For example, it is NO_y, rather than such components as NO_2, that is of primary interest in tracking the transport and deposition of tropospheric nitric acid on a regional basis.

Several NO_y measurement techniques have been proposed. In general, all rely on the reduction of the NO_y-species to NO followed by detection of the NO. A ground-based comparison of two of these techniques, the gold-catalyzed conversion of NO_y to NO in the presence of carbon monoxide and the reduction of NO_y to NO on a heated molybdenum oxide surface, has been done by Fehsenfeld et al. (1987). The instruments were found to give similar results for the measurement of NO_y concentration in ambient air under conditions that varied from clean continental background air to typical urban air, with NO_y ranging between 0.4 ppb and 100 ppb. However, it was found that when the molybdenum oxide converter was operated for extended periods (several hours) with NO_y concentrations > 100 ppb, the conversion efficiency dropped significantly. For this reason the gold-catalyzed converter was judged more reliable when used in a polluted environment.

MEASUREMENT TECHNIQUES
FOR CARBON MONOXIDE AND VOLATILE
ORGANIC COMPOUNDS

Unlike the NO_x measurement techniques described above, which are adequate, the techniques for measuring VOCs and their oxidation products do not meet current needs. The analysis of VOCs is complicated by the extreme complexity of the mixtures that can be present in the atmosphere. Over one hundred detectable VOCs can be present in air sampled from reasonably isolated rural sites (P.D. Goldan, pers. comm., Aeronomy Laboratory, NOAA, 1990). In urban locations this number is substantially greater (Winer et al., 1989). VOCs emitted by vegetation, estimated to account for 50% of the VOCs emitted into the atmosphere in the United States (Placet et al., 1990), are mainly highly reactive olefinic compounds. Moreover, it is believed that more than one-third of the natural compounds that are emitted are, as yet, unidentified. Air samples obviously can contain many different VOCs of natural and anthropogenic origin; the oxidation of each of these species creates a mixture that contains many additional oxidation products as well. This complex chemistry is discussed in detail in Chapters 5 and 6. It is clear that the analysis of VOCs and their oxidation products is a formidable task. The techniques that have been developed to measure these compounds, as well as several promising new methods, are described in the following sections.

The discussion of these measurement techniques is divided into five parts: (1) carbon monoxide (CO); (2) nonmethane hydrocarbons (NMHC), which are compounds other than methane (CH_4) that are composed entirely of hydrogen and carbon; (3) formaldehyde (HCHO), the simplest aldehyde; (4) other aldehydes and ketones; and (5) organic acids. These are the compounds that most strongly influence the photochemical production of tropospheric ozone, which have the most atmospheric variability (as opposed to CH_4), and for which measurement capability is a matter of the most concern. The term "volatile organic compound" refers to all the above compounds except CO, and also refers to other compounds, such as organic nitrates, peroxides, and radicals, that are discussed elsewhere in the report.

Carbon Monoxide

CO is ubiquitous in the atmosphere, and it has many sources, both natural (oxidation of methane and other natural VOCs and biomass burning) and anthropogenic (combustion processes). Its lifetime is long enough, on the order of a few months, that it is distributed globally, and its concentration ranges from 50 to 150 ppb in the remote troposphere. Concentrations above this background can indicate air masses that have had recent anthropogenic pollution input. Given the relatively unreactive nature of the gas and its high concentrations, it is expected to be one of the more easily measured trace atmospheric species.

Three techniques are used widely for measurement of CO in the troposphere: collection of grab samples followed by analysis using gas chromatography (GC); tunable-diode laser absorption spectrometry (TDLAS); and gas filter correlation, nondispersive infrared absorption spectroscopy (NDIR). The first two of these methods have been compared in rigorous sets of tests (Hoell et al., 1987b, and the references therein). In the earlier ground-based comparison of three GC techniques (one with direct injection of samples) and one TDLAS system, there was a high degree of correlation between the results of all four techniques in ambient measurements and for prepared mixtures of CO in ambient air (Hoell et al., 1987b). The general level of agreement was within 15%. However, a day-to-day bias between the techniques was observed to result in differences between techniques as large as 38%.

In the later airborne comparison (involving the two grab sample GC and the TDLAS systems), the techniques had been refined to the point that, at mixing ratios of 60-140 ppb, the level of agreement observed for the ensemble of measurements was well within the overall accuracy stated for each instrument (Hoell et al., 1987b). The correlation coefficient determined from the measurements taken from respective pairs of instruments ranged from 0.85 to 0.98, with no evidence of the presence of either a constant or proportional bias between any of the instruments. Thus, the reliability of the measurement of CO has been rigorously established.

Nonmethane Hydrocarbons

Much of the reactive carbon entering the atmosphere is in the form of nonmethane hydrocarbons (NMHCs). The standard approach for measurement of NMHCs is based on GC separation of individual hydrocarbons and the detection of each using a flame ionization detector (FID). Singh (1980) summarized the general procedures used in the analysis of ambient hydrocarbon samples. The GC column and temperature programming of the column are selected to give the desired resolution of the compound peaks. Over the years, separation and the integrity of compounds that pass through the columns have been improved by development of better column packing compounds for packed columns or coatings for open tubular columns.

The FID is a nonspecific hydrocarbon detector with a sensitivity that, in general, is linearly proportional to the number of carbon atoms in a VOC molecule (Ackman, 1968). Compound identification is usually established by compound retention in the GC column. In addition, mass spectrometric identification of given peaks can be made to confirm the elution time assignments or can be used to help in the identification of an unknown peak.

When very low concentrations of the VOCs are measured, it is necessary to concentrate the samples before they are injected into the column. This is done by concentrating a volume of air cryogenically or with a trapping matrix before injection onto the column. By temperature programming the trap, it is possible to separate the VOC compounds to be measured from compounds that comprise the bulk of the air sample: nitrogen, oxygen, water vapor, argon, and carbon dioxide. As a consequence of these improvements, VOCs with concentrations as low as 5 ppt in air have been measured with good resolution by GC-FID systems. However, there can be problems with using this method to measure reactive VOCs at low concentrations in air. Large amounts of compounds, particularly high-carbon-number compounds, can be retained by the trapping medium. In addition, reactions between the VOCs and oxidants, such as residual ozone that survives the collection procedures, may destroy some hydrocarbons and produce other compounds not originally in the sampled air. Hence, additional methods are required to reduce the oxidants to negligible concentrations before preconcentration without altering the concentration of the hydrocarbons to be analyzed.

Often, measurements of NMHCs in the field are done under circumstances that require maximum portability and low power consumption, and they are done in an environment adverse to the operation of a sensitive instrument. Consequently, many NMHC measurements are done by acquiring an air sample in a suitably prepared container and transferring it into the GC-FID. Sample containers have been made from glass, treated metal, and special plastics. Sampling procedures often require that the containers be purged before the air sample is obtained and that the sample be stored in the container above atmospheric pressure. Although much has been done to ensure the integrity of the compounds of interest in these containers, many of the difficulties attendant to this approach are associated with the stability of the sample during transport and storage. When samples of hydrocarbons are analyzed after having been stored for several days, substantial losses of the heavier hydrocarbons can occur (Holdren et al., 1979). This is significant because occasionally, several months may pass before hydrocarbon samples stored in these containers can be analyzed.

Thus far, there has been no rigorous comparison of the various versions of GC-FID systems and sampling containers. However, the limited comparisons that have been done indicate that the techniques could be satisfactory for measuring relatively high concentrations (>10 ppb carbons) of simple, light hydrocarbon compounds (five carbons or fewer).

Formaldehyde

The oxidation of NMHC forms the carbonyl compounds, the aldehydes and ketones (RCHO). Measurements of these compounds can test present understanding of the VOC oxidation mechanism. The oxidation of aldehydes and ketones can be an additional source of ozone and oxidizing free radicals, and the photolysis of aldehydes and ketones can be a primary source of radicals. The simplest aldehyde, formaldehyde (HCHO), is particularly important because it can be formed by the oxidation of methane. As a result, it is distributed throughout the troposphere. HCHO can also be emitted into the atmosphere as a direct product of hydrocarbon combustion (Lawson et al., 1990a). Thus the photolysis of HCHO could be a key process in the formation of tropospheric ozone.

Four techniques have emerged for the measurement of HCHO: TDLAS; enzymatic fluorometry (EF), which involves the absorption of HCHO from a sampled air stream into water followed by detection of the fluorescence from the reaction of the aqueous HCHO with β-nicotinamide adenine dinucleotide, as catalyzed by the enzyme formaldehyde dehydrogenase; a diffusion scrubbing fluorescence (DSF) technique, which involves the absorption of HCHO from a sampled air stream into water followed by detection of the fluorescence from the reaction of the aqueous HCHO with ammonia and acetylacetone; and a derivatization technique, which involves trapping HCHO on a substrate impregnated with 2,4-dinitrophenylhydrazine (DNPH) followed by extraction of the derivatized compounds and ultraviolet absorption analysis. Two studies have been reported that compare these techniques for the measurement of HCHO in ambient air. In the first, Kleindienst et al. (1988) compared the four techniques for the measurement of HCHO at the lower concentrations (<10 ppb) typically found in rural air. Because of its recent development, potential interferences for the DSF technique were not known in advance of the study and, as a consequence, the DSF technique was not involved in the ambient air measurements. In this study, no large systematic errors were observed in synthetic air mixtures with and without added interferants such as NO_2, SO_2, O_3, and H_2O_2, or, for the TDLAS, EF, and DNPH techniques, in ambient air where ambient concentrations of HCHO ranged from 1 to 10 ppb. Although reasonably low concentrations of HCHO were encountered during this comparison, no attempt was made to establish detection limits for these instruments.

In a more recent comparison, Lawson et al. (1990a) evaluated the four instruments in a reasonably polluted urban environment. In this evaluation two additional techniques were included, Fourier transform infrared spectrometry (FT-IR) and differential optical absorption spectrometry (DOAS). Because of their low sensitivity, these latter two techniques are not suitable for the measurement of HCHO in the nonurban atmosphere. However, because they are highly specific optical techniques that can measure higher concentrations of HCHO over limited paths in the free atmosphere, in this urban environment they provided independent measurements for comparison with the measurements made by the other techniques.

During the course of the 10-day study, a systematic diurnal variation was observed in the HCHO; it reached a maximum during the day and a minimum during the night. The average hourly ambient HCHO ranged from 4 to 20 ppb. Because reasonably high concentrations of HCHO were observed during the early morning rush hour, it was surmised that formaldehyde was being emitted directly into the atmosphere as a primary pollutant. Over the study period, the three spectroscopic techniques agreed to within 15% of the mean of these three methods. DNPH yielded values 15-20% lower than the mean of the spectroscopic techniques, whereas DSF yielded values 25% lower than the mean. Measurements obtained with the EF were found to be 25% higher than the mean. Measurements reported early in the study for DSF and EF were closer to the spectroscopic mean; problems developed in these instruments as the comparison progressed. The slight negative bias in the values obtained with DNPH was tentatively attributed to a negative ozone interference (ozone concentrations ranged from 0 to 240 ppb in this field study).

Other Aldehydes and Ketones

The measurement of higher molecular weight aldehydes and ketones has been performed with two different techniques, DNPH cartridges and GC-FID. DNPH has been the standard method for most field measurements of the carbonyls. For these compounds, the method has proven to have adequate selectivity. However, it suffers from low resolution and sensitivity when compared to GC-FID. Also, because DNPH involves liquid extraction of the compounds of interest from the cartridge, blank levels are a problem for measurements of carbonyl compounds at concentrations expected in the rural environment. GC-FID offers reasonable sensitivity and high resolution when capillary columns are used. This technique can achieve detection limits of <0.01-0.2 ppb in one liter of air, depending on the compound analyzed. In GC-FID analysis of ambient air, artifact formation of carbonyl compounds can arise in the cryogenic collection of an air sample. Thus far there have been no intercomparisons of these techniques for the measurement of the aldehydes and ketones.

Organic Acids

Although organic acids can be major components of atmospheric acidity, few measurements of these acids in the vapor phase have been reported (Norton, 1987; Farmer et al., 1987). The mechanisms for their formation are not well understood, and their role in tropospheric chemistry is uncertain. Organic acids can be removed from the atmosphere by deposition. Not enough is known about their chemistry and atmospheric distribution to predict how their oxidation will influence ozone formation.

There are many measurement techniques for collecting organic acids, but few tests have assessed their validity. Keene et al. (1986) compared techniques for collecting aerosol- and vapor-phase formic and acetic acid. The acids were collected in mist chambers, as cold plate condensates, in resin cartridges, in sodium-hydroxide-coated denuder tubes, in sodium-hydroxide-impregnated glass filters, in nylon filters, and on cellulose fibers impregnated with sodium or potassium. The study was limited to ambient air sampling. After collection, all of the samples were analyzed by ion chromatography. The mist chamber and denuder tube gave results that were statistically indistinguishable. The cold plate technique gave results that were in general agreement with the mist and denuder techniques but showed significant differences on some occasions. The nylon filters were found not to retain the acid vapors quantitatively. The sodium carbonate filters gave concentrations somewhat below those of the mist and denuder techniques. The resin cartridges, sodium-hydroxide impregnated glass filters, and the sodium- and potassium-impregnated cellulose filters gave concentrations substantially larger than those of the mist chamber and denuder tubes. Although this study was not able to establish a generally reliable method to measure organic acids, several correctable problems were identified with the techniques. The conclusion was that strong-base-coated filters and GC resin techniques suffer from interferences that can cause serious overestimation of the concentration of organic acids in the air samples.

MEASUREMENT TECHNIQUES FOR OXIDANTS

The oxidants discussed in this section are the hydroxyl radical (OH), the peroxy radical (HO_2), ozone, and hydrogen peroxide (H_2O_2). The processes responsible for the formation and loss of these highly reactive compounds and the roles these oxidants play in atmospheric chemistry are described in Chapters 5 and 6. The techniques that have been developed to measure these oxidants are described in the following sections.

The Hydroxyl Radical

Atmospheric oxidation is thought to be initiated by OH. However, to date, the verification of this chemistry has been derived solely from laboratory studies and from computer modeling of the chemistry. Neither the concentration of OH nor that of HO_2 has been measured in the atmosphere with instruments of established reliability to demonstrate whether the current mechanistic understanding of these fundamental processes is correct.

The requirements for such instruments are challenging. Although huge quantities of OH are generated during the sunlit hours by the photolysis of ozone, the very high reactivity of these oxidizing radicals implies that their atmospheric concentrations are small, typically less than $10^7/cm^3$ (~ 0.4 ppt) (Crosley and Hoell, 1985). Moreover, these free radicals can be lost by collision with the surfaces of instruments— for example, inside sampling inlets. Hence, although substantial effort has already been invested in the development of methods to measure OH, a definitive measurement in the troposphere is still to be done (Hoell, 1983; Crosley and Hoell, 1985; Platt et al., 1988; Smith and Crosley, 1990; Armerding et al., 1990; Hofzumahaus et al., 1990a). Techniques currently under development include in situ methods that use laser-induced fluorescence (Davis et al., 1981; Wang et al., 1981; Rodgers et al., 1985; Hard et al., 1986; Chan et al., 1990), a radioactive tracer technique (Campbell et al., 1986; Felton et al., 1990), long-path, laser absorption methods (Huebler et al., 1984; Perner et al., 1987; Dorn et al., 1988; Platt et al., 1988; Hofzumahaus et al., 1990b), and ion-assisted OH detection (Eisele and Tanner, 1990).

Peroxy and Organic Peroxy Radicals

Tropospheric measurements of peroxy (HO_2) and organic peroxy (RO_2) free radicals have been made with two different techniques, peroxy radical chemical amplification (PeRCA) and matrix isolation with electron spin resonance detection (MIESR). MIESR (Volz et al., 1988) relies on the cryogenic trapping of HO_2 radicals in a water matrix followed by the detection of the free radical using electron spin resonance. Problems with interference in the

ESR spectra have been overcome by using deuterium oxide (D_2O), instead of water, as the isolation matrix. This substitution has improved the signal-to-noise ratio and spectral resolution, allowing the identification of different free radical species during field measurements (Volz et al., 1988). The PeRCA technique relies on the oxidizing ability of the odd-hydrogen free radicals to convert NO and CO to NO_2 and CO_2 in a chain reaction (Cantrell and Stedman, 1982; Cantrell et al., 1984). The NO_2 produced in the chain reaction is measured using luminol chemiluminescence. Measurements of ambient concentrations of HO_2 free radicals in the atmosphere using this technique have also been reported (Cantrell et al., 1988).

Both techniques claim detection limits for HO_2 on the order of 1 ppt, which is sensitive enough for most ambient measurements. PeRCA has the distinct advantage of providing a continuous record of the total HO_2 radical concentration in an air mass. MIESR, on the other hand, provides an integrated measure of the HO_2 radical concentration, but it has the advantage of speciation. Calibration of both techniques under ambient conditions remains a research challenge. There have been no formal intercomparisons of these techniques, and their ability to measure HO_2 radical reliably is open to question.

In addition to these techniques specifically designed to detect HO_2, many of the in situ OH measurement techniques outlined in the preceeding section might be adapted to measure HO_2. Such adaptation requires that the HO_2 in the ambient air sampled by the instrument be titrated to OH, probably by reaction with NO before detection.

Ozone

Over the years several techniques have been developed to measure ozone. These include absorption of ultraviolet (UV) light, chemiluminescence, and chemical titration methods, particularly electrochemical techniques. Each has advantages for certain kinds of tropospheric ozone measurements.

The absorption of UV light by the ozone molecule provides a reasonably straightforward and accurate means to measure ozone. Most instruments rely on the 254 nm (nanometer) emission line of mercury (which happens to coincide with an absorption maximum of ozone) from a mercury discharge lamp as the UV light source. This technique, which has been incorporated into several high-quality commercially available instruments, is reliable, and interference that occurs because of the absorption of the UV light by molecules other than ozone can generally be ignored. Most high-quality, routine in situ measurements of ozone have been made with this technique.

The chemiluminescence produced by the reaction of ozone with nitric oxide forms the basis for a sensitive and specific ozone detection method. (The reactions of ozone with an unsaturated NMHC such as ethylene also have been used, but they are somewhat less sensitive.) Although chemiluminescence tends to be more complicated than the UV absorption method, it can make fast-response ozone measurements because of its greater sensitivity. For this reason chemiluminescence has been used to measure ozone fluxes that can be deduced from the correlation of ozone variation with atmospheric turbulence.

Electrochemical sondes measure the electrical conductivity of an electrolytic solution and rely on the conversion of chemicals in the solution by ozone in the sampled air, which alters the conductivity of the solution. A typical instrument, such as the electrochemical concentration cell (Komhyr, 1969), is composed of platinum electrodes immersed in neutral buffered potassium iodide solutions of different concentrations in anode and cathode chambers. When ozone-containing air is pumped into the cathode region of the cells, a current is generated proportional to the ozone flux through the cell. Sondes can be very lightweight and can therefore be lifted by small balloons; they are generally used to measure ozone profiles in the atmosphere. However, the measurements made by these instruments can suffer from positive or negative interference by compounds other than ozone, and this decreases the reliability of the instrument (Barnes et al., 1985).

Over the years, several formal and informal intercomparisons of these techniques have been made (Attmannspacher and Dutsch, 1981; Aimedieu et al., 1983; Robbins, 1983; Hilsenrath et al., 1986). Although the most recent and comprehensive of these studies was aimed at evaluating instruments used to measure stratospheric ozone, many of the findings can be applied to tropospheric ozone. The consensus to be drawn from these comparisons indicates that the best UV absorption instruments are probably reliable for measurement of tropospheric ozone with uncertainties of less than 3%. Chemiluminescence instruments should be equally good. The electrochemical sondes are susceptible to interference that reduces their intrinsic accuracy somewhat. However, it must be emphasized that all of these techniques when used in routine measurement likely will be subject to much larger uncertainties.

Hydrogen Peroxide

Several techniques are used to measure H_2O_2. These techniques have been subjected to an urban-area comparison (Kleindienst et al., 1988), and a second comparison is under way (Sakugawa and Kaplan, 1990). Kleindienst et al.

(1988) compared four techniques used to measure H_2O_2 in air: tunable diode laser spectrometry using infrared absorption to measure H_2O_2 (Slemr et al., 1986); continuous-scrubbing extracting gas-phase peroxides into aqueous solution that are analyzed by enzymatic fluorometry (Lazrus et al., 1986); diffusion-scrubbing followed by analysis using enzymatic fluorometry (Hwang and Dasgupta, 1986); and continuous-scrubbing extracting H_2O_2 into a luminol solution where it undergoes a chemiluminescence reaction (Zika and Saltzman, 1982).

The first three of these techniques were compared for the measurement of H_2O_2 (1) in synthetic air, sometimes spiked with common interferences; (2) in synthetic air containing UV irradiated mixtures of NMHC and NO_x; and (3) in ambient air. The luminol technique was not included in the last phase of the comparison; cf. Kleindienst et al. (1988). For the comparisons done in synthetic air and ambient air, the agreement was satisfactory–30% or better for the three techniques evaluated. These three techniques had detection limits for the measurement of H_2O_2 of approximately 0.1 ppb. In the tests done in synthetic air containing irradiated mixtures of NMHC and NO_x, the agreement among the techniques was not as good, suggesting the presence of some as-yet unidentified H_2O_2 interference. In this regard, one current concern is the degree to which H_2O_2 is contained on aerosols. It is not known how much of this aerosol H_2O_2 is measured by the various techniques.

The techniques using enzymatic fluorometry can also be used to measure organic peroxides. However, these techniques have yet to be tested through intercomparison.

CONDENSED-PHASE MEASUREMENT TECHNIQUES

The discussion to this point has centered on techniques to study gas-phase chemistry. Current understanding suggests that the production of ozone from precursors during the summer principally involves gas-phase processes. However, the formation and removal of ozone, particularly in seasons other than summer, may involve the condensed phase, including aerosols, fog droplets, or cloud droplets, and may depend on the chemistry that occurs in this phase. During the past 5 years, the potential importance of heterogeneous chemistry (chemistry occurring in more than one phase) has been demonstrated by research aimed at understanding the suppression of stratospheric ozone in polar regions in early spring (Geophys. Res. Lett., 1990).

There are review articles that discuss the role of multiphase chemistry in the troposphere (e.g., Charlson et al., 1991) and the techniques used to study these processes (e.g., Simoneit, 1986; Bidleman, 1988; Ayers and Gillett, 1990;

Turpin and Huntzicker, 1991). However, a review of the techniques that contribute to the measurement of condensed-phase processes as they pertain to the production and destruction of ozone in the troposphere is beyond the scope of this chapter. Inferences made from measurements using these techniques, as they pertain to the distribution of tropospheric ozone, would be premature. Nevertheless, there is evidence that the condensed phase may contain significant quantities of ozone precursors such as the NMHC (Simoneit, 1986; Sicre et al., 1987; Foreman and Bidleman, 1990; Pickle et al., 1990; Brorström-Lundén and Lövblad, 1991; Mylonas et al., 1991; Simoneit et al., 1991). The development of measurement techniques to study multiphase chemistry in the troposphere, the validation of these techniques, and the application of these techniques to atmospheric measurements should be encouraged.

LONG-TERM MONITORING
AND INTENSIVE FIELD MEASUREMENT PROGRAMS

The measurement techniques described here must be used in well-designed field studies to collect the data necessary to evaluate tropospheric ozone production. Particularly fruitful field studies have typically fallen into two categories: long-term monitoring of one or a few easily measured species and short-term intensive field campaigns measuring a wide suite of atmospheric species involved in chemical transformations. As a guide to planning future field studies, it is worthwhile to consider these two categories.

Long-Term Monitoring Programs

This approach is exemplified by the measurements of CO_2 that have been carried out continuously since 1957 to determine the global CO_2 background. These measurements clearly have established the increasing trend of CO_2 in the atmosphere and have defined its seasonal cycle. More recently, monitoring of ozone regionally and in specific locations has been instituted. The data set is still too short-term to reveal unambiguous trends in urban, suburban, or rural areas. Ozone precursors have not been monitored consistently over large areas, in part because of the difficulty and expense of such measurements. If resources can be found, it is realistic to begin such programs, because suitable instrumentation is now available for monitoring NO_x, NO_y, and NMHC. The use of atmospheric measurements to determine temporal trends in emissions will provide valuable checks for estimated emission reductions that inventories purport to show.

Designing the long-term monitoring programs will require careful thought. CO and ozone can be monitored with commercially available instruments. Because of the relatively long atmospheric lifetimes of these species, continuous measurements in appropriate locations can provide the necessary trend data. Investigators can rely on mixing in the atmosphere to provide representative samples. However, the important precursors of ozone are quite short-lived, and they undergo rapid chemical transformations in the atmosphere. Measurements at any location will reflect the intensity of proximate and distant sources, atmospheric mixing and transport, and the degree of chemical transformation. Hence, extraction of trends in emissions could be difficult.

The proper choice of sampling location and season could ease these difficulties. Measurements in late fall in suburban areas could be most fruitful because this season often provides low sunlight, temperatures, and precipitation, which will slow chemical transformations. Also during this season the lower troposphere is most stable, which will reduce the variability in atmospheric mixing. A suburban site could provide the best compromise between long transport times, which increases variability in degree of chemical transformation, and poor mixing due to proximity of sources, which increases variability due to specific source input. It is clear that monitoring of trends in emissions cannot easily be a part of a program designed to measure atmospheric transformation processes, because one obscures the other.

The simplest plan would be to concentrate on short-term, intense field studies in which the measurements are done by intercomparison-validated, high-quality instruments. The measurements could be made during one, or at most, a few periods each year rather than as an extended, routine program of measurements done throughout the year. It should be possible to adequately characterize the meteorology of each period to reduce the variability associated with varying transport processes. In this connection, it would also be fruitful to look at ratios of the species concentrations, because the ratios are much more independent of atmospheric mixing processes than are the concentrations themselves.

Intensive Field Studies

Understanding of ozone production outside urban areas has been greatly advanced by several field studies designed to elucidate atmospheric photochemical processes (for example, Dennis et al., 1990). The experience from these studies is that much more is learned from the simultaneous measurement of many photochemical processes and meteorological events than is possible from the separate measurement of each.

Intensive field studies have evolved to cover a wide range of the relevant variables. Two such studies, one carried out in an urban environment and the other in a rural location, are discussed by way of example.

A comprehensive urban study (Lawson, 1990 and the references cited therein), the Southern California Air Quality Study (SCAQS), was carried out in the South Coast Air Basin (SoCAB), which experiences the most severe air pollution in the United States. The SCAQS study was aimed at obtaining an integrated, basin-wide data base of the most important species contributing to air pollution in Los Angeles. By clarifying the VOC/NO_x/ozone relationships, these data will aid in improving the models used to design attainment strategies for ozone. The study, which was carried out in 1987, consisted of an intensive 11-day period in the summer and a second 6-day intensive period in the fall. The measurement suite included two highly instrumented ground sites (designated class "A" sites), 9 less instrumented support sites (class "B" sites), and 36 sites that make the routine measurements within the framework of the ongoing SCAQ Management District (SCAQMD).

The measurements made at the sites included surface meteorological parameters (temperature, dew point, wind speed and direction); gaseous species (H_2O_2, NO_x, O_3, CO, SO_2, HNO_2, and NO_3); and organic vapors (formaldehyde, RO radicals, organic acids, aldehydes, ketones, alcohols, C_1-C_{12} hydrocarbons, and PAN). Continuous measurements of aerosols included: particle sulfur, particulate matter, sulfate, organic carbon, elemental carbon and black carbon. Aerosol chemistry sampling included: mutagens, metals, organics, particulate matter, carbon, NO_3^-, SO_4^-, and polyaromatic hydrocarbons. Size resolved aerosol chemistry involved organic carbon, elemental carbon, and nighttime bromine and lead. These surface measurements were augmented by rawinsonde measurements, and the summer study was augmented by aircraft measurements.

The preliminary results of the study were presented at a symposium of the Air and Waste Management Association in 1989. Several tentative objectives and conclusions have been drawn for each phase of the study. Emissions inventories were checked using measurements capable of testing the reliability of emission factors used to generate the emissions inventories of NO_x, CO, and NMHC for point and area sources that are presently used in model calculations. It was found from tunnel studies that the emission factors being used to generate inventories for CO and NMHC for mobile sources may be too small. Also it was recognized that better inventories for ammonia emissions are required, because the formation of ammonium nitrate may be a significant source of aerosol formation and play a significant role in atmospheric denitrification. Transport was elucidated using tracer studies to determine: (1), in the summer, the relative impacts of elevated and ground level sources on the concentration of ozone and NO_2 during on- and off-shore flow; and (2), in the fall, the relative influence of elevated and ground level emission sources of NO_x on the ground level NO_x during stagnation events. Measurements of gas-phase pollutants and photochemistry elucidated the role of NO_x and NMHC in the formation of ozone and other oxidants.

The extensive data base obtained from this study has been archived and is being analyzed by various investigators. A primary goal of the SCAQS is to make this data base available to the scientific community with the aim of developing better models to simulate the atmospheric composition in urban areas.

A recent study (Parrish, 1990), Rural Ozone in a Southern Environment (ROSE), was organized at a ground site in Alabama. This study was part of the Southeastern Regional Oxidant Network (SERON) of the Southern Oxidants Study (SOS). Near the surface the chemical species measured included ozone and other oxidants (H_2O_2, organic peroxides, and HO_2 radicals), the major ozone precursors (NO, NO_2, and NMHC), their intermediate oxidation products (organic nitrates, HNO_3, NO_y, aldehydes, ketones, organic acids, and organic aerosols) and other primary pollutants (CO, SO_2, and sulfate aerosols). Meteorological parameters (e.g., temperature, relative humidity, wind speed and direction) also were measured at the surface. The concentration measurements were supported by measurements of biogenic emissions of NO_x and NMHC and of surface fluxes of CO_2 and ozone. The data from this surface site were supplemented by regional measurements of ozone, NO, NO_2, NMHC, organic nitrates, and NO_y from an aircraft. The vertical distributions of ozone and NMHC above the site were measured by the aircraft and by tethered balloons. The evolution of the planetary boundary layer, which controls much of the mixing in the lower troposphere, was monitored by balloon-launched radiosondes and by wind measurements by SODAR and boundary layer radar systems.

Other integrated, intensive field studies now planned or under way include the San Joaquin Valley Air Quality Study (SJVAQS)/Atmospheric Utility Signatures, Predictions, and Experiments (AUSPEX) (Roth, 1988; Ranzieri and Thuillier, 1990); the Lake Michigan Ozone Study (Bowne et al., 1990); and the Southern Oxidants Study (Chameides and Rogers, 1988). Nearly all present understanding of ozone formation in rural and remote environments has come from such integrated, intensive field studies, but much remains to be done. Previous studies have been limited in spatial coverage; sites in a wide variety of areas must be studied. Studies have been carried out only in the summer; future studies must compare atmospheric processes between seasons. The free radicals that drive atmospheric chemistry must be measured directly. Current measurements and computer models almost exclusively address homogeneous, gas-phase chemistry; the role of aerosols must also be investigated. These last two needs will require advances in instruments and in techniques.

In the planning and execution of field studies, chemical-dynamic models are useful for designing the measurement strategy, not just in interpreting the measurement data. In this way modelers can help identify crucial site characteristics, choose critical species and parameters to measure, and optimize the measurement schedule. In addition, the results of the field studies can best provide critical input to the modeling community that will lead to the improvement of computer models.

SUMMARY

Reliable techniques for measuring NO_x (oxides of nitrogen), even at the low concentrations found in rural and remote air, have been developed recently, and methods that measure nitric oxide (NO), nitrogen dioxide (NO_2), peroxyacetyl nitrate (PAN), and the total concentration of the nitrogen oxide (NO_y) have been compared. These intercomparisons indicate that methods are now available to measure these species throughout the troposphere. However, the validated techniques are of recent origin. In particular, the instruments used to measure NO_2 in most air-quality studies usually rely on heated surface converters to transform NO_2 to NO. All these methods will likely convert other NO_y species to NO as well, and hence can be subject to significant interference. Intercomparison of techniques used to measure HNO_3 shows significant variations among methods, and no definitive conclusions can be drawn about the reliability of individual techniques. Further development of HNO_3 measurement techniques is required before measurements of that compound can be considered reliable.

For VOCs, only the techniques that measure CO can be considered fully reliable. There has been no intercomparison of methods to measure non-methane hydrocarbons (NMHCs), so there is no way to judge the quality of the large body of data obtained using those techniques. As a general rule, however, measurements made at low NMHC concentrations (atmospheric mixing ratios <1 ppb of the compound) must be considered suspect. In addition, the heavier NMHCs (C_5 and larger) are subject to larger sampling uncertainties than are the lighter NMHCs. Sampling techniques for partially oxidized NMHCs—carbonyl compounds, including formaldehyde and organic acids—are under development. However, the data base of measured concentrations of those compounds using this emerging technology is limited.

The basic test of our understanding of oxidizing properties awaits the further development of techniques to measure the oxidizing radicals, hydroxyl radical (OH) and peroxy radical (HO$_2$). Much progress has been made toward the development of the necessary instrumentation, but reliable measurements are not yet available.

Reliable techniques to measure ozone are available, and the large data base obtained using these techniques is a valuable resource. However, the uncertainty of routine measurements using these techniques may be as much as 10%, which could limit the conclusions that can be drawn from the data. Intercomparisons have been made for techniques that measure H$_2$O$_2$. There is qualitative agreement among the techniques, but the data obtained on atmospheric concentrations of H$_2$O$_2$ are limited.

Much still must be learned about instrument reliability. An essential factor in establishing reliability is the existence of two or more field-worthy techniques that measure the species of interest. In addition, reliable or standardized calibration procedures must be available for these species. Even with those necessary conditions, the road to harmony can be long and twisting. For example, there is no one recipe for what to do when two or more methods disagree significantly.

Although it is arduous, time-consuming, and costly to develop individual instruments and track down discrepancies, it should be recognized that multiple techniques are essential (and are not wasteful duplication) and that intercomparisons are vital (and are indeed as much a part of doing atmospheric science as is gathering data to test a geophysical hypothesis). Without intercomparisons among different techniques, there is no assurance that what is measured is indeed correct.

Accurate and precise measurements of the trace species involved in ozone chemistry are needed to advance the understanding of the formation of high concentrations of ozone, to verify estimates of precursor emissions, and to assess the effectiveness of ozone control efforts. However, these species have not been adequately monitored. As a result, it is not known whether the lack of success of ozone control efforts is the result of failure to achieve targeted reductions in ozone precursors or failure to set appropriate targets. Also, questions remain about the relative importance of anthropogenic and biogenic VOCs, the extent to which ozone production is VOC-limited or NO$_x$-limited, and the role of VOC and NO$_x$ oxidation products in ozone formation.

To answer these questions, it is necessary to have reliable measurements of ozone, NO$_x$, VOCs, CO, and the oxidants that catalyze ozone production. Although reliable techniques for measuring many of these species have been available for several years, most of the data bases discussed in this report were not obtained using such techniques. Moreover, measurements made by inexperienced operators using sophisticated techniques may contain uncertainties that could mask important trends. Only measurements made by skilled operators with reliable instruments can ensure that the science on which emission controls are based is correct and that the effectiveness of these controls is adequately assessed.

8

Atmospheric Observations Of VOC, NO$_x$, and Ozone

INTRODUCTION

Chapter 6 contains a discussion of the central role of the VOC/NO$_x$ ratio (the ratio of volatile organic compounds to oxides of nitrogen) in determining the chemical character of the VOC-NO$_x$-ozone system. Two important points were made about the VOC/NO$_x$ ratio: First, the atmospheric boundary layer (defined here as a well-mixed layer extending from the surface to a height of about 1 or 2 km during the day) cannot be characterized by a single ratio because this ratio varies significantly with location and time of day. Second, ambient ratios often exceed by a substantial amount those calculated from emissions inventories. The goal of this chapter is to examine data gathered from atmospheric observations to determine if ambient VOC, NO$_x$, and O$_3$ concentrations follow a regular pattern as one moves from an urban or suburban area to a rural area and then to a remote area. By comparing these patterns with those observed in smog-chamber experiments, it may be possible to establish to what degree smog-chamber experiments, and the chemical models based on these experiments can be applied to the atmospheric VOC-NO$_x$-ozone system. By comparing patterns found in urban and suburban areas with those found in rural and remote locations, it may be possible to infer the relative effectiveness of controlling VOC versus NO$_x$ in different parts of the country.

The second point above—that discrepancies between ambient and emission-inventory-derived VOC/NO$_x$ ratios have major implications concerning the accuracy of emissions inventories—is addressed in Chapter 9.

It is useful to focus on four regions of the atmospheric boundary layer, each of which has a distinct mix of anthropogenic and natural VOC and NO_x emissions:

  • The urban-suburban atmosphere, which is the area most strongly affected by anthropogenic emissions
  • The rural atmosphere, which is somewhat less affected by anthropogenic emissions and more affected by natural emissions than is the urban-suburban atmosphere
  • The atmosphere over remote tropical forests, which is essentially free of anthropogenic VOC and NO_x emissions and strongly affected by natural emissions
  • The remote marine atmosphere, which is not only free of anthropogenic emissions, but also has relatively small biogenic sources of VOCs and NO_x

Because we are most interested in the conditions that foster episodes of high concentrations of ozone, our discussion concentrates on observations made during the daylight hours of the summer months. In the sections below, we first briefly examine the typical concentrations of ozone in these four regions and then turn to the more complex issues of NO_x and VOC concentrations and their variability.

OBSERVATIONS OF OZONE

Compared with those for NO_x and VOCs, the data base of ozone observations is fairly extensive, especially for urban and suburban areas. At most rural surface sites, ozone concentrations have been found to vary over a diurnal cycle with a minimum in the early morning hours before dawn and a maximum in the late afternoon (Figure 8-1). This pattern is believed to result from daytime photochemical production or downward transport of ozone-rich air from above, combined with ozone loss by dry deposition and reaction with nitric oxide (NO) at night, when photochemical production ceases and vertical transport is inhibited by an inversion of the normal temperature profile. In locations near large sources of NO, the nighttime minimum in ozone can be quite pronounced because of the rapid reaction between ozone and NO. In fact, in many urban areas the NO source is strong enough to cause the complete nighttime disappearance of ozone. A somewhat different pattern has been observed at high-altitude sites (i.e., sites located 1 km or more above the local terrain). At these sites, ozone often exhibits a shallow maximum rather than a minimum at night (Figure 8-1b). This diurnal pattern is thought to

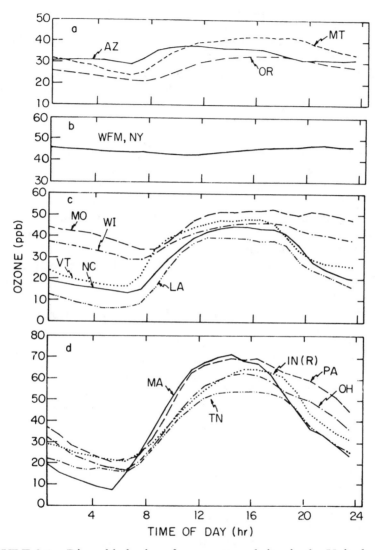

FIGURE 8-1   Diurnal behavior of ozone at rural sites in the United
States in July. Sites are identified by the state in which they are located.
(a) Western National Air Pollution Background Network (NAPBN); (b)
Whiteface Mountain (WFM) located at 1.5 km above sea level; (c) eas-
ter NAPBN sites; and (d) sites selected from the Sulfate Regional Air
Quality study. IN(R) refers to Rockport. Source: Logan, 1989.

reflect the contrasting dynamic conditions typically encountered at high altitude sites, with upslope flow bringing ozone-poor air from the boundary layer to the site during the day and downslope flow bringing ozone-rich air from the free troposphere to the site at night.

In addition to variations over a diurnal cycle, ozone concentrations at a given location also can vary significantly from one day to the next. It is not uncommon for the daily maximum ozone concentration at an urban site, for instance, to vary by a factor of two or three from day to day as local weather patterns change.

Despite the variable nature of ozone, the data base of ozone observations suggests a systematic pattern of decreasing daily maximum concentrations as one moves from urban-suburban locations to rural locations and then to remote locations. Table 8-1 shows that daily maximum ozone concentrations within the atmospheric boundary layer tend to be largest in the urban-suburban atmosphere, where 1-hour ozone concentrations most often exceed the National Ambient Air Quality Standard (NAAQS) concentration of 120 parts per billion (ppb), and maxima well above 200 ppb have been observed. Although the NAAQS can be exceeded in rural areas, ozone concentrations in these regions tend to be more moderate and rarely exceed 150 ppb. In remote locations, ozone concentrations tend to be quite low, typically ranging from 20 to 40 ppb.

Table 8-1   Typical Summertime Daily Maximum Ozone Concentrations

Region	Ozone, ppb
I Urban-suburban	100-400
II Rural	50-120
III Remote tropical forest	20-40
IV Remote marine	20-40

Sources: Cleveland et al. (1977); Hov (1984b); Gregory et al. (1988, 1990); Kirchoff (1988); LeFohn and Pinkerton (1988); Janach (1989); Logan (1989).

OBSERVATIONS OF NO_x

There is a sizable body of data on the concentrations of NO_x (the combined concentrations of NO and nitrogen dioxide (NO_2)) in the atmosphere, but caution must be exercised in drawing conclusions from these measurements. As noted in Chapter 7, most measurements of NO_x have been made by devices that convert NO_2 to NO, which is then measured by chemiluminescence. Comparison of these measurements with more specific techniques suggests that all surface converters that can convert NO_2 to NO also convert other reactive nitrogen oxide species, such as peroxyacetyl nitrate (PAN), to NO, thereby causing interference (Singh et al., 1985; Fehsenfeld et al., 1987, 1990; Gregory et al., 1990). Because PAN concentrations can vary considerably depending on other pollutant concentrations and the temperature, the potential error associated with PAN interference will depend strongly on the location, season, and altitude at which samples are taken. In urban locations, where the local NO sources are typically large, NO and NO_2 are probably the dominant constituents of the total reactive nitrogen or NO_y (NO_x + HNO_3 [nitric acid] + NO_3 [nitrate radical] + N_2O_5 [dinitrogen pentoxide] + HONO [nitrous acid] + PAN + other organic nitrogen compounds). Thus, in urban areas, interference from PAN and other oxides of nitrogen is believed to be relatively small. In rural and remote locations, however, the interference can be substantial. For this reason, all nonurban NO_x measurements made with surface converters must be considered upper limits (biased toward a high measurement); these measurements are so indicated in this discussion.

Urban NO_x

Given the dominant role of anthropogenic emissions in the budget of atmospheric NO_x and the fact that the sources of these emissions tend to be located in or near urban areas, elevated concentrations of NO_x are to be expected in these locations. Observations of NO_x support this expectation.

During the summer of 1986, NO_x measurements were made from 6:00 to 9:00 a.m. Daylight Saving Time (DST) at six locations in Philadelphia, Pennsylvania (Meyer, 1987). Four of the sites were downtown, where the average measured NO_x concentration ranged from 40 to 99 ppb. At two suburban sites, the average concentrations were 33 ppb (upwind of the core city) and 65 ppb (downwind of the core city). The average for the six-station network was 60 ppb. However, the NO_x concentrations at all locations exhibited a high degree of variability—standard deviations ranged from 37% to 50% of the average NO_x mixing ratio measured at the site.

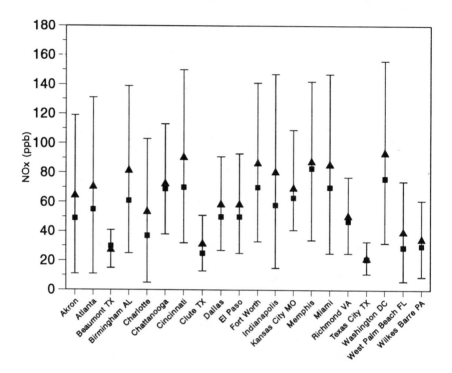

FIGURE 8-2a   NO$_x$ concentrations measured in urban locations in the
United States during the summer of 1984. All measurements were
made between 6:00 a.m. and 9:00 a.m. daylight savings time. The triangles are the averages for each site, the squares are the medians, and the
bars show the standard deviations of the averages. Adapted from
Baugues, 1986.

The range and variability found in the Philadelphia study's measurements
are reflected in measurements made in 29 other cities across the eastern and
southern United States during the summers of 1984 and 1985 (Baugues, 1986).
NO$_x$ measurements were made in 10 of the cities in both years. The measurements were made during the morning rush hour, 6:00 a.m. and 9:00 a.m.
DST. Figure 8-2a and 8-2b show the average and median NO$_x$ mixing ratios
and the standard deviations for each city. The average for all the cities studied varied between 18 ppb in Texas City, Texas (1985), and 114 ppb in
Cleveland, Ohio (1985).

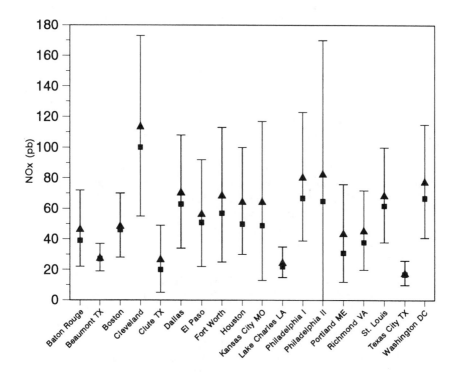

FIGURE 8-2b   NO$_x$ concentrations measured in urban locations in the United States during the summer of 1984. All measurements were made between 6:00 a.m. and 9:00 a.m. daylight savings time. The triangles are the averages for each site, the squares are the medians, and the bars show the standard deviations of the averages. Adapted from Baugues, 1986.

Because urban areas have concentrated sources of NO$_x$, urban measurements allow study of the rate of temporal and spatial decline of the NO$_x$ concentration with distance downwind of a source. Spicer et al. (1982) observed NO$_x$ concentrations in the Boston, Massachusetts, pollution plume a short distance from the city ranged from 27 to 131 ppb—concentrations similar to those typically found in surface measurements in urban areas. However, concentrations declined rapidly as the plume traveled away from the urban core. NO$_x$ concentrations in air masses 4-7 hours downwind of Boston were found to be 5-10 ppb. From this and similar plume studies (Spicer, 1977, 1982; Spicer et al., 1978; Spicer and Sverdrup, 1981) made over several cities in the United States, it has been estimated that the characteristic time for conversion of NO_x to other NO_y species is 4-20 hours.

Nonurban NO_x

Only during the past 10 years have techniques been available with sufficient sensitivity and range of detectability to measure NO_x in nonurban locales (NO_x concentrations below 1.0 ppb), and as a result the size and reliability of the data base needed to define nonurban NO_x concentrations are limited. Altshuller (1986) compiled and reviewed a series of NO_x measurements made at a number of rural sites in industrial regions of the United States; the results of these measurements are summarized in Table 8-2. Because of the proximity of these sites to urban and industrial sources, the NO_x concentrations usually exceeded 1 ppb and exhibited a high degree of short-term variability.

Measurements taken at more isolated rural sites in the United States are listed in Table 8-3. NO_x concentrations at these sites also can be dominated by anthropogenic NO_x emissions when meteorological conditions favor rapid transport of pollutants from urban and industrial centers to the site, but nevertheless tend to be significantly lower than concentrations measured at less-isolated rural sites (Table 8-2) and generally range from a few tenths to 1 ppb. Measurements of NO_x in the atmospheric boundary layer and lower free troposphere in remote maritime locations have generally yielded concentrations of 0.02-0.04 ppb (Bottenheim et al., 1986, Gregory et al., 1988, 1990; Montzka et al., 1989). Although the data base is still quite sparse, concentrations in remote tropical forests (not under the direct influence of biomass burning) appear to range from 0.02 to 0.08 ppb; the somewhat higher NO_x concentrations found in remote tropical forests, as compared with those observed in remote marine locations, could result from biogenic NO_x emissions from soil (Kaplan et al.,1988; Torres and Buchan, 1988).

A summary of the NO_x measurements made in the four areas considered here is presented in Table 8-4. It can be seen that, even more than is the case for ozone, NO_x concentrations decrease sharply as one moves from urban and suburban to rural sites in the United States and then to remote sites over the ocean and tropical forests. The striking difference of three orders of magnitude or more between NO_x concentrations in urban-suburban areas and remote locations is compelling evidence for the dominant role of anthropogenic emissions of NO_x over North America and suggests that NO_x concentrations in the United States would be significantly lower than their current concentrations in the absence of these emissions.

TABLE 8-2   Average Concentrations Measured at Nonurban Monitoring
Locations

Reference	Location	NO, ppb	NO$_2$, ppb	NO$_x$, ppb
Research	Fort McHenry, Maryland	ND[a]	6[b]	ND
Triangle	Dubois, Pennsylvania	ND	10[b]	ND
Institute,	McConnelsville, Ohio	ND	6[b]	ND
1975	Wilmington, Ohio	ND	6[b]	ND
	Wooster, Ohio	ND	6[b]	ND
Decker et	Bradford, Pennsylvania	2	3[b]	5[b]
al., 1976	Creston, Louisiana	4	2[b]	6[b]
	Deridder, Louisiana	1	3[b]	4[b]
Martinez	Montague, Massachusetts	2	3[b]	5[b]
and Singh,	Scranton, Pennsylvania	3	11[b]	14[b]
1979	Indian River, Delaware	3	5[b]	8[b]
	Research Triangle Park, North Carolina	10	13[b]	23[b]
	Lewisburg, West Virginia	1	5[b]	6[b]
	Duncan Falls, Ohio	1	8[b]	9[b]
	Fort Wayne, Indiana	3	7[b]	10[b]
	Rockport, Indiana	3	7[b]	10[b]
	Giles County, Tennessee	3	10[b]	13[b]
	Jetmore, Kansas	1	4[b]	5[b]
Pratt et al.,	Lamoure County, North	2.4	1.7[b]	4.1[b]
1983	Dakota	4.8	1.5[b]	6.3[b]
		3.3	2.8[b]	6.1[b]
		2.7	2.1[b]	4.8[b]
	Wright County, Minnesota	3.2	5.4[b]	8.6[b]
		3.0	6.7[b]	9.7[b]
		3.5	5.8[b]	9.3[b]
		2.9	4.7[b]	7.6[b]
Pratt et al.,	Traverse County, Minnesota	3.6	3.7[b]	7.3[b]
1983		4.8	3.6[b]	8.4[b]
		4.0	2.9[b]	6.9[b]
		2.0	2.2[b]	4.2[b]

| Parrish et al., 1986 | Scotia, Pennsylvania | 3.0[b] |
| Parrish et al., 1988 | Scotia, Pennsylvania | 3.1[b] |

[a]No data.
[b]Upper limit for NO_2 and NO_x

TABLE 8-3   Average Mixing Ratios Measured at Isolated Rural Sites and Coastal Inflow Sites

References	Location	NO, ppb	NO_2, ppb	NO_x, ppb
Kelly et al., 1980	Niwot Ridge, Colorado			0-2[a]
Kelly et al., 1982	Pierre, South Dakota[b]		1.2[a]	
Kelly et al., 1984	Schaeffer Observatory Whiteface Mountain, New York	≤0.2		1.1[a]
Bollinger et al., 1984	Niwot Ridge, Colorado			0.80
Fehsenfeld et al., 1987	Niwot Ridge, Colorado			0.56
Parrish et al., 1985	Point Arena, California			0.37

[a]Upper limit for NO_2 and NO_x
[b]Measurement site located 40 km WNW of Pierre.

TABLE 8-4   Typical Boundary Layer NO_x Concentrations

Region	NO_x, ppb
Urban-suburban	10-1000
Rural	0.2-10
Remote tropical forest	0.02-0.08
Remote marine	0.02-0.04

OBSERVATIONS OF NO_y

In addition to examining the measurements of atmospheric NO_x, it is instructive to consider the observed concentrations of atmospheric NO_y. The ratio of NO_x to NO_y reflects the chemical processing that occurs in an air mass after the initial introduction of NO_x. Thus this quantity is indicative of the oxidant formation that has occurred in the air mass.

Because urban areas have large sources of NO_x and because it takes several hours to convert NO_x to other NO_y compounds, NO_y concentrations in urban locations are generally dominated by NO_x. For this reason, the NO_y concentrations in urban and suburban locations should be approximately represented by the urban and suburban NO_x concentrations described above.

Because the ability to measure NO_y was developed only recently (see Chapter 7), the rural and remote NO_y data base is even more limited than that for NO_x. However, there are enough data to establish a rough indication of the NO_y distribution. During the summer of 1986, NO_y was measured at several rural sites in North America: Brasstown Bald Mountain, Georgia; Whitetop Mountain, North Carolina; Bondville, Illinois; Scotia, Pennsylvania; Egbert, Ontario; and Whiteface Mountain, New York. These were all rural sites in the industrial regions of the eastern United States or southern Canada. The NO_y concentrations recorded at these sites (c.f., Parrish et al., 1988), along with the period of the measurements at the various sites and their latitudinal locations, are shown in Figure 8-3. The concentrations covered a large range and were quite variable. In general, the low-elevation sites, where air from the atmospheric boundary layer was sampled, were closer to anthropogenic sources. These sites exhibited somewhat higher concentrations of NO_y than did the mountain sites, which were more remote and where the samples usually were from the free troposphere. However, the average NO_y concentrations observed at all the sites were quite similar; median values ranged from 3 to

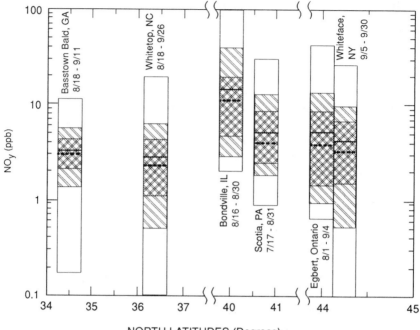

NORTH LATITUDES (Degrees)

FIGURE 8-3   NO$_y$ concentrations measured during the summer of 1986 at several rural sites in North America. The period of the measurements and the location of each site are indicated. Each rectangle encompasses the range of NO$_y$ concentrations recorded at a site. The 90% and 67% ranges of the data are shown as lightly and heavily shaded areas, respectively. The heavy solid line is the average, and the dashed line is the median concentration. Source: Parrish et al., 1988.

10 ppb. These concentrations are somewhat lower than NO$_x$ concentrations typically observed in urban and suburban locations, which range from 10 to 1000 ppb.

The contrast in NO$_y$ concentrations found in rural areas of the continental United States with those observed in the remote troposphere is illustrated in Figure 8-4. The measurement sites are Scotia (Parrish et al., 1988), a rural site in the eastern United States; Niwot Ridge, Colorado (Fahey et al., 1986; Parrish et al., 1988), an isolated inland site in the western United States; Point Arena, California (Parrish et al., 1985), a site on the West Coast that often receives maritime air from the Pacific Ocean; and Mauna Loa, Hawaii (Car

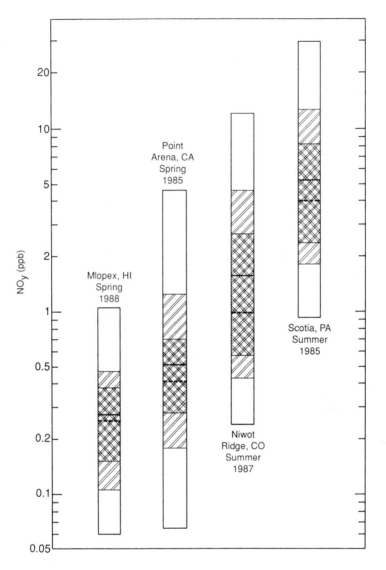

FIGURE 8-4   NO$_y$ measurements made at Mauna Loa (Carroll et al., in press), a remote site, and Point Arena, CA (Parrish et al., 1985), Niwot Ridge, CO (Parrish et al., 1988; Fahey et al., 1986b), and Scotia, PA (Parrish et al., 1988), three rural sites. Bars show range of measurements made. Dashed line = median; solid line = average; shaded area = central 68% of the data.

roll et al., in press), a remote maritime site. Two of the sites, Mauna Loa and Niwot Ridge, are at high elevations (10,000 feet or approximately 3 km), and thus the air sampled there is not necessarily representative of the boundary layer.

As was the case for NO_x, the observations summarized in Figure 8-4 show the progressive decrease in NO_y with increasing isolation from anthropogenic sources of NO_x. For example, the median NO_y mixing ratio decreases from 3.6 ppb at Scotia to 0.28 ppb at Mauna Loa. There is also a progressive decrease in the contribution of NO_x to NO_y as one moves toward more remote regions. On average, NO_x at Scotia accounted for 59% of the observed NO_y (Williams et al., 1987). At Niwot Ridge in 1987, NO_x accounted for 32% of the NO_y (Williams et al., 1987), and at Mauna Loa, NO_x accounted for only 15% of the NO_y (Carroll et al., in press). Because NO_y enters the atmosphere as NO_x, the decrease in the ratio of NO_x to NO_y as one moves to more remote sites can be understood in terms of the increasing chemical conversion of NO_x to organic nitrates (principally PAN) and to inorganic nitrates (principally HNO_3) with increasing distance of the site from major anthropogenic sources. This is why accurate measurements of NO_x at rural and remote locations must be free of interference from other NO_y species.

OBSERVATIONS OF VOCs

Determining the variation in VOC concentrations from one area to another is a much more complex task than is tracking variations in NO_x and ozone. There are scores of different VOCs in the atmosphere—some primary pollutants and others secondary—and the rate at which these compounds react in the atmosphere varies by orders of magnitude. We simplify the task somewhat by focusing on a subset of all the VOCs in the atmosphere, the nonmethane hydrocarbons, and by focusing primarily on midday. This subset of VOCs is composed of primary pollutants (i.e., those species that are emitted directly by sources and are not subsequently transformed in the atmosphere), and thus their concentrations probably most closely reflect local emissions. We exclude methane because its low reactivity precludes it from playing a significant role in urban and regional ozone formation. We focus on data collected at midday because this is the photochemically active period and also typically the time of the most intense vertical mixing; data from this period are most likely to reflect the balance between emissions and photochemical oxidation and least likely to be affected by inhomogeneities caused by local sources. On the other hand, because anthropogenic VOC concentrations are usually largest during the early morning and biogenic emissions tend to be most intense during the afternoon, VOC data from midday tend to overemphasize the contribution of biogenic compounds to the overall VOC mix. For this reason, some VOC data from other periods of the day are also included in our discussion.

The VOC data sets analyzed here are listed in Table 8-5. The analysis includes data gathered from 11 different areas: 5 urban-suburban areas in the United States, 2 rural sites in the eastern United States, 3 sites in the tropical forests of Brazil, and a region of the North Atlantic Ocean. From these 11 areas, 43 separate data sets have been formed for analysis. Each consists of a lengthy list of VOC species in the C_2 through C_{10} range and the concentrations of these species, as measured through analysis of whole-air samples using gas chromatography-flame ionization detection (Greenberg and Zimmerman, 1984; Westberg and Lamb, 1985; Christian and Riley, 1986; Rasmussen and Khalil, 1988). The concentration for each VOC species in each data set is an average of at least three measurements, and in many cases more than ten measurements were made at different times over the period shown in Table 8-5 for that particular data set.

Each data set has been assigned a data code: a Roman numeral, which shows the region; a capital letter, to indicate the area, and a number, to show the specific location and period of data collection. Thus, the data code for the Georgia Tech site in Atlanta is I.A1. The data code for the measurements made in Glendora (a suburb of Los Angeles) from 1200-1600 during the period of 8/12-8/17/86 (when moderately high temperatures, averaging 31 °C, were encountered) is denoted by I.C1, while the equivalent data set for the period from 8/18 - 8/20/86 (when extremely high temperatures, averaging 37 °C, were encountered) is denoted by I.C2. In some cases, a lower-case letter is also used to denote analyses for the same site during the same period of days but at different hours of the day. Thus data codes I.C2a, I.C2b, and I.C2c are for the Glendora site for 8/17 to 8/20/86 but for time intervals from 0800-1200, 1200-1600, and 1600-2000, respectively. In the case of the Baton Rouge data, a separate analysis was carried out for each hour of the day between 0600 and 1800 for each of the two sites considered here. A lower-case letter is used in the code to indicate each hourly measurement; the code I.E1a is used to denote the LSU data gathered at 0600, I.E1b is for data gathered at 0700, and so on.

The information listed in Table 8-5 is probably among the highest quality speciated VOC data available, but the data are not without problem areas and potential flaws. For instance, in all of the data sets, a significant fraction (generally about 10% by mass) of the VOCs in the air samples could not be identified. Furthermore, although many of the species identified in one data set appeared in others, there were some notable exceptions. In the case of

TABLE 8-5 Speciated VOC Data Analyzed

Region	Area	Site location	Data code[a]	Time period		Comments	Data source
				Days	Hours		
Urban-suburban	Atlanta	Georgia Tech	I.A1	7/13-8/03/81	1100-1400	"Midtown" site p-cymene identified	W. Lonneman
		Fulton County Health Dept.	I.A2	7/13-8/03/81	1100-1400	"Downtown" site p-cymene identified	
		DeKalb Comm. College	I.A3	7/13-8/03/81	1100-1400	"Perimeter" site p-cymene identified; α-pinene identified	
	Detroit	Downtown	I.B1	7/20-8/03/81	1200-1500	Avg. T = 31 °C	W. Lonneman

Los Angeles	Glendora	I.C1	8/12-8/17/86	1200-1600	Avg. T = 37 °C	D. Lawson., R. Rasmussen, and W. Lonneman
	Glendora	I.C2a		0800-1200		
	Glendora	I.C2b	8/18-8/20/86	1200-1600		
	Glendora	I.C2c		1600-2000		
	Claremont	I.C3	8/27-8/29/87	1400-1700	High isoprene days	
Columbus	Fort Hayes	I.C3	8/20-8/29/87	1200-1600	"Downtown" site	W. Lonneman
	WVCO Radio Tower	I.D2	7/20-8/14/80	1200-1600	"Suburban" site	

Region	Area	Site location	Data code[a]	Time period		Comments	Data source
				Days	Hours		
	Baton Rouge	La. State University	I.E1 a...m	7/18-7/26/89	0600-1800	"Downtown" site, analysis for each hour of period	M.O. Rodgers
		Pride	I.E2 a...m	7/27-8/20/89	0600-1800	"Suburban" site analysis for each hour of period	
Rural	Pennsylvania	Scotia	II.A	Summer, 1988	0800-1700	$N=200$ samples	D. Parrish
	Georgia	Brasstown Bald Mtn.	II.B	Summer, 1988	0800-1700	High altitude site $N=30$ samples	M.O. Rodgers

Remote tropical forest	Brazil	Ducke Forest	III.A	Dry season, 1985	24 hr	10km North of Manaus, $N=81$ samples	
		Amazon	III.B	September 1980	24 hr	Samples collected after biomass burning $N=12$ samples	
		Lago Calado	IIIC.	Wet season, 1984	24 hr	50km north of Manaus, $N=9$ samples	P. Zimmerman, et al., 1988b

Region	Area	Site location	Data code[a]	Time period Days	Time period Hours	Comments	Data source
Remote marine	N. Atlantic Ocean	26-34°N 72-80°W	IV.A	2/3, 2/8, 2/10/86	1300-1500		P. Zimmerman and J. Greenberg

[a]Each data set has been assigned a data code: a Roman numeral (which shows the region); a capital letter (to indicate the area); and a number (to show the specific location and period of data collection). Source: Parrish et al., 1988

[b]Personal communications in April 1990: W. Lonneman (EPA), D. Lawson (California Air Resources Board), R. Rasmussen (Oregon Graduate Center), M.O. Rodgers (Georgia Institute of Technology), D. Parrish (National Oceanic and Atmospheric Administration), P. Zimmerman and J. Greenberg (National Center for Oceanic Research).

the urban data sets, isoprene was generally the only biogenic VOC identified. In the Atlanta data, however, relatively large concentrations of *p*-cymene appear, and in the Detroit data small concentrations of α-pinene were reported. Although these differences could reflect the actual chemical variability of the atmosphere, the possibility that they are caused by analytical problems with one or more of the measurements cannot be ruled out. All of the VOC data analyzed here were gathered from sites located at or near the earth's surface. Because virtually all VOC sources also are located near the surface and because concentrations tend to decrease as VOCs disperse in the atmosphere at a rate proportional to their reactivity, the data analyzed here could be somewhat biased in favor of reactive as opposed to less reactive VOCs. It is difficult to assess the magnitude of this bias, but the limited measurements of Zimmerman et al. (1988b) and Rasmussen and Khalil (1988) over tropical forests suggest that it is not especially large. (When averaged over the atmospheric boundary layer, it is less than a factor of two.) Furthermore, as noted earlier, we have focused on midday data in an attempt to minimize bias.

ANALYSIS OF VOC DATA SETS

The analysis of the data sets was done with two contrasting methods. The first, which is the simplest and the one probably most often adopted, is a concentration-based method in which the various species are ranked in importance according to their concentrations (on a carbon-atom basis), and data sets from different locations and times are compared according to the total VOC concentration (the sum of the concentrations of all the individual VOCs).

However, because the concentration-based method does not account for the different reactivities of the various VOC species, it can be misleading about the involvement of the various species in ozone formation. Recall from Chapter 5 that the rate of ozone production from a given VOC is essentially a function of three factors: the species' atmospheric concentration, its rate of reaction with OH (its OH-reactivity), and the number of ozone molecules produced each time the species is oxidized (its mechanistic reactivity). Although the concentrations and OH-reactivities of VOCs can vary by orders of magnitude from one species to another, the mechanistic reactivities of the VOC species generally found in the atmosphere are fairly uniform, varying only by factors of two or three from one species to another (see Table 5-5). The product of a VOC's concentration and its rate of reaction with OH will determine its relative role in an air mass as an ozone precursor. A species with a large concentration will not necessarily be an important precursor if it is unreactive; conversely, another with a small concentration can be important if it is extremely reactive. (An example is methane, typically the most abundant VOC in the atmosphere but of negligible importance in producing ozone on urban or regional scales over the contiguous United States because of its extremely low reactivity.) An air mass can have a large total VOC concentration but a low ozone-producing capacity if the VOCs present are relatively unreactive.

To account for the combined effect of OH-reactivity and concentration, we have adopted a second, OH-reactivity-based method. In this method, we define a propylene-equivalent concentration, Propy-Equiv(j), for each VOC species j. This equivalent concentration is given by

$$Propy\text{-}Equiv(j) \ = \ Conc(j) \ \frac{k_{OH}(j)}{k_{OH}(C_3H_6)}, \qquad (8.1)$$

Conc(j) is the concentration of species j in ppb of carbon (C); $k_{OH}(j)$ is the rate constant for the reaction between species j and OH; and $k_{OH}(C_3H_6)$ is the rate constant for the reaction between OH and propylene. Propy-Equiv(j) is a measure of the concentration of species j on an OH-reactivity-based scale normalized to the reactivity of propylene and is literally the concentration (in parts per billion carbon) required of propylene to yield a carbon oxidation rate equal to that of VOC species j. Thus if a VOC species has an atmospheric abundance of 10 ppbC and is twice as reactive as propylene, its Propy-Equiv is 20 ppbC; if the species is half as reactive as propylene, its Propy-Equiv is 5 ppbC.

Because the OH-reactivity-based method accounts for a species' rate of reaction as well as its atmospheric concentration, it provides a more accurate picture than does the concentration-based method of the relative contribution of each VOC species to the photochemical production of ozone at the specific time and location of the measurement. To the extent that the measurements from a given site are representative of the average concentrations throughout the air mass, the OH-reactivity-based method provides a more rational basis for assessing the relative importance of the various VOCs in the air mass to ozone formation as well as for comparing the VOC concentrations in different air masses. The reader should note that the use of propylene's reactivity as the normalization factor in the above formulation is completely arbitrary, and equivalent results would be obtained if the reactivity of any other species had been chosen. In addition, even though Equation 8.1 considers only VOC reactions with OH and does not account for ozone reactions, this has little effect on our conclusions; calculations using an alternate formulation to account for ozone as well as OH reactions with VOCs yield results essentially identical to those obtained from Equation 8.1. A ranking of the 35 most important species from data set I.A1 is presented in Table 8-6. In Part A of the table, the ranking was made using the concentration-based method, and in Part B the OH-reactivity-based method was used. (The OH rate constants used to calculate the Propy-Equiv concentrations in Table 8-6 and below were obtained from Middleton et al. [1990].) Although the 35 most important compounds obtained from both methods include a complex list of alkanes, alkenes, and aromatics ranging in concentration from 1 to 20 ppbC, the relative ranking of the species on the two lists is quite dissimilar. The two highest ranking species using the concentration-based method are *i*-pentane and *n*-butane, two relatively unreactive compounds; the third highest ranking species is toluene, a moderately reactive compound. Isoprene, a highly reactive species normally associated with biogenic emissions, is ranked fifteenth using this method, with an average concentration of 4.6 ppbC; it constitutes only about 2% of the total VOCs present during the sampling period. The total VOC concentration for the data set, obtained by adding the concentrations of each individual VOC, is about 200 ppbC.

In contrast to the above results, the highest ranking species obtained with the OH-reactivity-based method are isoprene and *p*-cymene, two biogenic compounds, and *m*- and *p*-xylene, highly reactive aromatics associated with evaporative emissions. (An unusual feature of the data sets from Atlanta is the high concentration for reported *p*-cymene, a reactive aromatic thought to be emitted biogenically.) Whereas *i*-pentane, *n*-butane, and toluene are the most abundant species in the data set, their rankings on an OH-reactivity-based scale are eleventh, eighteenth, and sixth, respectively. Furthermore, although the total VOC concentration for data set I.A1 is about 200 ppbC, the total Propy-Equiv concentration is only about 105 ppbC, indicating that the mix of VOCs in data set I.A1 is on average less reactive than propylene. (This conclusion is totally consistent with the data given in Chapter 5 concerning the incremental reactivities of an "all-city" urban mix and representative alkanes, alkenes, and aromatic VOCs.) These results illustrate the importance of accounting for a species' OH-reactivity in assessing its role in the photochemistry of an air mass. For instance, it might be concluded using a concentration-based approach that biogenics are unimportant in ozone formation in the Atlanta area because their ambient concentrations are only a few percent based scale indicates that a significant fraction of the total organic carbon being cycled through the atmosphere at the Atlanta site originated from biogenic VOCs and thus that these species could be quite important.

TABLE 8-6   Top 35 and Total VOCs Measured at Georgia Tech Campus, Atlanta, 1100-1400, 7/13/81 - 8/03/81 (dataset I.A1)

Species sorted by concentration		Species sorted by OH-reactivity	
Species	Concentration, ppbC	Species	Propy-Equiv., ppbC
1 *i*-pentane	19.8	1 isoprene	17.6
2 *n*-butane	16.9	2 *p*-cymene	10.4
3 toluene	14.7	3/*m*&p-xylene	7.1
4 *p*-cymene	11.0	4 2-methyl 2-butene	5.9
5 *n*-pentane	9.4	5 1,3,5-trime-benzene	4.3
6 benzene	8.8	6 toluene	3.5
7 *m*&*p*-xylene	7.6	7 *m*&*p*-ethyl toluene	3.4
8 2-me-pentane	5.9	8 1,2,4-trime-benzene	3.4
9 cyclohexane	5.4	9 *t*-2-pentane	3.1
10 2-me-hexane	5.2	10 iso-butene	3.1
11 ethane	5.0	11 *i*-pentane	3.1
12 undecane	4.9	12 *t*-2-butene	2.9
13 propane	4.8	13 *c*-2-butene	2.7
14 *i*-butane	4.8	14 undecane	2.3

15	isoprene	4.6
16	acetylene	4.3
17	n-hexane	3.8
18	m&p-eth-toluene	3.6
19	3-me-pentane	3.4
20	ethylene	3.0
21	me-cyclopentane	2.9
22	ethylbenzene	2.8
23	o-xylene	2.8
24	3-me-hexane	2.6
25	2,3-dime-pentane	2.5
26	1,4-dieth-benzene	2.4
27	iso-butene	2.2
28	2,2,4-trime-pentane	2.2
29	1,2,4-trime-pentane	2.2
30	i-butyl-benzene	2.2
31	2-me-2-butene	1.8
32	1,3,5-trime-benzene	1.8
33	cyclopentane	1.6

15	1,4-diethyl benzene	2.3
16	c-2-pentene	2.2
17	cyclohexane	1.8
18	n-butane	1.7
19	1,2,3-trime-benzene	1.7
20	o-xylene	1.6
21	2-methyl 1-butene	1.6
22	propene	1.5
23	2-methyl hexane	1.4
24	n-pentane	1.4
25	2-methyl pentane	1.2
26	o-ethyl toluene	1.0
27	ethylene	1.0
28	1-pentene	0.9
29	2,3-dimethyl pentane	0.8
30	ethylbenzene	0.8
31	methyl cyclopentane	0.8
32	n-hexane	0.8
33	3-methyl hexane	0.7

Species sorted by concentration		Species sorted by OH-reactivity	
Species	Concentration, ppbC	Species	Propy-Equiv., ppbC
34 propene	1.5	34 3-methyl pentane	0.7
35 *i*-propyl-benzene	1.5	35 *i*-butyl benzene	0.5
Total[a]	197	Total[a]	105.0

[a]Includes all measured VOCs and comprises more than the 35 species presented in the table.

The total VOC concentrations and total Propy-Equiv concentrations for each of the midday data sets included in our analysis are compared in Figure 8-5. (Note that data sets III.A, III.B, and III.C from the remote tropical forest actually represent diurnally averaged concentrations rather than midday concentrations. Because VOC concentrations in forests generally peak at midday, the values indicated in Figure 8-5 are underestimates of the midday concentrations to be expected in tropical forests.) The results obtained from the concentration-based method are fairly predictable. On average, the urban-suburban areas have the highest total VOC concentrations, ranging from somewhat less than 100 ppbC to slightly more than 500 ppbC. Lower total VOC concentrations in the range of a few tens of ppbC are found in the rural and remote areas, where anthropogenic influences are small to negligible. The highest total VOC concentration obtained for the remote tropical forest was from data set III.B. Note in Table 8-5 that this data set was gathered during a period following extensive biomass burning; the relatively large abundance of VOCs measured during this period perhaps indicated the sizable effect biomass burning can have on a region's air quality.

A very different pattern emerges in Figure 8-5 from the results obtained with the OH-reactivity-based method. In the urban-suburban areas, the total Propy-Equiv concentration is always less than the total VOC concentration, indicative of the large amounts of relatively unreactive VOCs typically present in the urban atmosphere. In rural areas and in the remote tropical forests, on the other hand, where emissions of highly reactive biogenic VOCs are the largest, the total Propy-Equiv concentration is always larger than the total VOC concentration. As a result, we find using the OH-reactivity-based meth-

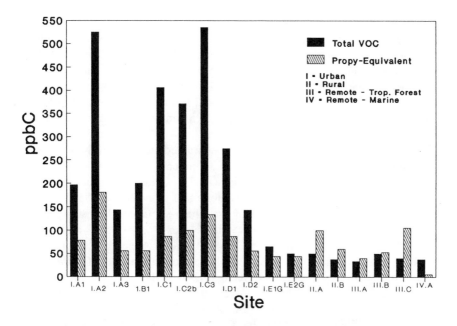

FIGURE 8-5   Total nonmethane VOC concentrations and propylene
equivalents (Propy-Equiv) concentrations measured at urban-suburban,
rural, and remote sites from Table 8-5. The urban-suburban data were
gathered during the midday, the rural data are averages from 0800 to
1700 local standard time (LST). The remote, tropical forest data are
diurnally averaged, and the remote marine data were gathered from
1300-1500 (LST).

od that the concentrations of VOCs in rural regions of the United States and
in the remote tropical forests of Brazil are similar to those found in urban
centers of the United States; the total Propy-Equiv concentrations from these
three regions all range from 50 to 250 ppbC. In fact, the total Propy-Equiv
concentration calculated for the tropical forests during the wet season (data
set III.C) is actually larger than the total Propy-Equiv concentrations obtained
from the majority of the midday data sets from the urban-suburban regime.
The only data set from a remote regime that yielded a significantly smaller
total Propy-Equiv concentration than that of the urban-suburban data sets was
IV.A. This data set, however, was gathered in the remote marine atmosphere,
where biogenic VOC emissions are negligible and anthropogenic influences
are minimal.

SOURCE APPORTIONMENT

To gain insight about the relative importance of various kinds of sources in generating the VOCs observed in the atmosphere, the VOC data sets from the urban, suburban, and rural United States were further analyzed by assigning each species' Propy-Equiv concentration to an emission or source type. As described below, two broad categories of sources were considered for this analysis: biogenic and anthropogenic, with the anthropogenic sources further divided into mobile and stationary sources.

Biogenic VOCs were assumed to be characterized exclusively by isoprene and all of the terpenoid compounds (eg., α-pinene, β-pinene) normally associated with vegetative emissions (Chapter 9). Evidence in support of this assumption in the case of isoprene, the dominant biogenic VOC in all the data sets analyzed here (and for most of the urban-suburban data sets the only biogenic VOC identified), is presented in Figures 8-6, 8-7, and 8-8. Figure 8-6, which is based on data gathered from the Los Angeles, California, and Baton Rouge, Louisiana, areas, shows that although reactive VOCs normally associated with mobile sources, such as 2-methyl-2-pentene, cyclohexene, trans-2-pentene, and cis-2-butene (Middleton et al., 1990), are strongly correlated with one another, isoprene shows a weak negative correlation with these compounds. Figure 8-7 illustrates that the ratio between isoprene and a mobile-source VOC exhibits a minimum during the night and early morning hours, when biogenic emissions are suppressed, and a maximum in the late afternoon, when isoprene emissions are at their peak. Figure 8-8 illustrates that although the variability in urban isoprene concentrations is large, it exhibits a temperature dependence consistent with the laboratory-measured temperature dependence of biogenic isoprene emissions.

All VOCs not assigned to the biogenic category were assumed to originate exclusively from anthropogenic sources. These anthropogenic VOCs were further divided into mobile and stationary sources for all but the Los Angeles sites using the 1985 NAPAP (National Acid Precipitation Assessment Program) speciated VOC inventory for the United States (J. Wagner, EPA, and M. Saeger, Alliance Technologies, pers. comm., April 1990). For the Los Angeles sites, we used a speciated VOC inventory prepared by the California Air Resources Board (CARB) to simulate VOC emissions during an August day in the Los Angeles area (T. McGuire and P. Allan, pers. comm., California Air Resources Board, June 1990). If an anthropogenic VOC had a Propy-Equiv concentration of 10 ppbC and if, in the inventory, the source of this VOC was 50% from mobile sources and 50% from stationary sources, 5 ppbC of its Propy-Equiv concentration would be assigned to mobile sources and the other 5 ppbC to stationary sources. Any species that appeared in the ambient measurements but did not appear in the inventory was assumed to come exclusively from stationary sources. By adding up the contributions from all

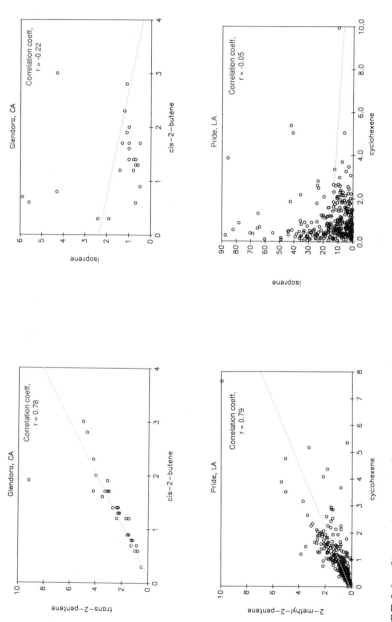

FIGURE 8-6 Observed atmospheric concentrations of trans-2-pentene, cis-2-butene, cyclohexene, 2-methyl-2-pentene, and isoprene. The data from Pride, a suburb of Baton Rouge, were obtained from M.O. Rodgers (personal communication). The data from Glendora, a suburb of Los Angeles were obtained from R. Rasmussen and D. Lawson (personal communication).

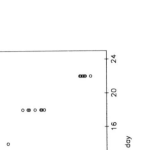

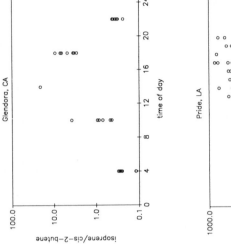

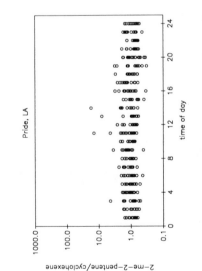

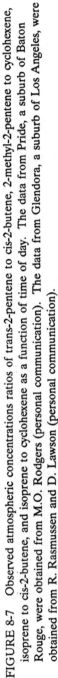

FIGURE 8-7   Observed atmospheric concentrations ratios of trans-2-pentene to cis-2-butene, 2-methyl-2-pentene to cyclohexene, isoprene to cis-2-butene, and isoprene to cyclohexene as a function of time of day. The data from Pride, a suburb of Baton Rouge, were obtained from M.O. Rodgers (personal communication). The data from Glendora, a suburb of Los Angeles, were obtained from R. Rasmussen and D. Lawson (personal communication).

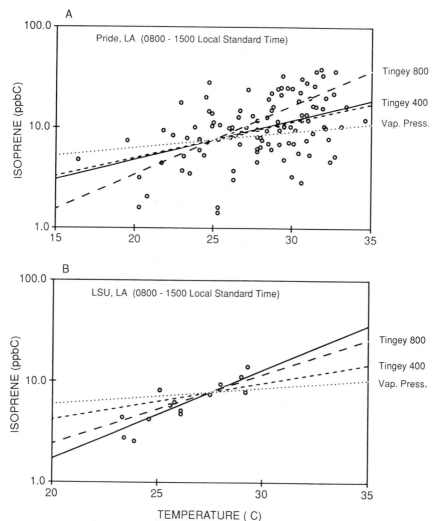

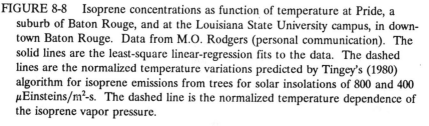

FIGURE 8-8   Isoprene concentrations as function of temperature at Pride, a
suburb of Baton Rouge, and at the Louisiana State University campus, in downtown Baton Rouge. Data from M.O. Rodgers (personal communication). The
solid lines are the least-square linear-regression fits to the data. The dashed
lines are the normalized temperature variations predicted by Tingey's (1980)
algorithm for isoprene emissions from trees for solar insolations of 800 and 400
μEinsteins/m^2-s. The dashed line is the normalized temperature dependence of
the isoprene vapor pressure.

VOCs, a total Propy-Equiv concentration for mobile sources and for stationary sources was obtained.

The resulting Propy-Equiv concentrations from mobile, stationary, and biogenic sources for the urban-suburban midday data sets as well as the rural data sets are shown in Figure 8-9. In Figures 8-10 and 8-11, we illustrate how the contributions of the various sources to the total reactivity (Propy-Equiv concentrations) at single locations (downtown Baton Rouge and Glendora) vary as a function of time of day. Because the distinction between biogenic and anthropogenic VOCs in the urban-suburban and rural data sets is straightforward, the sum of the reactivity of VOCs apportioned to mobile and stationary sources in the figures is probably a fairly reliable estimate of the total anthropogenic contribution to the reactivity of VOCs concentrations at the various sites. However, the apportionment of these VOCs between mobile and stationary sources should be viewed only as a rough estimate because of the sizable uncertainties in the inventories used to make the apportionment (see Chapter 9). Furthermore, because isoprene was the only biogenic species measured in the urban samples and because it generally constitutes 30-50% of the total biogenic emissions in an area (Chapter 9), the total reactivity of VOCs assigned to biogenic sources should be viewed as a lower limit in these cases.

With the exception of the Baton Rouge data sets, mobile and stationary sources make the largest contributions to the total reactivity of VOCs at the urban sites. In Baton Rouge, biogenic sources contribute most to the total reactivity, and in the eastern U.S. rural data sets, biogenic sources dominate over anthropogenic sources. The anthropogenic and biogenic contributions to the total reactivity vary significantly over the course of the day. In Baton Rouge and Glendora, anthropogenic VOCs peak in the early morning hours; biogenic VOCs tend to peak during the mid- and late afternoon. In Los Angeles, the contribution from biogenic VOCs never equals the contributions from anthropogenic sources. In Baton Rouge, however, a very different pattern emerges; biogenic VOCs surpass the contributions from mobile and stationary sources by 1,000 hours and remain dominant for the rest of the daylight period.

Although mobile and stationary sources make the largest contributions to the reactivity of VOCs at most of the urban-suburban sites, the contribution from biogenic sources to these areas should not be discounted as negligible. In most of the urban-suburban data sets, a significant fraction of the total reactivity was found to arise from VOCs of biogenic origin. Except for downtown Detroit, Michigan, and Columbus, Ohio, where extremely low isoprene concentrations were reported, midday biogenic VOC concentrations in the urban-suburban data sets accounted for a Propylene-Equiv (reactivity) of at

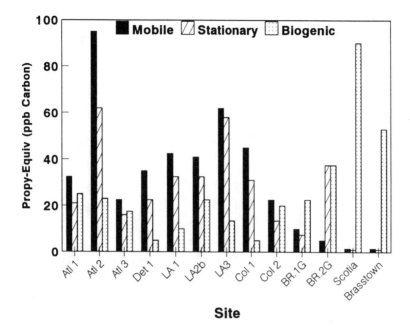

Site

FIGURE 8-9   Total nonmethane VOC in propylene-equivalent concentrations in units of ppb carbon observed at urban-suburban sites (midday) and rural sites (daylight hours) and apportioned by source category. Atl is Atlanta, Det is Detroit, LA is Los Angeles, Col is Columbus, BR is Baton Rouge, Scotia is in Pennsylvania, Brasstown is Brasstown Bald in Georgia. Because of uncertainties, the apportionment between source categories should be viewed as an estimate. For instance, except for the Los Angeles sites, the splitting of anthropogenic VOCs between mobile and stationary source VOCs was based on a national rather than a local inventory. The assignment of biogenic VOCs is a lower limit because isoprene was generally the only biogenic VOC identified in the speciated data.

least 10 ppbC—and in many cases, more than 20 ppbC. Even in Glendora, in an area not generally noted for large biogenic emissions, midday isoprene concentrations were about 10 ppbC Propy-Equiv during a moderately hot period (see Table 8-5) and more than 25 ppbC during a particularly hot 3-day period. Furthermore, given that isoprene, which typically amounts to 30-50% of the total VOC emissions from vegetation (Chapter 9), was generally the only VOC of biogenic origin identified in the urban-suburban data sets, it is possible that the actual contribution of biogenic emissions to the reactivity of

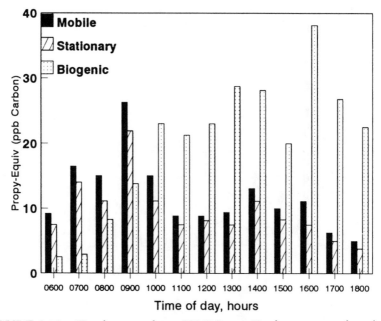

FIGURE 8-10   Total nonmethane VOC Propy-Equiv concentrations in units of ppb carbon observed at the Louisiana State University campus as a function of time of day and apportioned by source category. Sampling period was July 18-26, 1989.

VOCs in urban areas is considerably larger than we estimate here.

It is also important to note that the biogenic contribution is a background VOC blanket that cannot be removed from the atmosphere through emissions controls. If VOC emissions from anthropogenic sources are reduced in the future, this background will be a larger and more significant fraction of the total reactivity of VOCs. Our analysis suggests that in many cities, even if anthropogenic VOC emissions are totally eliminated, a background of reactive biogenic VOCs will remain; on hot summer days, this background can be equivalent to 10 or 20 ppbC, or perhaps more, of propylene. Without control of NO_x emissions, this VOC background should be able to generate ozone concentrations that exceed the NAAQS concentration of 120 ppb (Chameides et al., 1988).

Another interesting aspect of our results relates to the relative contributions of mobile- and stationary-source VOCs. In the NAPAP and CARB inventories, stationary-source VOC emissions are estimated to be significantly larger than are mobile-source emissions. For instance, in the CARB invento-

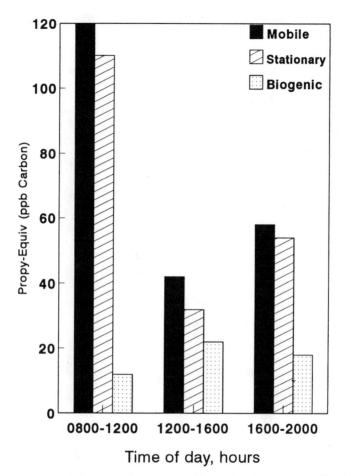

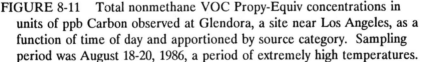

FIGURE 8-11   Total nonmethane VOC Propy-Equiv concentrations in
units of ppb Carbon observed at Glendora, a site near Los Angeles, as a
function of time of day and apportioned by source category. Sampling
period was August 18-20, 1986, a period of extremely high temperatures.

ry, daily stationary-source emissions in the Los Angeles area are estimated to
total 1,881,000 kilograms (kg); mobile source emissions total only 732,000
kg/day. By contrast, our analysis based on the ambient concentrations of
mobile-source and stationary-source VOCs indicates that mobile sources
contribute as much as or perhaps somewhat more than stationary sources
(Figures 8-9, 8-10, and 8-11). Moreover, as illustrated in Figure 8-12, this
finding appears to be essentially independent of whether the NAPAP invento-

ry or the CARB inventory is used in the analysis. The fact that the apportionment of the observed anthropogenic VOCs yields a relatively larger role for mobile sources than implied by the emissions inventories suggests that mobile emissions could have been underestimated in these inventories.

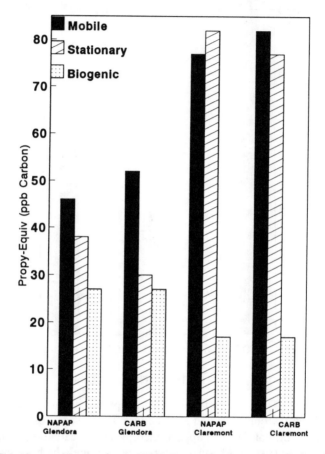

FIGURE 8-12   Nonmethane VOC Propy-Equiv concentrations in units of ppb Carbon apportioned by source category using the 1985 National Acid Precipitation Assessment Program (NAPAP) speciated VOC inventory for the nation and the California Air Resources Board (CARB) speciated VOC inventory for the Los Angeles area during an August day. Results are shown for Glendora (data set I.C2b) and Claremont (data set I.C3). Data sets are described in Table 8-5.

SUMMARY OF VOC, NO$_x$, AND OZONE OBSERVATIONS

The ranges of VOC, NO$_x$, and ozone concentrations measured in the four atmospheric boundary layer regions are summarized in Figure 8-13. The format for this figure was chosen to resemble the traditional ozone isopleth diagram (Chapter 6). However, although the x-axis in the ozone isopleth typically adopts a concentration-based scale (total VOC concentration), an OH-reactivity-based scale (Propy-Equiv) is adopted in Figure 8-13.

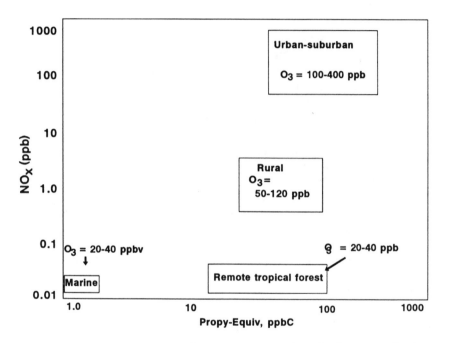

FIGURE 8-13   VOC, NO$_x$ and ozone concentrations in the atmospheric boundary layer at four locations. VOC is shown as Propy-Equiv concentrations in units of ppb carbon.

The position of the four regions in the diagram shows a strong relationship between observed ozone and NO$_x$ concentrations but little or no consistent relationship between ozone and VOC reactivity. Although ozone and NO$_x$ concentrations increase substantially as one moves from the tropical forest to rural areas and then to urban and suburban regions, VOC reactivity as measured in Propyl-Equiv remains essentially the same in all three. Similarly, although VOC reactivity increases by more than an order of magnitude from the remote marine region to the tropical forest, ozone concentrations remain the same; the NO_x concentrations in these two areas also are quite similar.

SUMMARY

An analysis of observed concentrations of ozone, oxides of nitrogen (NO_x), and volatile organic compounds (VOCs) in remote, rural, and urban-suburban areas implies the following:

• A large gradient in NO_x concentrations exists between remote, rural, and urban-suburban areas. This gradient suggests that anthropogenic sources dominate the NO_x atmospheric budget in the continental United States and that the greatest domination usually occurs in urban centers.

• On an OH-reactivity-based scale, the VOC concentrations observed at surface sites in the remote tropical forests of Brazil, in rural areas of the eastern United States, and in urban-suburban areas of the United States are comparable. All three regions tend to exhibit total VOC concentrations equivalent to 50-150 ppbC of propylene.

• In urban-suburban and rural areas of the United States, VOCs from mobile and stationary sources contribute about equally to the total ambient VOC reactivity. Comparison of this observation with VOC inventories suggests that the inventories have underestimated the contribution of mobile sources.

• In urban-suburban areas of the United States at midday, biogenic VOCs can account for a significant fraction of the total ambient VOC reactivity. Under some conditions in Atlanta and Los Angeles, isoprene alone was found to be present in near-surface air at concentrations equivalent to 25 ppbC of propylene on an OH-reactivity scale. In Baton Rouge, isoprene concentrations equivalent to 40 ppbC of propylene were observed.

• In rural areas of the eastern United States, biogenic VOCs contribute more than 90% of the total ambient VOC reactivity in near-surface air and dominate over anthropogenic VOCs.

• As one moves from remote forests to rural areas in the eastern United States and then to urban and suburban areas in the United States, ozone concentrations are found to correlate with NO_x but not with VOC reactivity. This suggests that, in the gross average, NO_x and not VOCs is the limiting factor in ozone photochemical production.

These conclusions have been drawn from an analysis of a limited data base. It is hard to establish that the data are representative, particularly in the case of the VOC measurements, and it is likely that there are specific urban, suburban, and rural areas in the United States that do not follow the trends implied by the data analyzed here. A more representative analysis would be possible with a more complete data base, including a more comprehensive set of VOC measurements that more accurately establishes the horizontal and vertical variability of these species in the troposphere. In addition, one cannot exclude the possibility that the VOC analysis has been significantly biased by the inability of current techniques to identify and quantify the concentrations of all VOCs in the atmosphere. For these reasons, an important research priority for the coming decade should be the development and application of accurate and reliable techniques for the measurement of VOCs that react to form ozone.

9

Emissions Inventories

INTRODUCTION

According to the air quality management approach in environmental regulation, emission limits are set according to the stringency needed to achieve a desired concentration of an atmospheric pollutant. Such an approach is based on an understanding of the quantitative relationship between atmospheric emissions and ambient air quality. The task of evaluating this relationship is straightforward for primary pollutants, such as sulfur dioxide (SO_2) or carbon monoxide (CO), whose ambient concentrations are directly related to emissions because the pollutant of interest in the atmosphere is the pollutant that is emitted. For many large emission sources of SO_2, it is possible to measure simultaneously emissions and ambient air quality in the affected region. With CO, which is emitted mostly by mobile sources—cars and trucks—data on the actual emission rates by source (real-time) are not available, and the source contribution is much more ubiquitous, but real-time ambient measurement is possible.

The air quality management approach for secondary pollutants, such as ozone, introduces issues additional to those raised for primary pollutants. These issues result from the added complexity introduced by the coupled chemical relationship between ozone production and precursor emissions. One class of the primary emitted precursors—the oxides of nitrogen (NO_x), which have attributes similar to those described above for CO—is measurable in the ambient air, and subject to limitations in real-time source monitoring. Point source NO_x emissions make up approximately 57% of the national

251

(California's Inspection/Maintenance program) inventory, and 82% of the point sources emit 5000 tons or more annually; 43% of the NO_x inventory is generated by mobile sources (EPA, 1989a). Volatile organic compounds, in contrast, are less well characterized from both a real-time emissions and ambient monitoring perspective.

This chapter provides an overview of the anthropogenic emissions inventory: how it is compiled, what the major contributing sources are, and where uncertainties lie. There is a similar overview of the inventory of biogenic emissions, and finally a review of efforts to evaluate the accuracy of emissions inventories.

COMPILATION OF EMISSIONS INVENTORIES

In 1971, the U.S. Environmental Protection Agency (EPA) established the National Emissions Data System (NEDS) on sources of airborne pollutants. This system was to summarize annual cumulative estimates of source emissions by air quality control region, by state, and nationwide for the Clean Air Act's five criteria pollutants: particulate matter, sulfur oxides, nitrogen oxides, VOCs, and carbon monoxide. At that time the developers did not envision the evolving demands on emissions inventories that have become common with the advent of increasingly sophisticated air quality models. The original intent to compile annual national trends in the emissions of VOCs, NO_x, SO_2, CO, and particulate matter has been expanded and amended by the need for chemical speciation of VOCs, consideration of additional chemical species, more detailed information on spatial and temporal patterns of inventoried species, and techniques to project trends in emissions.

An estimate of emissions of a pollutant from a source is based on a technique that uses "emission factors," which are based on source-specific emission measurements as a function of activity level (e.g., amount of annual production at an industrial facility) with regard to each source. For example, suppose one wants to sample a power plant's emissions of SO_2 or NO_x at the stack. The plant's boiler design and its Btu (British thermal unit) consumption rate are known. The sulfur and nitrogen content of fuel burned can be used to calculate an emissions factor of x kilograms (kg) of SO_2 or NO_x emitted per y megagrams (Mg, or metric tons) of fuel consumed.

EPA has compiled emission factors for a variety of sources and activity levels (such as production or consumption), reporting the results since 1972 in "AP-42 Compilation of Air Pollutant Emission Factors," for which supplements are issued regularly (the most recent was published in 1985) (EPA, 1985). Emission factors currently in use are developed from only a limited sampling of the emissions source population for any given category, and the values reported are an average of those limited samples and might not be statistically representative of the population. As illustrated in Figure 9-1 (Placet et al., 1990), 30 source tests of coal-fueled, tangentially fired boilers led to calculations of emission factors that range approximately from 5 to 11 kg NO_x per Mg of coal burned. The sample population was averaged and the emission factor for this source type was reported as 7.5 kg NO_x per Mg coal. The uncertainties associated with emission factor determinations can be considerable. They are discussed later in this chapter.

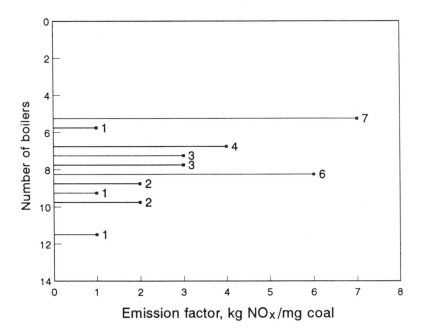

FIGURE 9-1   Results of 30 NO_x-emissions tests on tangentially-fired boilers that use coal. An average of 7.5 kg NO_x/Mg coal was obtained. Source: Placet et al., 1990.

The formulation of emission factors for mobile sources, the major sources of VOCs and NO_x, is based on rather complex emission estimation models used in conjunction with data from laboratory testing of representative groups of motor vehicles. Vehicle testing is performed with a chassis dynamometer, which determines the exhaust emission of a vehicle as a function of a specified ambient temperature and humidity, speed, and load cycle.   The specified testing cycle is called the Federal Test Procedure (FTP) (EPA, 1989b). Based on results from this set of vehicle emissions data, a computer model has been developed to simulate for specified speeds, temperatures, and trip profiles, for example, the emission factors to be applied for the national fleet average for all vehicles or any specified distribution of vehicle age and type. These data are then incorporated with activity data on vehicle miles traveled as a function of spatial and temporal allocation factors to estimate emissions. The models used to estimate mobile source emissions have been developed primarily by EPA; California has developed its own model. Recent versions of the EPA and California mobile source emission factor models are MOBILE4 (EPA, 1989b) and EMFAC7E (CARB, 1986; Lovelace, 1990), respectively.

The basic approach in estimating emissions therefore is derived from a simple calculation that requires an estimate of an activity level, an emissions factor, and, if the source has a pollution control device, a control factor:

Emission =
activity level × emission factor × control factor

Although obtaining an estimate of the activity level can be simple and as direct as monitoring fuel use or power plant load for a specified period, it also can be quite complex and indirect, requiring spatial aggregation or disaggregation of estimated activity measures, which may depend on the source type or category and its emission rate. Essential data elements compiled as part of the National Acid Precipitation Assessment Program (NAPAP) point source emissions file are presented in Table 9-1; data elements related to area source compilations are presented in Table 9-2.

ANTHROPOGENIC EMISSIONS INVENTORIES

The 1985 NAPAP emissions inventory prepared by EPA for NAPAP (EPA, 1989a), includes emissions from the U.S. and Canada for 1985. It was developed to provide information for assessment and modeling objectives of the national program. The inventory listed emissions of CO, SO_2, NO_x, VOCs, total suspended particulate matter, ammonia (NH_3), primary sulfate (SO_4^{-2}), hydrogen chloride (HCl), and hydrogen fluoride (HF). Of specific interest to this report are the NO_x, VOC, and CO emissions, which are summarized by source category in Figure 9-2 and by state in Figure 9-3. The specifics of the compilation are discussed by EPA (1989a) and are only briefly reviewed in this report.

The U.S. emissions data were derived primarily using existing methodologies previously developed by EPA (Zimmerman et al., 1988a; Demmy et al.,

TABLE 9-1   Types of Point Source Emissions Data for NAPAP

Plant Data

State, county, air quality control region, and UTM[a] zone codes

Point Data

Point identification number
Standard industrial classification (SIC) code
UTM coordinates
Stack, plume data (height, diameter, temperature, flow rate)
Points with common stack
Boiler design capacity
Control equipment (devices and control efficiencies by pollutant)
Operating schedule (season, hr/day, days/week, weeks/year)
Emissions estimates for criteria pollutants (method)

Process Data

Source classification code (SCC)
Operating rates (annual, maximum hourly design)
Fuel content (sulfur, ash, heat, nitrogen)

---

[a]UTM, Universal Trans Mercator, a type of map projection. Source:
Placet et al., 1990.

1988) and use the NEDS point and area source inventory as a starting point
for modification and refinement in the development of the 1985 inventory.

Anthropogenic VOCs

Forty percent of anthropogenic VOC emissions result from transportation,
according to the 1985 NEDS and NAPAP emissions inventories (see Figure
9-2); light-duty cars and trucks make up the largest contributing fraction.
Solvent emissions, which are distributed across a broad group of sources,

TABLE 9-2   Types of Area Source Emissions Data for NAPAP

Source Category Data

Stationary (residential, commercial, institutional, and industrial fuel emissions less than 25 tons per year)

Mobile (highway and off-highway vehicles, locomotives, aircraft, marine)

Solid waste (on-site incineration and open burning)

Miscellaneous (gasoline marketing and evaporation of solvents used by consumers, unpaved roads and airstrips, construction, wind erosion, forest fires, agricultural and managed burning, structural fires, orchard heaters)

Other (publicly owned treatment works; hazardous-waste treatment, storage and disposal facilities; fugitive emissions from petrochemical operations; synthetic organic chemical manufacturing and bulk terminal storage facilities; process emissions from bakeries, pharmaceutical, and synthetic fiber manufacturing; oil and gas production fields; and cutback asphalt-paving operations)

Activity Level Data

Fuel use (by gross vehicle weight and type, by state and county)

Vehicle miles of travel (VMT, by road type and speed, by state and county)

Surrogate geographic and economic data (population, dwelling units, vehicle registration, manufacturing employment, commercial employment, solvent user category employment)

[a]Cutback asphalt refers to asphalt that is thinned with volatile petroleum distillates, such as kerosene. Adapted from Placet et al., 1990.

contribute 32% of total VOC emissions; the remaining 28% result from other sources such as industrial manufacturing activities and fuel combustion.

An independent analysis by the Congressional Office of Technology Assessment (OTA, 1989), reported that 94 cities exceeding the ozone National Ambient Air Quality Standard (NAAQS) during 1986 to 1988 generated 44% of VOC emissions nationwide. In these cities 48% of VOC emissions were from mobile sources, and an additional 25% came from the evaporations of organic solvents and the application of surface coatings. Because these three categories alone account for about 75% of the total estimated VOC emissions in cities where the NAAQS was not met, significant attention to the quality and accuracy of the estimates seems warranted.

VOCs emitted from motor vehicles are mainly hydrocarbons that result from the incomplete combustion of fuel or from its vaporization. These contributions are generally categorized and reported as exhaust and evaporative emissions. Within the exhaust emissions category are included the unburned and partially burned fuel and lubricating oil in the exhaust and gases that leak from the engine. The evaporative emissions category includes fuel vapor emitted from the engine and fuel system that can be attributed to several sources: vaporization of fuel as a result of the heating of the fuel tank, vaporization of fuel from the heat of the engine after it has been turned off (hot-soak emissions), vaporization of fuel from the fuel system while the vehicle is operating (running losses), fuel losses due to leaks and diffusion through containment materials (resting losses), and fuel vapor displacement as a result of filling fuel tanks (refueling losses) (EPA, 1990f).

Only recently has it been recognized that running losses are not treated adequately in EPA's emissions estimating (MOBILE) models (Black, 1989; EPA, 1989b). The magnitude of running-loss emissions will depend on ambient temperature, gasoline volatility, operating cycle, and engine and emission control system design. An OTA study (1989) reported that, using preliminary emission factors provided by EPA for fleet average running losses, for ambient temperatures of 79°F (26°C) and gasoline volatility of 11.5 pounds per square inch (psi), MOBILE4 model estimates of VOC emissions were 1.5 grams per mile (g/mi). In assuming ambient temperatures of 87 °F and gasoline volatility of 11.7 psi, however, the resulting estimate of VOC emissions increased to 2.9 g/mi, a 93% change. Because of the difference in these estimates it has been suggested that past emissions inventory compilations underestimated mobile source VOC contributions by as much as 30% on hot summer days (OTA, 1989).

The use of organic solvents in the dry cleaning industry, in metal degreasing, in cutback asphalt paving, and in a variety of consumer and commercial product manufacturing contributed about 15% of the total VOC emissions in the 1985 national inventory. These sources are difficult to inventory, because almost half of their emissions are estimated to come from facilities that emit less than 50 tons (45 Mg) annually.

The emissions of VOCs from surface-coating-related industries contributed

Total NO$_x$, 5.67 Teragrams nitrogen/year

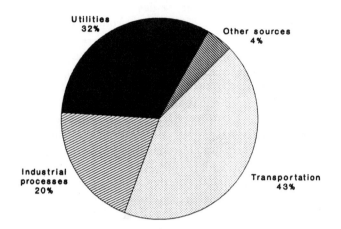

FIGURE 9-2a   NAPAP 1985 national emissions inventory for NO$_x$ and VOCs by source category.   Source: Placet et al., 1990.

about 9% of the total in the 1985 national inventory. The sources include automotive, furniture and appliance manufacturing, printing, and metal and plastic fabrication industries. As with stationary-source solvent evaporation, this source category is not well quantified, and the combined contributions of the two categories are expected to have significant uncertainties.

VOC Speciation by Source Category

The evolving knowledge of the chemical reaction mechanisms that provide quantitative information on the VOC-NO$_x$ relationship to ozone production has highlighted the need for compound-specific information on the chemical composition of the compounds in the VOC emissions inventory. The methods developed to generate compound-specific information are similar to those

Total VOCs, 20.02 Teragrams/year

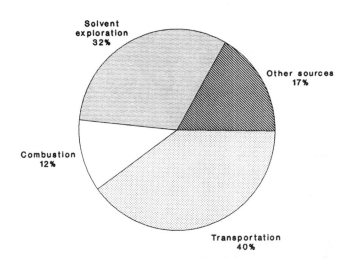

FIGURE 9-2b   NAPAP 1985 national emissions inventory for NO_x and VOCs by source category.   Source: EPA, 1989c.

used for emission factors.

Each point or area VOC emissions source, identified by its source classification code (SCC), has an associated VOC speciation profile, which provides a weight-percent breakdown of the individual compounds that contribute to total VOC mass emissions from the source. Speciation profiles are derived typically from compilations of detailed gas chromatographic analyses of the VOC emissions from sources in representative categories. The data sets used to derive the speciation profiles are quite limited and within a given source category can be highly variable. The current profile data base (Shareef et al., 1988), although the most comprehensive to date, suffers from major uncertainties in estimating compound-specific emissions. Shareef et al. (1988) used a rating procedure considered in EPA's AP-42 emission factor analysis technique (mentioned above) to assign subjective data quality rankings to the speciation profiles used in the NAPAP 1985 inventory. They used a scale of

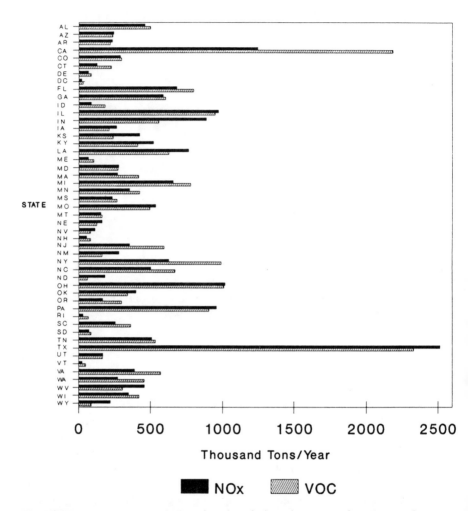

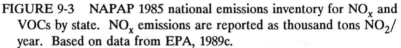

FIGURE 9-3   NAPAP 1985 national emissions inventory for NO_x and VOCs by state. NO_x emissions are reported as thousand tons $NO_2/$ year. Based on data from EPA, 1989c.

A through E, for the highest to the lowest quality, respectively, and their subjective analysis indicates that about 50% of the national VOC emissions of the most reactive chemical species would fall into the class B quality rating. The scale does not have quantitative error estimates associated with the letter ratings. Uncertainty issues associated with the chemical speciation will be discussed further in a later section.

Anthropogenic NO_x

Nitric oxide (NO) is formed by high-temperature chemical processes during combustion of fossil fuels, from both the nitrogen present in fuel and from oxidation of atmospheric nitrogen. Detailed inventories are available for Canada, the United States, and western Europe that describe the spatial patterns of NO_x emissions from combustion of fossil fuels and from industrial processes (Lubkert and Zierock, 1989; Placet et al., 1990). Table 9-3 lists several estimates (cf. Placet et al., 1990) of NO_x emissions associated with fossil fuel combustion in the United States. Between 40 and 45% of all NO_x emissions in the United States are estimated to come from transportation, 30-35% from power plants, and about 20% from industrial sources. About half the NO_x emissions associated with transportation come from light-duty gasoline trucks and cars and approximately one-quarter are from heavy-duty gasoline and diesel vehicles.

TABLE 9-3   Estimated Annual U.S. NO_x Emissions from Anthropogenic Sources Obtained from Recent Inventories

Source category	Emissions (teragrams of nitrogen/year)			
	NAPAP Inventory[a]	EPA Trends[b]	MSCET[c]	EPRI[d]
Electric utilities	1.8	2.1	1.9	2.2
Nonutility combustion	1.1	1.0	1.1	1.3
Transportation	2.4	2.7	2.3	2.4
Other sources	0.3	0.2	0.2	0.4
Total	5.6	6.0	5.5	6.3

[a]EPA, 1989a.
[b]EPA, 1990a.
[c]MSCET, Month and State Current Emissions Trends; Kohout et al., 1990.
[d]EPRI, Electric Power Research Institute, Heisler et al., 1988. Amounts are for 1982. The other amounts presented in the table are for 1985.
Source: Placet et al., 1990.

Emissions have been measured from typical sources for most of the important combustion and industrial processes. These are used to derive emission factors, which are then applied to statistics for fuel consumption and industrial production by various sectors of the economy. The states are responsible for providing estimated emissions for all major point sources and for area sources by county (EPA, 1981). These estimates are subsequently disaggregated to provide emissions data of greater spatial resolution for models. For mobile-source emissions, measured emissions were also used to calculate emission factors.

There are many uncertainties involved in the estimates. For example, NO_x emissions from vehicles vary with temperature and vehicle speed. Although the emission factors take this into account, different emission factor models use different assumptions about vehicle speed and ambient operating temperature.

Although the NAPAP NO_x inventory for 1985 is considered the most reliable to date, some researchers have tried to estimate its uncertainty. Placet et al. (1990) published a statistical analysis of the measured differences in the emission factors used to construct the annual NO_x emissions inventory for power plants in the United States. That study indicated an uncertainty of approximately 10% in the NAPAP emissions estimates for these large point sources. Such a detailed analysis was not possible for the other sources due to insufficient data. A subjective estimate of the level of uncertainty can be obtained from the relative differences in the various estimates of the emissions for the source categories listed in Table 9-3. It should be noted that the source estimates compiled in Table 9-3 have not generally been compiled independently of one another, and have a considerable amount of shared resource data. Therefore estimates of uncertainty by this comparative approach must be viewed as a lower limit. In addition, such estimates of uncertainty would be larger if made for smaller spatial domains (less than the area of the U.S.) and shorter time periods (less than a year). Finally, in assessing the value of this approach for estimating the uncertainty in an emissions inventory, it must be recognized that using a statistical estimate of uncertainty assumes that there are no systematic biases in the emission factors that are used to construct the inventories. Moreover, this uncertainty applies to an annual emissions estimate constructed for a particular year, 1985. It does not indicate the variability in estimates that could be expected from year to year based on market/economic and climatic factors.

ESTIMATES OF BIOGENIC EMISSIONS

In 1960, F.W. Went (1960) first proposed that natural foliar emissions of VOCs from trees and other vegetation could have a significant effect on the chemistry of the earth's atmosphere. Since that study, numerous investigators have focused on biogenic VOCs. Researchers have investigated the speciation of natural VOCs (Rasmussen and Went, 1965; Rasmussen, 1970, 1972), their rate of emission (Tingey et al., 1979, 1980; Zimmerman, 1979, 1981; Arnts and Meeks, 1981; Tingey, 1981; Lamb et al., 1986, 1987; Zimmerman et al., 1988b; Monson and Fall, 1989), and the fate and distribution of these compounds in the atmosphere (Arnts and Meeks, 1981; Duce et al., 1983; Hov et al., 1983).

Measurements of wooded and agricultural areas coupled with emission studies from selected individual trees and agricultural crops have demonstrated the ubiquitous nature of VOC emissions and the variety of organic compounds that can be emitted. The VOCs emitted by vegetation are primarily in the form of highly reactive, and therefore relatively short-lived, olefinic compounds. One compound typically emitted by deciduous trees is isoprene (C_5H_8); conifers typically emit terpenes such as α-pinene and β-pinene (Rasmussen and Went, 1965). However, foliar VOC emissions are by no means limited to these compounds; often 50% or more of the VOC mass emitted by vegetation is made up of other VOCs, and a significant fraction of the mass can be composed of VOCs whose structure and chemistry have not yet been determined. Table 9-4, taken from a study of VOC emissions from vegetation in California's Central Valley (Winer et al., 1989), illustrates the complex mix of organic species that can be emitted.

Not only do the isoprene and terpenoid emissions vary considerably among plant species, but the biochemical and biophysical processes that control the rate of these emissions also appear to be quite distinct (Tingey et al., 1979, 1980). Isoprene emissions appear to be a species-dependent by-product of photosynthesis, photorespiration, or both; there is no evidence that isoprene is stored within or metabolized by plants. As a result, isoprene emissions are temperature and light dependent; essentially no isoprene is emitted without illumination. By contrast, terpenoid emissions seem to be triggered by biophysical processes associated with the amount of terpenoid material present in the leaf oils and resins and the vapor pressure of the terpenoid compounds. As a result, terpene emissions do not depend strongly on light (and they typically continue at night), but they do vary with ambient temperature. The dependence of natural isoprene and terpenoid emissions on temperature can result in a large variation in the rate of production of biogenic VOCs over the course of a growing season. An analysis of emissions data by Lamb et al. (1987) indicates that an increase in ambient temperature from 25 to 35°C can result in a factor of 4 increase in the rate of natural VOC emissions from isoprene-emitting deciduous trees and in a factor of 1.5 increase from terpene-emitting conifers (Figure 9-4). This analysis aptly illustrates the fact that, all other factors being equal, natural VOC emissions are generally highest on hot summer days, when the weather also favors ozone generation.

TABLE 9-4   Compounds Identified[a] as Emissions from the Agricultural and Natural Plant Species Studied

Isoprene

MONOTERPENES

Camphene
Δ^3-Carene
d-Limonene
Myrcene
cis-Ocimene
trans-Ocimene
α-Phellandrene
β-Phellandrene
α-Pinene
β-Pinene
Sabinene
α-Terpinene
γ-Terpinene
Terpinolene
Tricyclene or α-Thujene (tentative)[b]

SESQUITERPENES
β-Caryophyllene
Cyperene
α-Humulene

ALCOHOLS
p-Cymen-8-ol (tentative)[b]
cis-3-Hexen-1-ol
Linalool

ACETATES
Bornylacetate
Butylacetate (tentative)[b]
cis-3-Hexenylacetate

ALDEHYDES
n-Hexanal
trans-2-Hexenal

KETONES[c]
2-Heptanone
2-Methyl-6- methylene-1,7-octadien-3-one (tentative)[b]
Pinocarvone (tentative)[b]
Verbenone (tentative)[b]

ETHERS
1,8-Cineole
p-Dimenthoxybenzene (tentative)[b]
Estragole (Tentative)[b]
p-Methylanisole (tentative)[b]

ESTERS
Methylsalicylate (tentative)b

n-ALKANES
n-Hexane
$C_{10} \rightarrow C_{17}$

ALKENES
1-Decene
1-Dodecene
1-Hexadecene (tentative)[b]
p-Mentha-1,3,8-triene (tentative)[b]
1-Pentadecene (tentative)[b]
1-Tetradecene

AROMATICS
p-Cymene

---

[a]Unless labeled "tentative," identifications were made on the basis of matching full mass spectra and retention times with authentic standards.

[b]Tentative identifications were made on the basis of matching the mass spectra (and retention order when available) with published spectra (Adams et al., 1989).

[c]Acetone was tentatively identified from the gas chromatography-flame ionization detection analysis on the GS-Q column (see text for details).

Source: Winer et al., 1989

Biogenic VOC Inventories

As with anthropogenic emissions, an accurate inventory of natural VOC emissions is needed to help quantify their part in fostering episodes of ozone pollution. Also similar to the case of anthropogenic emissions inventories, natural VOC inventories typically are developed first by determining an emission factor (actually an emissions rate per unit area) for specific kinds of sources (in the case of biogenic emissions, the source types are biomes such as deciduous forests, coniferous forests, mixed forests, grasslands, and agricultural lands) and then extrapolating to an integrated region-wide emissions rate or inventory, using landuse statistics to determine the extent of each source type or biome within the region.

Three field methods have been used to determine VOC emission factors for individual biomes. The most widely used was developed by Zimmerman (1979, 1981): A teflon bag is placed around a live branch of known biomass and the rate of emissions per unit biomass of the species is determined from the rate at which VOCs are found to accumulate in the bag (Figure 9-5). These data can then be coupled with a detailed biomass survey of the immediate area to deduce an emissions rate per unit area for the specific biome. Table 9-5 gives the emission factors for mixed forests derived by Zimmerman

(1979) from a study (one of the first comprehensive studies of its kind) done in the Tampa-St. Petersburg area of Florida during the summer.

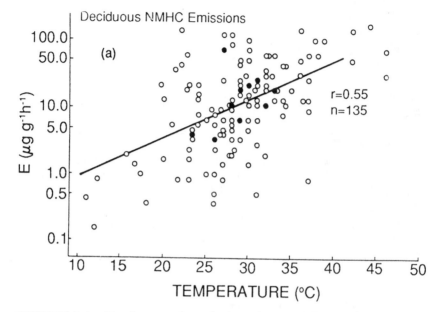

FIGURE 9-4   Total nonmethane hydrocarbon emissions (NMHC) (a) from deciduous trees and (b) from conifers. Multiple data points (points based on more than one measurement) are shown as closed circles. Single data points are shown as open circles. The solid line is the best fit of the data based on least-square regression. r is the Pierson correlation coefficient. n is the number of data points. These are total NMHC emissions as a function of temperature, in units of micrograms (μg) of NMHC per g of dry leaf biomass per hour. Source: Lamb et al., 1987.

Although the enclosure method is relatively simple to implement and interpret, it has some distinct disadvantages. The placement of an enclosure on the plant may disturb the leaf surface or alter the plant's physical environment (for example, the temperature and humidity) and thus cause anomalous emission rates. There are also uncertainties associated with the large extrapolation that must be made to derive an emission rate for an entire forest from the rates measured for a fixed number of branches and trees. For this reason other approaches have been used, and although they have their own drawbacks, they can at least serve as an independent check on the enclosure technique.

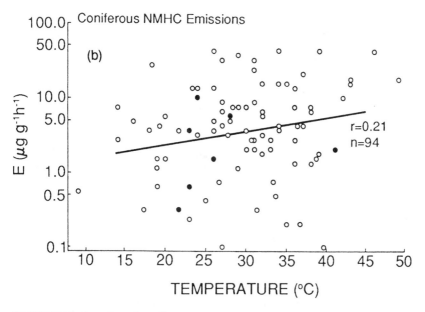

FIGURE 9-4   (continued)

One such method is the micrometeorological approach (Lamb et al., 1985; Knoerr and Mowry, 1981), which uses measurements of the vertical gradient of natural VOCs near the earth's surface; this measured gradient, coupled with micrometeorological data such as the local wind and temperature profile, can be used to estimate a local emission rate. Another technique (Lamb et al., 1986) uses the release of a tracer, such as sulfur hexafluoride (SF_6), coupled with downwind measurements of the tracer and VOC concentrations. The natural VOC emission rate is then calculated from the tracer release rate multiplied by the ratio of the VOC concentration to the tracer concentration measured at the downwind location. These two methods integrate data from trees throughout a region, but they are much more complex to carry out and interpret than the enclosure method; they also require specific meteorological conditions with a well-defined and relatively homogeneous wind profile and, in the case of the micrometeorological approach, cannot be carried out in regions with complex terrain (such as mountainous areas). Despite the inadequacies of each method and the spatial and temporal inhomogeneity inherent in biogenic emissions, the emission factors yielded by the different approaches have all been reasonably consistent; emissions rates obtained by these approaches agree within a factor of two or three (Lamb et al., 1987).

Once the emission factors for all the biomes in a region have been deter-

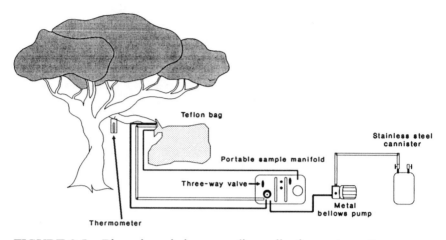

FIGURE 9-5   Biogenic emissions sampling collection system. Source: Zimmerman, 1979.

mined, a biogenic VOC inventory for the region can estimated by combining these results with landuse data on the amount of the earth's surface covered by the biomes in the region. This information can come from a variety of sources, including the Geoecology Data Base (Olson, 1980) or digitized satellite images, such as those available from LANDSAT. Typical of this approach is the work of Lamb et al. (1987), who estimated county-by-county the emissions of biogenic VOCs into the atmosphere over the entire continental United States; the results of this study are illustrated in Figures 9-6 and 9-7.

It is interesting to note that the characteristics of the biogenic emissions change from one region of the country to another. The South, with its dense vegetation and high temperatures, has relatively high natural VOC emissions. Furthermore, although the inventories for the other regions of the nation are dominated by α-pinene and other nonmethane VOCs, with relatively small emissions of isoprene, isoprene emissions comprise almost 50% of the total biogenic inventory for the South.

TABLE 9-5   Emission Factors, $\mu g/m^2$-hr

	Day		Night	
Compound	ER	S^2	ER	S^2
TNMHC[a]	5969.490	2532520.0	1745.940	439545.0

	Day		Night	
Paraffins	573.827	21846.0	411.475	10462.6
Olefins	4953.520	2139170.0	1024.990	376886.0
Aromatics	423.550	11713.9	298.629	5638.1
Methane	830.981	75674.0	798.813	71427.9
α-Pinene	462.466	134816.0	348.456	69977.9
β-Pinene	428.301	168643.0	311.320	82895.6
d-Limonene	53.234	2243.0	36.998	1085.5
Isoprene	3539.270	1372810.0	0.000	0.0
Myrcene	7.806	1030.3	5.426	498.1
Unknown Terpenes	-4.658	614.625	-2.832	285.2
21A	2.196	0.0	1.531	0.0
18	0.170	0.668	0.115	0.313
20	0.081	0.153	0.055	0.072
21	90.549	62997.8	61.404	31599.1
22	4.063	170.0	2.749	79.9
23	1.261	6.479	0.880	3.156
24	2.179	35.7	2.179	35.7
27	0.756	1.593	0.525	0.767
28	1.427	1.879	0.991	0.907
29	6.719	132.4	3.997	55.8
Δ^3-Carene	102.137	13875.9	71.374	6695.7
26A	10.842	2253.4	7.522	1083.7
29A	21.210	227.0	14.714	109.0

[a]TNMHC, total nonmethane hydrocarbons
Source: Zimmerman (1979)

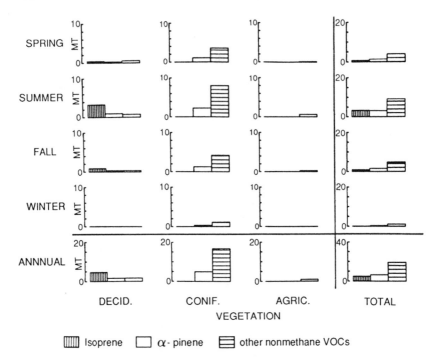

FIGURE 9-6   Nonmethane VOC emissions in Montana by season and
source type. Source: Lamb et al., 1987.

Placet et al. (1990) estimated that biogenic emissions integrated over the
entire continental United States add approximately 30,000,000 tons (27.4 Tg)
of carbon per year to the atmosphere. This rate is comparable to the estimated input of VOCs from anthropogenic sources discussed earlier in this chapter. Because of their extensive land area and high biomass density, forests are
by far the major contributor to the biogenic VOC inventory; agricultural crops
contribute only a few percent.

In addition to being a significant source of VOCs in the continental United
States as a whole, biogenic emissions can also represent a significant source
of VOCs in urban airsheds. For example, in Atlanta, Georgia, where almost
60% of the metropolitan area is wooded, anthropogenic and biogenic VOC
emissions have been estimated to be roughly comparable; each contributes 10-
60 kg C/km^2-day (Chameides et al., 1988). This is qualitatively consistent
with the analysis of VOC concentration data presented in Chapter 8, which
also implies a significant role for biogenic VOCs in a variety of urban areas,
including Atlanta.

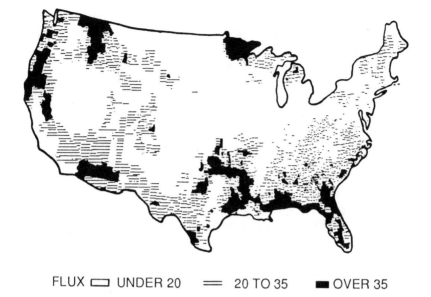

FLUX ☐ UNDER 20   ═══ 20 TO 35   ■ OVER 35

FIGURE 9-7   Average nonmethane VOC flux (kg/hectare) during the summer in the United States. Source: Lamb et al., 1987.

Uncertainties in Biogenic Inventories

The magnitude of biogenic VOC emissions, coupled with the high reactivity and ozone-forming potential of these compounds (see Chapter 5), suggests that they can have a significant effect on ozone formation that should be accounted for in any ozone control strategy. However, biogenic emissions inventories for yearly or seasonal periods are not yet well-quantified and could be in error by as much as a factor of two or three (Placet et al., 1990). Moreover, because of the extreme variability in emissions that can occur over the growing season, much larger errors can be incurred when annual or even seasonal inventories are applied to a given single- or multiple-day episode of high concentrations of ozone.

The uncertainties in biogenic inventories arise from a variety of causes. In addition to the experimental uncertainties associated with the biogenic emission measurements, there are also significant uncertainties associated with the interpretation of these measurements as well as with the related biomass and land use data used to derive the total emission rates from the experimentally measured data.

Probably the largest source of uncertainty in biogenic VOC inventories involves the empirically derived models used to calculate emission factors for individual biomes and the adjustments that must be made to account for variable environmental conditions throughout a region of interest. A large amount of variation in biogenic emissions has been reported (cf. Placet et al., 1990). The challenge is to develop a statistically robust algorithm to account for the variability in emissions caused by environmental factors. Although the mechanisms responsible for biogenic emissions of VOCs are complex and many environmental factors must play a role, temperature and light intensity are the only factors included in current models, even though they account for only 30-50% of the variability observed in emissions rates (Lamb et al., 1987). The remaining variability, which is caused by some combination of environmental as well as biologic factors, is not accounted for in current biogenic emissions models and must therefore be removed by averaging to obtain a "representative" emission factor. However, because the variability is so great, the averaging method can affect the resulting emission factors; it has been found, for instance, that emission factors formed by using a geometric mean of the data—which tends to minimize the influence of members of the sample population found to be gross or large emitters—is typically a factor of two or three lower than are emission factors formed from an arithmetic mean of the data (Placet et al., 1990). It is not now clear which averaging procedure is most appropriate. To address this problem, a good deal more laboratory and field work will be needed to illuminate the mechanisms responsible for biogenic emissions and thus account for the variability of the emissions data so that more accurate and robust models can be developed.

Another complication arises from the application of the temperature- and light-dependent algorithm to actual environmental conditions. Although observed ambient temperatures and light levels might be appropriate for leaves at the top or edge of a forest canopy, they are not appropriate for leaves within a canopy. To account for the reduced temperatures and light intensities inside a canopy, temperatures and light intensities observed at the top of the forest must be adjusted by a model that simulates the flow of light and heat through the forest canopy. The current generation of these models (such as that documented by Pierce and Waldruff, 1991) are highly simplified treatments of the meteorology within a forest canopy and as such might introduce some as yet undefined error into the resulting emission factors.

Another significant source of uncertainty in biogenic VOC inventories is associated with their speciation. As noted above, in addition to isoprene, α-

pinene, and β-pinene, plants emit a wide variety of other VOCs. Furthermore, a significant fraction of the mass of the natural emissions is in forms that have not yet been identified. It also should be noted that for the VOCs that have been identified, very little is known of the specific chemical mechanisms by which they are oxidized in the atmosphere (see Chapter 5). To accurately assess the chemical effect of these emissions, the complex mix of hydrocarbons emitted by different species of vegetation and the atmospheric chemistry of these hydrocarbons must be better elucidated.

Natural Sources of NO_x

Biomass burning generates NO_x and other chemically and radiatively important species, including carbon dioxide (CO_2), carbon monoxide (CO), methane (CH_4), VOCs other than CH_4, and nitrous oxide (N_2O). In 1980, Seiler and Crutzen published the first estimate of the amount of CO_2 emitted from biomass burning. Since then, considerable progress has been made toward quantifying the compounds emitted by burning vegetation (Crutzen et al., 1979, 1985; Greenberg et al., 1984; Andreae et al., 1988).

Biomass burning is a significant source of NO_x in tropical and subtropical regions. Most is associated with agriculture and the clearing of land for agricultural use. By contrast, in the United States, most of the NO_x associated with biomass burning comes from wildfires. The sporadic and unpredictable nature of wildfires precludes their assembly into a meaningful inventory. In any event, although the occasional large forest fire can be a significant regional source of NO_x, on average, wildfires are believed to be a minor source of NO_x in the United States (Logan, 1983).

The release of energy generated by lightning produces the extremely high temperatures required to convert nitrogen (N_2) and oxygen (O_2) to nitric oxide (NO). For the purposes of this discussion, it is useful to discriminate between two types of lightning: intracloud (IC) and cloud-to-ground (CG) discharges (Kowalczyk and Bauer, 1982; Boruki and Chameides, 1984). IC lightning is more frequent but less efficient at converting N_2 and O_2; it accounts for approximately one-third of the NO_x generated. However, because IC lightning occurs at relatively high altitudes, removal processes are reasonably slow, and IC-lightning-generated NO_x is transported over long distances. The remaining two-thirds of lightning-produced NO_x is generated by CG discharges. These NO_x emissions, which are generated at lower altitudes, and thus are similar to those from anthropogenic sources, are more readily removed from the atmosphere and therefore less available for long-distance transport.

Lightning shows strong spatial and seasonal variations. Equatorial regions have the highest concentrations; about 60% of the earth's lightning occurs between 20° south and 20° north (Hameed et al., 1981). Less than 10% occurs over North America, and of that, about 75% occurs from May to September (Logan, 1983).

Quantitative estimates of NO_x from lightning usually are derived from a combination of three factors: the frequency of lightning flashes (a flash consists of one or more strokes closely spaced in time), the energy dissipated per flash, and the NO_x produced per energy unit dissipated. A thorough review of this calculation procedure resulted in an estimate of global emissions from lightning of 2.6 Teragrams of nitrogen (Tg-N)/ year within a range of 0.8 to 7.9 Tg-N/year (Boruki and Chameides, 1984). Albritton et al. (1984) used a different approach to estimate the amount of NO_x production attributable to lightning. They used nitrate deposition data from remote oceanic and polar sites where NO_x from IC lightning is believed to be the dominant nitrate precursor. The values obtained were approximately 5.5 Tg-N/year globally and 0.5 Tg-N/year in North America.

Based on the literature cited above, an estimate of the NO_x-generated lightning over North America has been calculated by Placet et al. (1990). They estimated seasonal emissions of NO_x from lightning over the United States as a function of latitude and longitude (Table 9-6) by using the lightning distribution of Turman and Edgar (1982) and assumed total global NO_x formation from lightning to be 6 Tg-N/year. The estimates shown in Table 9-6 are reported for 10° × 10° grid cells that extend from 30° north to 60° north, and from 80° west to 120° west. The estimates indicate that the total NO_x produced by lightning over North America is 0.36 Tg-N/year, that NO_x production peaks in the summer, and that the largest rate of production is in the Southeast.

TABLE 9-6   Production of NO_x by Lightning over the United States as a Function of Season, Tg-N

Latitude	Longitude	Flux (0.001 Tg-N)
Winter (January, February, and March)		
30-40	80-90	4.8
30-40	90-100	4.4
30-40	100-110	3.8
30-40	110-120	3.3
40-50	80-90	3.5
40-50	90-100	3.2
40-50	100-110	3.0
40-50	110-120	2.7
50-60	80-90	2.8
50-60	90-100	2.6
50-60	100-110	2.5
50-60	110-120	2.4
Total		**39**
Spring (April, May, June)		
30-40	80-90	18.2
30-40	90-100	15.3
30-40	100-110	12.3
30-40	110-120	9.4
40-50	80-90	10.1
40-50	90-100	8.7
40-50	100-110	7.2
40-50	110-120	5.7
50-60	80-90	6.0
50-60	90-100	5.3
50-60	100-110	4.6
50-60	110-120	3.9
Total		**107**

Latitude	Longitude	Flux (0.001 Tg-N)
Summer (July, August, and September)		
30-40	80-90	25.9
30-40	90-100	21.6
30-40	100-110	17.2
30-40	110-120	12.9
40-50	80-90	14.0
40-50	90-100	11.8
40-50	100-110	9.7
40-50	110-120	7.5
50-60	80-90	7.9
50-60	90-100	6.9
50-60	100-110	5.8
50-60	110-120	4.7
Total		**146**
Fall (October, November, and December)		
30-40	80-90	12.6
30-40	90-100	10.7
30-40	100-110	8.8
30-40	110-120	6.8
40-50	80-90	7.3
40-50	90-100	6.4
40-50	100-110	5.4
40-50	110-120	4.5
50-60	80-90	4.6
50-60	90-100	4.2
50-60	100-110	3.7
50-60	110-120	3.2
Total		**78**

U.S. Total: 0.37 Tg-N/year

Source: Placet et al., 1990

These results can be compared with estimates of NO_x generated by lightning during a specific period over an area of the eastern United States (Placet et al., 1990). The estimates were done for a case study between July 1, 1988, and Sept. 30, 1988, for an area between latitudes 30° north and 40° north and longitudes between 80° west and 90° west. Their calculation used "visible" lightning flashes recorded for this area during the period (a total of 8,723,258 flashes), as taken from the State University of New York at Albany, Lightning Data Base. The assumptions made for the calculation were as follows:

(1) Each recorded cloud-to-ground flash produces 4×10^{10} joules of energy (see Boruki and Chameides, 1984).

(2) Each joule yields 9×10^{16} molecules of NO (see Boruki and Chameides, 1984).

(3) An additional amount of NO is generated by "unseen" cloud-to-cloud flashes (as estimated by Bauer, 1982).

This calculation yields a total nitrogen source for this area and period of 0.0073 Tg-N as compared with the corresponding estimate of 0.026 Tg-N from Table 9-6. The difference, almost a factor of four, provides a gauge of the uncertainty and the variability that can be expected for the estimates of NO produced by lightning.

NO_x is produced in soils by the microbial processes of nitrification and denitrification and by several chemical reactions that involve nitrite (Galbally and Roy, 1978; Galbally, 1989). The local variables that frequently influence NO_x emissions are soil temperature, moisture content, soil nutrient level, and vegetation cover (Johansson, 1984; Johansson and Granat, 1984; Slemr and Seiler, 1984; Galbally et al., 1985; Anderson and Levine, 1987; Galbally et al., 1987; Williams et al, 1987; Johansson et al., 1988; Kaplan et al., 1988; Williams et al., 1988). The wider scale factors that influence NO_x emissions to the atmosphere include climate (through temperature and rainfall), plant growth and decay, clearing of forests, biomass burning (Anderson et al., 1988), and fertilization.

Several observations can be made about the measurements of soil emissions of NO_x. NO is the principal nitrogen species emitted by soils. Overall, the source of NO_2 is a small fraction ($< 10\%$) of the NO_x emissions. The published results cited above show a large range of variability in fluxes. The averages reported by the various investigators range from 0.034 ng (nanograms)-N/m^2-second (Williams and Fehsenfeld, 1991) for a tidal marsh to 60 ng-N/m^2-second (Williams and Fehsenfeld, 1991) for a recently fertilized corn field. For the set of 26 measurements reported, there was an average emission of 8.1 ng-N/m^2-second and a standard deviation of 13.7 ng-N/m^2-second.

Various approaches have been used to estimate the annual soil emission of NO_x globally and for North America. The earliest meaningful estimate was based on the field measurements of NO_x emissions in Australian pasture land during the spring and autumn (Galbally and Roy, 1978). Extrapolation of these measured rates to all seasons and land areas yielded a global estimate of about 9 Tg-N/year. A refinement of this method, which included field measurements for two other ecosystem types and covers the growing season, resulted in a much smaller estimate of 1 Tg-N/year (Johansson, 1984), with an estimated range of uncertainty of 0.2 to 3.2 Tg-N/year. Another study (Slemr and Seiler, 1984) used relatively high measured flux rates to estimate about 11 Tg-N/year. In view of recent measurements of high rates of emissions from soils in the tropics (Duffy et al., 1988; Johansson and Sanhueza, 1988; Johansson et al., 1988), the current consensus favors larger values for global emissions—about 12 Tg-N/year. In view of the extreme variability of emission rates, the limited data base, the uncertainty about the controlling biogenic processes, and the lack of detailed global geoecologic data, this value must be considered as an estimate that is accurate within a factor of 10. It gives a range for global NO_x emissions from soils of 3.3 to 18 Tg-N/year. Based on the fraction of global land area occupied by North America, an approximate value for soil emission of NO_x of 0.6 Tg-N/year for North America can be inferred.

However, given the increasing number of measurements of NO_x emissions from soil by type, a somewhat more detailed estimate of NO_x soil emissions for the continental United States is now feasible. Despite variability and uncertainties, various distinct patterns emerge. The four local variables that frequently influence NO_x emission are soil temperature, soil moisture content, soil vegetation cover and soil nutrient level (Johansson, 1984; Johansson and Granat, 1984; Slemr and Seiler, 1984; Galbally et al., 1985; Anderson and Levine, 1987; Galbally et al., 1987; Williams et al., 1987; Johansson et al., 1988; Kaplan et al., 1988; Williams et al., 1988). One variable that can be accounted for systematically is soil temperature. Williams et al. (1992a) deduced a simple relationship between soil temperature and NO_x emissions for different measurement sites. This relationship greatly reduces the variation of data taken at individual sites by different investigators. By separating the results of the NO_x soil emissions measurements according to the biome associated with those measurements, it is possible to identify biome-characteristic emissions patterns. With temperature effects taken into account, the NO_x emissions measured for the biomes are relatively consistent. Making allowances for seasonal temperature variations, it is possible to estimate the seasonal NO_x soil emissions for those biomes. Finally, using data which characterize biomes for the United States, it is possible to identify areas of the United

States that correspond to the biomes included in the study of soil temperature and NO_x emissions.

The biomes within the United States are classified as forests, grasslands, wetlands, and agricultural lands. The forests, grasslands (including grazing lands), and wetlands make up 53.5% of land area in the United States; agricultural lands make up 16.0%. Agricultural lands are further subdivided into four crop types, reflecting the farming practiced—particularly the amount and kinds of fertilization. The crop types are corn, cotton, wheat, and unfertilized harvested feed crops (soybeans, alfalfa, and hay). The growing season is assumed to extend from May 1 to August 31. Outside the active growing season, the NO_x emissions from these agricultural lands are assumed to be the same as for the grasslands. Excluded from the inventory is the remaining 30.6% of the total land area of the United States. This includes largely arid lands and deserts where limited NO_x flux measurements have been made. Because of the very low moisture content of the soils, these lands are not expected to be major sources of NO_x.

The estimates of NO_x emissions from soil for the 10 EPA regions are listed in Table 9-7. This inventory estimates annual emissions in the continental

Table 9-7   Annual NO_x Emissions from Soil by EPA Regions

Region	Annual emissions, Tg-N/year	Percentage of region inventoried
1-3	0.015	76.83
4	0.033	79.58
5	0.079	75.57
6	0.055	50.69
7	0.082	74.42
8	0.067	64.38
9	0.012	27.23
10	0.003	60.55

Total = 0.346 Tg-N/year

Placet et al., 1990

United States at 0.36 Tg-N/year. Given uncertainties, it is estimated that the annual flux of NO_x from the soils in the United States is between 0.11 and 1.1 Tg-N/year.

The prevailing conditions for the high fluxes appear to be high temperatures and bare soil or open crops. The lowest fluxes are observed in flooded soils or under cool conditions and in the presence of extensive vegetation cover of the soil surface. Indeed, the presence of heavy vegetation covering the soils and the subsequent deposition of NO_2 on the surface of the vegetation, particularly when wet, could prevent the escape of NO_x above the vegetation canopy. Thus, even in relatively remote unpolluted regions, forests represent a net sink for NO_x. This is expected to hold for most forested regions of the United States. Hence, forested regions in the United States will represent net sinks of NO_x even when there are modest emissions of NO_x from the soils of the forest floor.

ACCURACY OF EMISSIONS INVENTORIES

Knowledge of the precision and accuracy of emissions estimates for VOCs and NO_x is essential to quantify the relationship of these precursor species to ozone air quality. The desired accuracy and precision of the emissions inventory is governed by its intended application. The analysis of emissions and air quality trends for primary emitted pollutants has performance criteria different from those for air quality management applications and each is likely to stress different elements of the methods of estimating emissions inventories. For example, an urban photochemical airshed model is used in an air quality management application. The input data for that model require oxidant precursor emissions data that have accurate and precise spatial and temporal distributions and chemical speciations. Such requirements are significantly more rigorous than those needed to specify the overall annual emissions of precursors within an urban region in an emissions trend analysis.

Because of extensive costs and lack of proven measurement techniques to test the accuracy and precision of emissions data, the uncertainty of emissions inventories is estimated. In lieu of field verification and direct intercomparison testing of the methods, estimates of uncertainty have been developed for the data inputs for the modeling algorithms used in estimating emissions from various sources.

Under the aegis of NAPAP, Placet et al. (1990) have summarized results of an uncertainty analysis of the emissions models that support the NAPAP 1985 emissions inventory. Sources of emission uncertainty identified in the estimation process include variability in each of the individual components contributing to the estimate, systematic errors associated with the emissions factor, variability in the accuracy and precision of data required to generate emissions estimates, and the limited number of methods available to treat uncertainties in emissions estimates. A list of the potential sources of variability is provided in Table 9-8.

TABLE 9-8   Sources of Emission Variability

Emission Factor Variability

Temporal variability as a result of operating conditions
Inherent variations within source category due to equipment configuration
Location specific factors: ambient temperature, wind speed, or fuel characteristics
Variability in Btu and sulfur content of fuel

Annual Activity Level Variability

Temporal variability in fuel consumption or production levels
Variability in activity values disaggregated using surrogate activity levels

Variability of Allocation Factors

Temporal variability due to changes in demand or production distribution, maintenance, and random outages
Variability in chemical speciation due to process changes and product variability

Variability of Emission Control Efficiency

Variability in removal efficiency due to changes in operating conditions, equipment failures, and fuel characteristics.

Adapted from Placet et al., 1990

Estimates of uncertainty in emissions data have been derived mainly from subjective judgments of the variability of the various components identified in

Table 9-8 combined with subjective estimates of uncertainties associated with the emissions estimation parameters (for example, fuel consumed, vehicle miles travelled, and number of vehicles produced). Statistical methods have been used to characterize uncertainties by aggregating the uncertainty values of components in Equation 9-1 assuming independence among components. The validity of this assumption has allowed the application of Goodman's formula in the treatment of combined uncertainties in the emissions uncertainty estimation process.[1] Although applying Goodman's formula provides a methodological approach to aggregating the various uncertainties associated with components contributing to the emissions estimate, the lack of quantitative estimates of these uncertainties has limited the credibility of the approach. Estimates of coefficients of variation for the National Annual NO_x Emissions Inventory by economic sector are given in Table 9-9, along with the 90% relative confidence intervals that can be inferred from the coefficients of variation. The confidence intervals are presented as a measure of uncertainty. However, they do not account for systematic error or nonrepresentative samples.

TABLE 9-9   90% relative confidence intervals (RCI) for national annual NO_x emissions

Sector	National Annual NO_x Emissions[a] (Tg)	Coefficient of Variation	90% RCI
Lower variability case[b]			
Electric Utilities	6.09	0.07	0.11
Industrial Combustion	2.25	0.15	0.25
Industrial Process	0.84	0.25	0.41
Residential/	0.62	0.30	0.49
Commercial	8.03	0.02	0.03
Transportation			
All Other	0.81	0.40	0.66

[1]An illustrative example showing the application of Goodman's formula to two power plants with varying CVs (coefficient of variation; CV = standard deviation/mean) with respect to their activity levels and emission factors can be found in Placet et al. (1990).

Sector	National Annual NO_x Emissions[a] (Tg)	Coefficient of Variation	90% RCI
Aggregate	18.64	0.04	0.06
Higher variability case[b]			
Electric Utilities	6.09	0.09	0.15
Industrial Combustion	2.25	0.20	0.33
Industrial Process	0.84	0.25	0.41
Residential/ Commercial	0.62	0.30	0.49
Transportation	8.03	0.10	0.16
All Other	0.81	0.40	0.66
Aggregate	18.64	0.06	0.10

[a]Emission values are for 1985 and were taken from a preliminary version of the report by Kohout et al. (1990). The values were slightly modified after their use in this uncertainty analysis; however, slight changes in the emission values do not strongly affect the uncertainty estimates.

[b]The lower and higher variability cases refer to alternative uncertainty assumptions regarding low NO_x burners and the variability of emissions from the industrial combustion and transportation sectors.

Source: Placet et al., 1990.

The development and refinement of methods for estimating emissions is a fundamental element in the implementation of the regulatory approach to air quality management. Demonstrating the accuracy of emissions estimates has been the Achilles' heel in evaluating the effectiveness of air quality management control strategies and is a fundamental flaw in the regulatory process.

MOTOR VEHICLE EMISSIONS

Accurately estimating emissions from the population of motor vehicles in an area is one of the most difficult problems in the entire emissions inventory process. The problem is compounded because motor vehicle emissions typically account for about 40% of the total VOC and NO_x emissions in any region. Moreover, motor vehicle emissions result not only from tailpipe exhaust but also from evaporation of fuel from various locations in the fuel tank-engine system, and total emissions depend critically on the mode of operation of the vehicle, its state of repair, and the ambient temperature. As noted earlier, the motor vehicle emissions inventory for a region is a compilation of the emissions and driving cycle of each vehicle. (A driving cycle provides a specific pattern of activity—for example, acceleration and speed profiles and trip duration—for defined segments of the motor vehicle population.) Although this section addresses only vehicle emissions and not driving cycles, it is important to note that determination of a representative cycle for an area is an essential aspect of the overall motor vehicle emissions inventory and is generally overlooked, due to expense and a lack of proven techniques.

Dynamometer testing of new and in-use vehicles has been performed over the years on limited numbers of vehicles to assess pre- and postcontrol tailpipe emissions, emissions control deterioration factors, and inspection and maintenance programs (Bonamassa and Wong-Woo, 1966; Black and High, 1977,1980; Sigsby et al., 1987). Dynamometer testing is the key source of emissions rate data used in the development of exhaust emissions factors for use in mobile-source emissions models. As a tool for measuring mobile-source exhaust emissions under operating conditions, dynamometer testing cannot be widely used because of the expense. Sigsby et al. (1987) reported one example of a dynamometer test of 46 state-owned vehicles, model years 1975-1982. In that test, only 17% of the vehicles met their exhaust emission standards for CO, VOCs, and NO_x.

Ashbaugh et al. (1990) reported the results of the 1989 California Random Roadside Inspection Survey Program, which has been carried out each year since 1983 to evaluate the effectiveness of California's Inspection/Maintenance program (I/M, called Smog Checks). The 1989 survey was done at 60 urban locations and involved about 4500 vehicles. It was found, for the no-load 1000 rpm idle test performed; that 10% of the vehicles were responsible for about 60% of the exhaust idle CO, and that another (not necessarily the same) 10% were responsible for about 60% of the exhaust VOCs. The results showed only a weak relationship between vehicles that emit large amounts of CO and those that emit large amounts of VOCs, based on the idle test.

Tunnel Studies

Ambient measurements of VOCs, CO, and NO_x in parking garages and roadway tunnels have been used to determine emissions from motor vehicles within these somewhat controlled environments. This approach serves to reduce the uncertainties associated with atmospheric chemistry and transport of emissions from their source to the location where they are measured. Because the concentrations of compounds found in these contained environments are large, many of the uncertainties associated with measurement of ambient concentration are circumvented, such as the inability of an instrument to measure a low ambient concentration accurately.

Tunnel studies were conducted as early as 1970 (Lonneman et al., 1974) and have been carried out from time to time since then (Pierson et al., 1978; Gorse and Norbeck, 1981; Hampton et al., 1983; Gorse, 1984; Lonneman et al., 1986). The studies have been done in the Lincoln Tunnel, which connects New York and New Jersey, and the Allegheny Mountain Tunnel on the Pennsylvania Turnpike, among others. Most recently, a tunnel in Van Nuys, California, a suburb of Los Angeles, was sampled (Ingalls, 1989; Ingalls et al., 1989) as part of the Southern California Air Quality Study (SCAQS).

The SCAQS tunnel study consisted of measurements of CO, VOCs (excluding methane), and NO_x in a tunnel and a parking garage to evaluate emissions factors and to assess methods for evaluating total vehicle emissions. From the measurements it was possible to deduce emissions factors appropriate to California vehicles and to compare these with emissions factors that were being used for the California Air Quality computer program (EMFAC7C).

According to the SCAQS tunnel study, measured CO and VOC emissions rates in the tunnel were a factor of 2.7 ± 0.7 and 3.8 ± 1.5 higher, respectively, than predicted by CARB's EMFAC7C model for estimating emissions from on-road vehicles; NO_x emissions rates agreed reasonably well with model predictions. The EMFAC7C emission factors were calculated with the assumption that all the vehicles in the tunnel had reached their stable operating temperature. The actual fraction of vehicles in the cold operating mode was not known. Inclusion of cold operating mode, (contributing increased CO and VOC emissions) and running losses (contributing increased VOC emissions) would reduce the discrepancies between measured and predicted CO and VOC. Pierson (1990) has considered the extent to which inclusion of both cold operation and running evaporative losses could reconcile the discrepancies between measured and predicted CO and VOC emissions. He showed that even if 100% of the vehicles in a tunnel were operating in the cold mode, the VOC data could not be explained on this basis. Thus, no admixture of cold and hot start can account for the discrepancies between experiment and predictions using EMFAC7C. With regard to running losses, Pierson (1990) considered methane and the sum of C_4 hydrocarbons (a subset of VOCs) from the tunnel experiment, together with published abundances of methane and total C_4 hydrocarbons in exhaust and evaporative emissions, and concluded that running-loss evaporative emissions could have been no more than about 30% of the hydrocarbon mass.

Pierson et al. (1990) have analyzed results from past tunnel studies, including the SCAQS tunnel study, to compare the studies' measured emissions with those predicted by mobile-source emissions models. The instances in which on-road CO and VOC measurements have been directly compared with model predictions are only the 1979 and 1981 Allegheny Tunnel, 1983 North Carolina, and the 1987 Van Nuys (SCAQS) tunnel experiments. Aside from the SCAQS tunnel study, a clear underprediction of absolute CO or VOC emission rates exists in only one-fourth of the cases (Pierson, 1990). Underprediction of CO/NO_x or VOC/NO_x, however, is found without exception. CO/NO_x is underpredicted consistently by a factor of about two. The inescapable conclusion from the comparison between on-road data and the results from emissions models is that vehicles on the road have higher CO/NO_x and VOC/NO_x ratios, or "run richer," than the models predict.

Roadside Measurements

Three approaches to monitoring roadway vehicle emissions for the purpose of evaluating emissions factors and emissions models have been considered. Two methods use ambient measurements near roadways to estimate vehicle emissions indirectly; a third uses remote sensing to monitor vehicle emissions directly on the road.

The first approach uses field measurements designed on the principle of conservation of mass. The experimental design requires measuring the vertical concentration profiles of chemical constituents and appropriate meteorological parameters (for example, wind speed and direction, temperature, and turbulence) upwind and downwind of the road, the source region. Such measurements allow the determination of the net mass flux (in units of mass/area-time) due to vehicle emissions on the roadway. This technique has been used (Bullin et al., 1980; Hlavinka and Bullin, 1988) to determine CO emissions factors along an interstate highway, and the results have been compared with the MOBILE3 emissions model.

The use of an added inert tracer, sulfur hexafluoride (SF_6), released by a pace car in traffic to determine dilution effects is another technique (Cadle et al., 1976; Zweidinger et al., 1988) that has been applied to study on-road vehicle emissions. By using measurements of ambient concentrations of the tracer and the vehicle source gases, with knowledge of the tracer emission rate, it is possible to estimate the emissions rate of the composite vehicle fleet on the road.

The third and by far most promising development in roadway vehicle emissions testing has evolved from new real-time in situ remote-sensing instruments that measure CO emissions from thousands of vehicles during normal on-road operation (Bishop et al., 1989; Bishop and Stedman, 1990; Lawson et al., 1990a; Stephens and Cadle, 1991). The technique uses a collimated infrared source placed to direct a beam across the road about 0.33 meters above the road surface to an infrared detector filtered appropriately for the spectral absorption regions of interest (CO and CO_2). The equipment also monitors the reference source. The Bishop et al. (1989) instrument uses a rotating gas filter correlation cell with one detector to alternately sample infrared intensity at two separate spectral regions; the Stephens and Cadle (1991) instrument uses three separate detectors to monitor the attenuated infrared beam over the spectral regions specified. A limited intercomparison of the two techniques (Stephens and Cadle, 1991) showed good agreement, with 294 direct measurement comparisons of CO/CO_2.

Field tests of the remote-sensing technology have been performed in Denver, Colorado (Bishop and Stedman, 1990), and in the Lynwood area of Los Angeles (Lawson et al., 1990a). The Denver study sampled CO emissions from 117,000 vehicles at two locations in the area. The study reported that half of the CO emissions produced by the sampled vehicle population were from 7.0-10.2% of the vehicles, depending on location and time of day. More than 70% of the vehicles were measured to be emitting less than 1% CO. During the testing period of the state's oxygenated fuels program, a statistically significant decrease of $16 \pm 3\%$ in average CO emissions was observed at the two locations and was independent of initial emissions for the 25% highest emitters in the fleet and insignificant for the remaining 75%.

The Lynwood study was a pilot to assess the accuracy of the remote-sensing technique and to demonstrate its application in performing roadside vehicle inspections. Specially equipped vehicles were used to make on-board exhaust CO measurements in a series of blind and double- blind tests. The remote-sensing technique performed with an accuracy of $\pm 10\%$. The remote sensing of 2771 vehicles on La Cienega Boulevard in the Lynwood study showed that 10% of the sample population was responsible for 55% of the total CO emissions, averaged on a basis of grams CO per gallon of fuel burned. The 10% all registered CO emissions values greater than 4%. In a limited sample of 60 vehicles, it was found that most vehicles for which the remotely sensed CO reading was greater than 2% failed on-site roadside inspections. The roadside data set was compared with data from the biennial smog-check tests (California's Inspection/Maintenance program) for the same vehicles. It was observed that CO and exhaust VOCs from high-emitting vehicles were much higher than when the vehicles received their routine inspections. Furthermore, for the high-emitting vehicles in the data set, the length of time since the biennial smog check had little influence on the cars' emissions in the roadside inspection.

Measurements from roadside tests, tunnel studies, and remote sensing of in-use vehicles present consistent and compelling evidence that vehicles on the road have substantially higher CO and VOC emissions than current emissions models predict. Moreover, data from roadside tests and remote sensing indicate that the problem is with a relatively small number of high-emitting vehicles; 10% of the vehicles are responsible for almost 60% of the CO. Another 10%, but not necessarily the same vehicles, are responsible for almost 60% of the VOCs. Although new cars are required to emit no more than 0.04 g/mile of VOCs, the vehicle fleet has an average nearly two orders of magnitude higher. The higher-than-predicted CO/NO_x and VOC/NO_x ratios in tunnel and roadside tests suggest richer operation of on-road vehicles than predicted or than observed in the vehicle dynamometer tests that serve as the model inputs. Issues that must be addressed to make the emissions models more realistic include (Gertler et al., 1990):

- Representativeness of volunteer fleets used in dynamometer testing
- Speed correction factors
- Running loss evaporative emissions
- Limitations in sampling statistics

Also, the FTP is designed to provide vehicle certification using a dynamometer, which is not a real-world driving simulation, so results from the FTP must be used with great caution when extrapolating to actual on-road conditions.

ATMOSPHERIC MEASUREMENTS
VERSUS EMISSIONS INVENTORIES

The distribution of chemical compounds in the atmosphere is governed by source characteristics and locations, air mass transport patterns, and the nature of the processes through which individual species are removed. All of these processes acting in concert must be understood in detail if the atmospheric measurements of the compounds are to be used to extract information concerning any one of them. This section reviews existing atmospheric measurements and their potential usefulness in addressing questions about the accuracy of emissions estimates for various sources.

The relative composition of VOCs in ambient air has been used as an indicator of different source contributions to the atmosphere by Stephens and Burleson (1967, 1969), Altshuller et al. (1971), and Mayrsohn and Crabtree (1976). Lonneman et al. (1968), Pilar and Graydon (1973), and Ioffe et al. (1979) discussed the use of the ratio of toluene-to-benzene concentration measured in ambient air to discern the contribution of automobile exhaust to VOCs measured in urban air. Whitby and Altwicker (1978) reviewed the use of acetylene as a tracer of urban pollution and discussed the use in numerous studies of compound ratios to acetylene as applied to source determination. A critique of the approaches that are used to deduce these and other quantities from atmospheric measurements has been published by McKeen et al. (1990).

The ozone precursors, VOCs (excluding CH_4) and NO_x, are introduced into the atmosphere from a variety of natural and anthropogenic sources. The anthropogenic emissions of these compounds are concentrated, usually around urban areas, and emissions are reasonably uniformly distributed throughout the year. The natural sources, by contrast, tend to be widely dispersed, highly variable, and seasonal, with peak emissions usually occurring in the summer.

In the case of the VOCs the principal compounds emitted from natural sources, the terpenoid compounds, are not emitted by anthropogenic sources. As a consequence, if the measurement of the VOCs in the atmosphere also determines the identities of the compounds, the attribution of these compounds to natural or anthropogenic sources is straightforward.

Urban Measurements

Because most of the concern to date about elevated ozone concentrations has focused on urban pollution, most measurements have been made in urban areas, which have large anthropogenic sources of ozone precursors. The concentrations of these compounds in the urban atmosphere are large and easily measured. However, because the measurement sites are embedded in a complex matrix of different sources, the measurements are highly variable, and there is little prospect of successfully estimating the strength of a particular source, such as automobiles, or a particular compound, such as NO_x, from the atmospheric measurements of that compound. The aim rather has been to establish the relative emissions of the various classes of ozone precursors—NO_x, VOCs and CO—from the combination of all the sources of those compounds in the urban area, and interpretations have focused on the relative concentrations of the compounds.

Many of the uncertainties associated with source analyses can be reduced by examining a well-defined plume from a concentrated source, undertaking measurements when removal processes are at a minimum, or by making measurements with more reliable and sensitive measuring techniques.

Measurements made in the free atmosphere are much less definitive in establishing the relative importance of emission sources. Most of the measurements carried out in the atmosphere were not designed to determine source strength, distribution, or allocation, but rather to establish concentrations of the compounds for model simulations of ozone production or to elucidate the photochemical processes that control ozone production.

The range and variability found in VOC/NO_x is reflected in measurements made in 29 cities across the eastern and southern United States during the summers of 1984 and 1985 (Baugues, 1986). VOC/NO_x was determined in 10 of the cities both years. The measurements were made during the morning rush hours, between 6:00 a.m. and 9:00 a.m. By measuring in the early morning, the effects of daytime photochemistry on the relative concentration of the compounds would be minimized. However, because most of these samples were taken in urban centers, the contribution of auto exhaust emissions relative to emissions from heavy vehicles or from stationary industrial sources may very well be overemphasized.

The results of these measurements are shown in Figures 9-8 and 9-9. The figures show the average and median VOC/NO_x, along with the standard deviation of the ratios. Excluding one set of measurements (West Orange, Texas), which was judged to be inaccurate, the average VOC/NO_x in the 29 cities studied was found to vary between 7.9 (Cleveland, Ohio; 1985) and 69 (Beaumont, Texas; 1985). In general, the largest ratios were found in the cities in Tennessee and Texas. These cities have industrial petrochemical sources, and higher ratios of VOC to NO_x are found there than are found from the typical mix of source categories in other urban areas.

Baugues (1986) used acetylene as a mobile-source tracer to establish the percentage of the VOC emissions associated with mobile sources. Tunnel studies show that acetylene accounts for approximately 3.7% of the VOC emissions from mobile sources (Whitby and Altwicker, 1978; Baugues, 1986). The results from the Baugues study of such comparisons for seven cities are listed in Table 9-10.

In all cases, the percentage of VOCs from mobile sources is much greater than the percentage of the mobile-source emissions used in each State Implementation Plan. Again, this difference is attributed to higher emissions of VOCs from operating vehicles than are estimated from emission factors deduced from measurements made on test vehicles.

Because of the choice of urban locations for these studies, little attention was given to emissions from natural sources. However, when biogenic emissions were measured, they usually represented a very small fraction of the

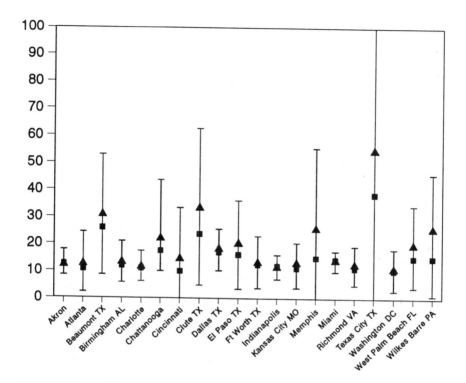

FIGURE 9-8   VOC/NO$_x$ ratios measured in urban locations in the United States during the summer of 1984. (VOCs do not include methane.) These measurements were made during the morning rush hour, 6:00 a.m. to 9:00 a.m. The triangles are the averages of the measurements made at each site and the bars show the standard deviation. The squares are the median. Adapted from Baugues, 1986.

total. Measurements by Baugues (1986) determined the concentrations of the VOCs found in the urban sites. The medians of the mixing ratios (in parts-per-billion carbon) of isoprene and α-pinene for 10 cities that were studied during the summers of 1984 and 1985 are listed in Figure 9-10. For comparison with the anthropogenic VOCs, the percentage concentrations of the natural VOCs to the total are plotted in Figure 9-11 for these 10 cities. The natural VOC fraction ranged between 0.23% (Indianapolis, Indiana, 1984) and 1.9% (West Orange, Texas, 1985) of the total VOCs. (Beaumont, Texas recorded 9.9% of the total NMHCs as natural in 1985, but these data are considered questionable). Baugues (1986) presented the data as support for the

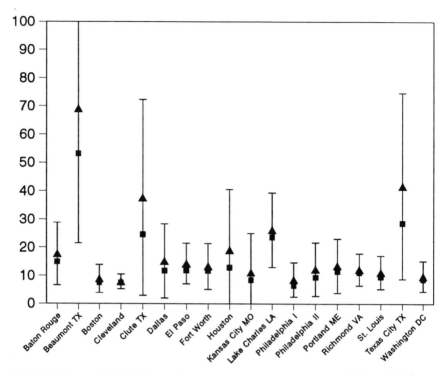

FIGURE 9-9   VOC/NO$_x$ ratios measured during summer 1985. (VOCs
do not include methane.) These measurements were made during the
morning rush hour, 6:00 a.m. to 9:00 a.m. The triangles are the average
of the measurements made at each site and the bars show the standard
deviation. The squares are the median. Adapted from Baugues, 1986.

contention that anthropogenic emissions are the predominant sources for the
ambient VOCs concentrations recorded in urban areas.

In assessing the implications of this finding, several factors must be considered. First, as discussed in Chapter 5, the natural VOCs are, in general, more
reactive than are the anthropogenic VOCs. As a consequence, any chemical
processing of the VOCs that occurs before early-morning samples are taken
will tend to increase the concentrations of the anthropogenic VOCs relative
to those of the more reactive natural VOCs. Second, because of their greater
reactivity, the natural compounds tend to be discriminated against in the
sampling method (container sampling) used to collect the data. This discrimi-
nation tends to reduce the measured concentrations of natural VOCs relative

TABLE 9-10.   Comparison of mobile-source contribution deduced from emissions inventory data with estimates deduced from ambient measurements.   The VOC inventory data for highway sources were taken from the State Implementation Plans.

City	Percentage of 1980 emissions inventory	Estimated from measurements	Year
Boston, Massachusetts	46	82	1985
Philadelphia, Pennsylvania	32	50 69	1984 1985
Washington, D.C.	66	87 96	1984 1985
Cincinnati, Ohio	41	50	1984
Cleveland, Ohio	45	70	1985
Houston, Texas	26	39	1985
St. Louis, Missouri	27	63	1985

Source: Baugues, 1986

to the anthropogenics. Third, because of the temperature and light dependence, the emissions of the natural VOCs will tend to increase during the daytime and peak during the periods of greatest photochemical activity. Finally, the concentration of samplers near street-level in downtown areas tends to discriminate against natural VOCs, whose sources tend to be concentrated in areas outside the core city. These measurements, therefore, might not be a reliable gauge of the importance of the natural compounds in the portion of the atmosphere responsible for the ozone sampled at these urban locations.

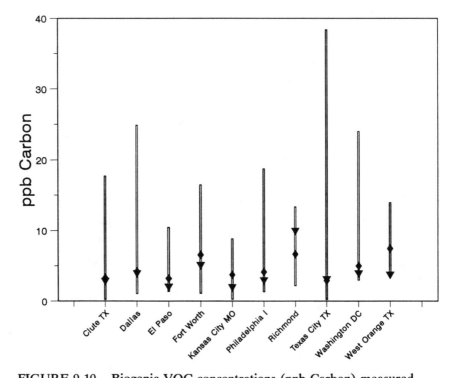

FIGURE 9-10   Biogenic VOC concentrations (ppb Carbon) measured during the summers of 1984 and 1985. The medians of the concentrations of isoprene and α-pinene are listed for 10 cities studied during both summers. The 1984 measurements are the triangles; the 1985 measurements are the diamonds. Bars represent the range of concentration. Adapted from Baugues, 1986.

Rural Measurements

As air masses leave the urban areas, they mix with air masses from other urban centers that can contain a somewhat different mix of ozone precursors. To this mixture are added the emissions from more isolated industrial sources or power plants, with their characteristic emission patterns. In addition, there is a relatively large input of natural compounds; in particular, the emissions of VOCs from forests. Photochemical processing and deposition serve to reduce the concentrations of the more reactive and polar compounds relative to the less reactive and nonpolar compounds. Finally, as the concentrations of the compounds are reduced through dilution, chemical destruction, and

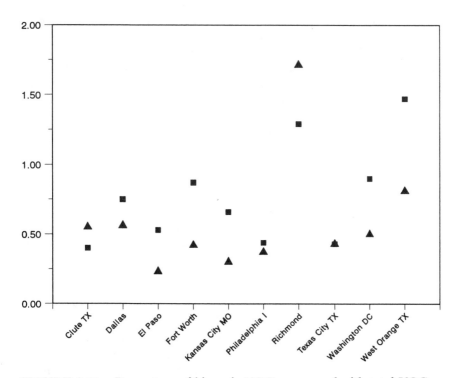

FIGURE 9-11   Percentage of biogenic VOCs compared with total VOC measured during the summers of 1984 and 1985. (The VOCs do not include methane.) The medians of the percentage contribution of isoprene and α-pinene to the total VOCs for 10 cities studied during both summers are listed. The 1984 measurements are the triangles; the 1985 measurements are the squares. Adapted from Baugues, 1986.

deposition, the difficulties associated with measurement increase the chance of sizable errors in measurement. This makes the extraction of useful information for emissions inventories extremely difficult. However, large data sets measured with high-quality instruments afford some interesting insights.

As is the case in the urban studies, atmospheric measurements made in rural areas are not very reliable for establishing the relative importance of emissions sources. Most of the measurements are not designed to determine source strength, distribution, or allocation, but rather to establish air concentrations of the compounds for model simulations of ozone production or to indicate the photochemical processes that control ozone production. In addition, there have been no long-term studies of rural areas that reveal the trends in the emissions of these compounds.

Dodge (1989) suggested that aged urban air reaching rural locations has a ratio of VOC to NO_x of 17:1. To this would be added local emissions dominated by stationary rather than mobile sources with VOC/NO_x ratios of 2:1. In those simulations natural VOCs also were added in varying amounts to test sensitivities of the mechanism to the compounds. This scenario would suggest that the VOC to NO_x ratio in rural areas depends strongly on the photochemical processing of the compounds and on the effect of urban versus local anthropogenic or natural sources.

During the summer of 1986, D.D. Parrish (pers. comm., NOAA, 1990) observed a daytime VOC/NO_x ratio of 7.2:1 at a rural site in western Pennsylvania. Of the total non-methane hydrocarbons, 45% were identified as natural, with the dominant natural compound being isoprene. The bulk of the remaining compounds were alkanes (39%) and aromatics (8%), presumably anthropogenic. Neglecting the biogenic hydrocarbons, the VOC/NO_x ratio was approximately 4:1. This is much lower than the ratios typically registered in the urban locations.

A principal-components analysis was carried out on a similar data set obtained at this site in 1988 (M.P. Buhr, pers. comm., University of Colorado, 1990). This analysis indicated that the atmospheric concentrations of alkanes, aromatics, and CO correlated with each other, suggesting they are dominated by anthropogenic, probably mobile, sources. NO_x on the other hand correlated most strongly with SO_2, suggesting that NO_x was most strongly associated with stationary anthropogenic sources.

By contrast, in 1987 similar measurements were carried out at a rural site in the Colorado mountains (D.D. Parrish, pers. comm., NOAA, 1990). In this case VOC/NO_x was 15.2:1. Of the total VOCs, 28% were identified as natural, principally terpenes. Neglecting the biogenic compounds, VOC/NO_x was approximately 11:1. This is very similar to the ratios typically registered in the urban locations. The bulk of the remaining compounds were alkanes (58%) and aromatics (9%), presumably of anthropogenic origin. A principal-components analysis was carried out on a similar data set obtained at this site in 1989 (M.P. Buhr, pers. comm., University of Colorado, 1990). This analysis indicates that the anthropogenic VOCs and NO_x correlated with each other, suggesting they are dominated by an anthropogenic urban source, the Denver metropolitan area.

It must be recognized that these rural measurements are strongly biased. The measurements were made at the surface in forest clearings. Hence, they tend to amplify the concentration of the natural VOCs relative to anthropogenic VOCs. A much lower relative concentration of these compounds could be expected on average throughout the boundary layer.

In addition to the VOCs, CO was measured at the rural locations in Pennsylvania and Colorado, as well as a suburban site near Boulder. The lifetime of CO in the atmosphere is long enough that there is a global background concentration, which is known to vary systematically with season because of photochemical processing. For Northern Hemisphere midlatitudes, the CO concentration varies between 127 parts per billion in winter and 84 ppb in summer (Seiler et al., 1976). For this reason the amount of CO relative to NO_x or VOC measured in the atmosphere will be greater than the relative amounts contained in the primary emissions depending on the degree of photochemical processing that has occurred between the point that the emissions entered the atmosphere and the location where this air mass is sampled. Thus, when the concentrations of NO_x and the NMHCs are large near anthropogenic sources, the ratio of NO_x or NMHCs to CO will approach the ratio of compounds expected for that source. Likewise, in so far as NO_x, the NMHCs, and CO are derived from anthropogenic sources, when NO_x or the NMHCs become quite sparse the concentration of CO will approach background levels.

Figure 9-12 shows the mixing ratio of CO versus NO_y (NO_x + PAN + HNO_3 + ...) measured near Boulder (Parrish et al., 1991). NO_y is the more conserved atmospheric reservoir of the primary NO_x emissions because NO_y comprises NO_x as well as its oxidation products. The curves on the graph show the expected relationship between CO and NO_x if the CO concentration is equal to the wintertime CO background mixing ratio, 127 ppb, along with the CO emitted by sources that have a CO/NO_x ratio of 5:1, 10:1, or 20:1. During this period, when photochemical processing is reduced, the data indicate that the sources of anthropogenic emission influencing the atmospheric concentrations of these compounds have a CO/NO_x ratio of 10:1 to 20:1. This can be compared with a CO/NO_x ratio for the Denver metropolitan area of 7.3:1 estimated by the 1985 NAPAP inventory or 8.2:1 estimated by the Colorado Department of Health (R. Graves, pers. comm., 1990).

ATMOSPHERIC MEASUREMENTS VERSUS EMISSIONS INVENTORIES

Fujita et al. (1990) compared CO/NO_x and VOC/NO_x derived from ambient measurements taken between 7:00 a.m. and 8:00 a.m. at eight sites in the South Coast Air Basin of California during the summer phase of the 1987 South Coast Air Quality Study (SCAQS) (Lawson, 1990) with corresponding ratios derived from the day-specific, gridded emissions inventory for the same

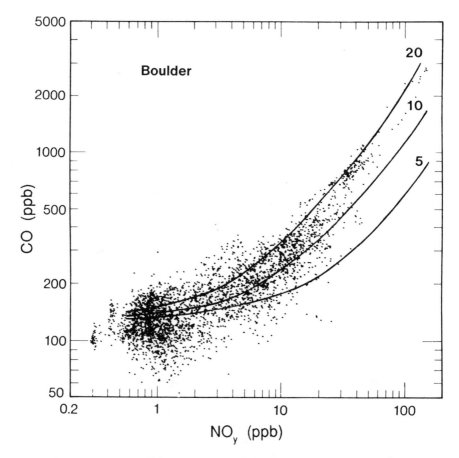

FIGURE 9-12   Correlation between CO and NO_y measured at a suburban site in Boulder. NO_y comprises NO_x and its oxidation process. The
individual 5-min averages of the measurements are shown. The background CO concentration for this season was taken to be 130 ppb. The
curves indicate the various emission ratios of CO and NO_y being added
to this background CO concentration. Source: Parrish et al., 1991.

period. Similar ambient CO/NO_x ratios at each of the monitoring locations
measured at the same times suggested a common emissions source. Both the
ambient CO/NO_x and VOC/NO_x ratios were about 60-80% higher than
corresponding ratios derived from emissions inventories (Figures 9-13 and 9-
14). The accuracy of the VOC speciation was evaluated by comparing the

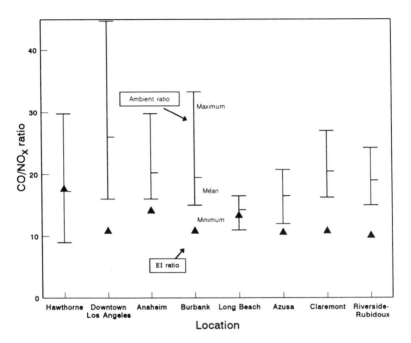

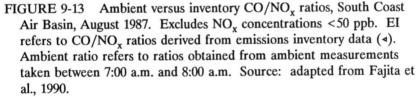

FIGURE 9-13   Ambient versus inventory CO/NO_x ratios, South Coast Air Basin, August 1987. Excludes NO_x concentrations < 50 ppb. EI refers to CO/NO_x ratios derived from emissions inventory data (◄). Ambient ratio refers to ratios obtained from ambient measurements taken between 7:00 a.m. and 8:00 a.m. Source: adapted from Fajita et al., 1990.

composition of emissions at the monitoring sites with the observed ambient composition.

Figure 9-15 is a comparison of VOC/NO_x derived from ambient measurements and the emissions inventory for seven cities (Morris, 1990); the mismatch is believed to be a nationwide phenomenon.

SUMMARY

For 2 decades, EPA has compiled inventories of emissions of volatile organic compounds (VOCs), oxides of nitrogen (NO_x), and other airborne pollutants. As sophisticated air quality models have been developed, emissions inventories have been expanded to provide detailed information on

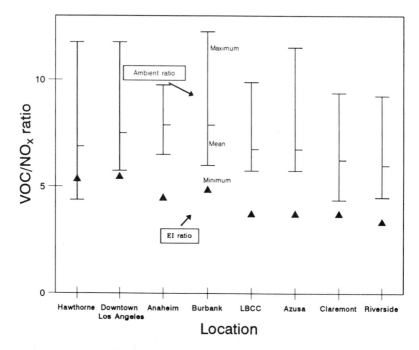

FIGURE 9-14   Ambient versus inventory VOC/NO$_x$, South Coast Air
   Basin, August 1987. Excludes NO$_x$ concentrations <50 ppb. EI refers
   to VOC/NO$_x$ ratios derived from emissions inventory data (◄). Ambient ratio refers to ratios obtained from ambient measurements taken
   between 6:00 a.m. and 9:00 a.m.

chemical speciation of VOCs, spatial and temporal patterns of inventoried
species, and projected trends in emissions. The development of sound ozone
control strategies requires knowledge of the precision and accuracy of emissions estimates. However, estimates of the uncertainty in emissions data have,
for the most part, been highly subjective. The methods used to estimate emissions have not been adequately checked by intercomparison or field measurements.

   Ambient monitoring data from many urban and rural areas of the United
States, along with data from roadside motor vehicle emissions tests, tunnel
studies, and remote sensing studies of on-road vehicle exhaust, show that
current inventories underestimate anthropogenic VOC and carbon monoxide
(CO) emissions by large margins. The motor vehicle portion of the emissions
inventory has been demonstrated conclusively to underestimate VOC and CO
emissions. Moreover, roadside tests and remote-sensing data indicate that

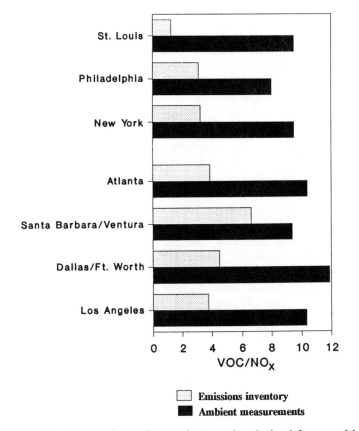

FIGURE 9-15   Comparison of VOC/NO_x ratios derived from ambient measurements and emissions inventories for seven cities. Source: Morris et al., 1990.

approximately 10% of the vehicles on the road contribute at least 50% of the CO and VOC emissions. The VOC bias in current emissions inventories is a serious impediment to progress in designing effective ozone reduction strategies.

These findings have substantial national implications for strategies to control VOCs. A rigorous program to resolve differences between on-road emissions and those predicted by emission models must be undertaken immediately. Resolution of the on-road versus model-predicted vehicle emission rates will have to account for the mass emission rate differences as well as the emission rate ratios that imply rich on-road operation. Issues such as super-emitters, speed-dependent emission rates, and off-cycle operation must be considered.

Biogenic VOC emissions appear to be of comparable magnitude to anthropogenic VOC emissions in the United States as a whole. Biogenic emissions can also be a significant source of VOCs in urban airsheds. For yearly or seasonal periods, these emissions are not well quantified. Moreover, because of the large variability in emissions that can occur over the growing season, much larger errors can be incurred when annual or seasonal inventories are applied to a given single- or multiple-day episode of high ozone concentrations. Because natural VOC emissions tend to be highly reactive and to increase during the day, past measurements of these emissions may have understated their importance relative to anthropogenic VOC emissions. Much research is needed to improve the methods used to calculate biogenic VOC emissions.

If VOC emissions have been underestimated as much as the studies discussed in this chapter suggest, then VOC emission reductions in many areas of the United States will be less effective than was previously believed (see Chapters 6 and 11). Hence a major upward revision in VOC emissions inventories could force a fundamental change in the nation's ozone reduction strategy, which has been based primarily on VOC control.

10

Ozone Air-Quality Models

INTRODUCTION

To predict compliance with the ozone air-quality standard at some future date it is necessary to know how ozone concentrations change in response to prescribed changes in source emissions of precursor species: the oxides of nitrogen (NO_x) and volatile organic compounds (VOCs). This assessment requires an air-quality model, which in the case of ozone prediction is often called a photochemical air-quality model. The model in effect determines the emission reductions needed to achieve the desired air-quality standard, such as the National Ambient Air Quality Standard (NAAQS) for ozone.

Air-quality models are mathematical descriptions of the atmospheric transport, diffusion, and chemical reactions of pollutants. They operate on sets of input data that characterize the emissions, topography, and meteorology of a region and produce outputs that describe that region's air quality. Mathematical models for photochemical air pollution were first developed in the early 1970s and have been developed, applied, and evaluated since that time. Much of the field's history is described in reviews by Tesche (1983), Seinfeld (1988), and Roth et al. (1989).

The air-quality model is theoretically the ultimate integrator of one's knowledge of the chemistry and physics of the ozone-precursor system. Photochemical air-quality models can be used to demonstrate NAAQS attainment or to educate planners about the emissions controls needed to head toward attainment. Whether or not they are actually used in determining abatement strategies, models are essential for examining of the complex interactions among emissions, meteorology, and atmospheric chemistry.

A practical model consists of four structural levels:

- The conceptual formulation; that is, a set of assumptions and approximations that reduce the actual physical problem to an idealized one, which, within the limits of present understanding, retains the most important features of the actual problem.
- The basic mathematical relations and auxiliary conditions that describe the idealized physical system.
- The computational schemes (numerical procedures) used to solve the basic equations.
- The computer program or code that actually performs the calculations.

The term "model" has been used to apply collectively or separately to all four levels. Models of a particular process, or a group of interacting processes, are called component models or modules. The basis for air-quality models is the atmospheric diffusion equation, which expresses the conservation of mass of each pollutant in a turbulent fluid in which chemical reactions occur (Seinfeld, 1986).

For at least a decade, the Environmental Protection Agency (EPA) has offered guidelines on the selection of air-quality modeling techniques for use in State Implementation Plan (SIP) revisions, new source reviews, and studies aimed at the prevention of significant deterioration of air quality. EPA guidelines (EPA, 1986b) identify two kinds of photochemical model: The urban airshed model (UAM) is the recommended model for modeling ozone over urban areas and EKMA (empirical kinetic modeling approach) is identified as an acceptable approach.

As noted in Chapter 3, the 1990 Clean Air Act Amendments specify that three-dimensional, or grid-based, air-quality models, such as UAM, be used in SIPs for ozone nonattainment areas designated as extreme, severe, serious, or multistate moderate (EPA, 1991b). Grid-based models use a fixed Cartesian reference system within which to describe atmospheric dynamics (Seinfeld, 1988). The region to be modeled is bounded on the bottom by the ground, on the top by the inversion base or some other height that characterizes the maximum extent of vertical mixing, and on the sides by east-west and north-south boundaries, unless the coordinates are rotated. This space is then subdivided into a three-dimensional array of grid cells. The horizontal dimensions of each cell are usually a few kilometers for urban applications. Some older grid-based models assumed only a single, well-mixed vertical cell extending from the ground to the inversion base; current models subdivide the region into layers. Vertical dimensions can vary, depending on the number of vertical layers and the vertical extent of the region being modeled. A compromise generally must be reached between the better vertical resolution afforded by the use of more vertical layers and the associated increase in computing time. Although aerometric data, such as the vertical temperature profile, that are needed to define the vertical structure of the airshed are generally lacking, it is important to use enough vertical layers so that NO_x emissions from tall stacks are not overdiluted computationally. There are practical and theoretical limits to the minimum horizontal grid cell size. Increasing the number of cells increases computing and data acquisition effort and costs. In addition, the choice of the dimension of a grid cell implies that the input data—information about winds, turbulence, and emissions, for example—are resolved to that scale. In practice, most urban models use horizontal grid cell of a few kilometers, whereas regional models use horizontal grid cells of tens of kilometers.

There is a need for a set of directives about how to specify the size of the modeling domain, the horizontal grid spacing to be used, the vertical extent of the modeling region, and the number and resolution of vertical layers. These directives must be based on the exercise of models having a wide range of spatial resolutions and on the comparison of model performance against a wide variety of high-quality field data. It has been found that increasing the horizontal grid spacing in a photochemical air-quality model will generally result in a reduction in the peak ozone concentration. It also is important to provide adequate vertical resolution—the order of five vertical layers or more for urban-scale applications (EPA, 1991b). The minimum amount of meteorological and air-quality data must be prescribed as modeling inputs. The choice of the size of the modeling domain will depend on the resolution available in the data, including the distribution of emissions in the region, the weather conditions, and, to some extent, the computational resources available. The spatial resolution of the concentrations predicted by a grid-based model corresponds to the size of the grid cell. Thus, effects that have spatial scales smaller than those of the grid cell cannot be resolved. Such effects include the depletion of ozone by reaction with nitric oxide (NO) near strong sources of NO_x like roadways and power plants.

Several grid-based photochemical air-quality models have been developed to simulate ozone production in urban areas or in larger regions. They differ primarily in their treatment of atmospheric processes and in the numerical procedures used to solve the governing system of equations. Table 10-1 lists grid-based photochemical air-quality models in current use or under development. As noted in Chapter 3, photochemical air-quality models are used in determining the emissions controls needed to attain the ozone NAAQS. But

TABLE 10-1 Photochemical Air Quality Models

Model	Developer	Status	Reference
UAM (Urban Airshed Model) Carbon Bond-II Chemistry	Systems Applications, Inc.	Wide use Documented	Reynolds et al. (1973, 1989) Reynolds (1977) Ames et al. (1985)
UAM Carbon Bond-IV Chemistry	Systems Applications, Inc.	Limited	
UAM Carbon Bond-IV Chemistry	Radian Corporation	In testing Code available	Tesche et la. (1988a,b) Tesche and McNally (1989)
CIT (California Institute of Technology) CIT Chemistry	California Institute of Technology and Carnegie Mellon University	Research applications Not documented	McRae et al. (1982) McRae and Seinfeld (1983) Russell and Cass (1986)
CIT SAPRC (Statewide Air Pollution Research Center) Chemistry	Carnegie Mellon University	Limited use	

CALGRID SAPRC Chemistry	Sigma Research	In review Documented	Yamartino et al. (1989)
ROM Carbon Bond-IV Chemistry	EPA	Operational (code and documentation due in 1990)	Lamb (1983)
RADM (Regional Acid Deposition Model)	National Center for Atmospheric Research and SUNY Albany	Operational	Chang et al. (1987)
ADOM (Acid Deposition and Oxidant Model)	ENSR and Ontario Ministry and Environment	Operational	Venkatram et al. (1988)

first, the validity of the model must be demonstrated by its ability to simulate adequately a base-year episode of high concentrations of ozone. Then, using the same meteorology as in the base-year episode, the emissions are hypothetically reduced to the point at which the peak 1-hr ozone concentration in the region does not exceed 120 ppb.

The remainder of this chapter is structured as follows. First the development of meteorological inputs to air-quality models is discussed, followed by a brief discussion of boundary and initial conditions. We then address how attainment of the ozone NAAQS is demonstrated using a grid-based modeling approach. Because urban grid-based photochemical air-quality models have received a great deal of attention in the literature (see, for example, Seinfeld, 1988), we do not devote significant coverage to them here. However, regional air-quality models, which are quite similar in structure to the earlier models used in urban applications, have not undergone the same degree of evaluation and application as their urban counterparts. For this reason, and because such regional models are currently being used to assess ozone abatement strategies in areas like the northeastern United States, we will review regional grid models in this chapter. The particular model on which we focus is EPA's Regional Oxidant Model (ROM). After the analysis of regional models, the general issue of evaluation of model performance is addressed.

METEOROLOGICAL INPUT
TO AIR-QUALITY MODELS

Grid-based air-quality models require, as input, the three-dimensional wind field for the episode being simulated. This input is supplied by a so-called meteorological module. Meteorological modules for constructing wind fields for air-quality models fall into one of four categories (Tesche, 1987; Kessler, 1988):

• Objective analysis procedures that interpolate observed surface and aloft wind speed and direction data throughout the modeling domain.
• Diagnostic methods in which the mass continuity equation is solved to determine the wind field.
• Dynamic, or prognostic, methods based on numerical solution of the governing equations for mass, momentum, energy, and moisture conservation along with the thermodynamic state equations on a three-dimensional, finite-difference mesh.
• Hybrid methods that embody elements from both diagnostic and prognostic approaches.

Objective analysis procedures are inexpensive and simple to use. Their disadvantage is that they contain no physics-based calculations, and the results are highly dependent on the temporal and spatial resolution of the observed wind speeds and directions. Results are often unsatisfactory in areas of the modeling domain where observations are either sparse or not representative of the physical geography. Areas of complex terrain, variations in land use, and ocean-land contrasts cannot be accounted for. Diagnostic procedures impose mass consistency on the flow field through appropriate equations, and can crudely include terrain blocking effects or estimates of upslope and down-slope flows if observed values are entered into the analysis. Diagnostic procedures have modest computational requirements and can require fewer observations than does objective analysis to produce a three-dimensional wind field. Without representative data, however, diagnostic models cannot simulate such features as sea and land breezes. Prognostic numerical prediction models are intended to simulate all relevant physical processes without requiring a significant amount of observed data. These models require specification of the large-scale flow, surface conditions, and the initial state of the atmosphere. Because prognostic models simulate the temperature field in addition to the wind field, it is possible to determine atmospheric stabilities and mixing-height fields from the output. However, the computations performed by prognostic models can be expensive, and they do not necessarily reproduce available observations. Recent developments in data assimilation techniques could overcome the latter problem by forcing models to be more consistent with the available local observations, provided these observations can be shown to be truly representative of the actual meteorological field. Also, with better computer systems becoming available, it is increasingly practical to use full numerical prediction models.

There are several hybrid models that use standard finite-difference techniques for horizontal advection but replace a rigid, vertical, finite-difference grid with one or more layers. The simplest example is a single-layer model in which the height and quantities of potential temperature and moisture, for example, are predicted in the boundary layer (Lavoie, 1972). This hybrid of numerical prediction and layer-averaged approaches is the basis of the EPA Regional Oxidant Model (ROM) (Lamb, 1983). An approach that is becoming more common is the use of the outputs from a prognostic model along with observed data as inputs to a diagnostic model.

An EPA five-city study has identified important limitations in the routinely available meteorological data bases and raises several questions (Scheffe, 1990): Is it possible to set minimum data requirements given the great diversity of the areas in the United States to which air-quality models will be applied? Must data be of sufficient spatial and temporal density to result in such narrow uncertainty bounds that performance evaluation for wind field generation routines is not needed, or is "sufficient data" defined by the amount needed to attain a specified level of uncertainty? An emerging issue in wind field modeling is that of performance evaluation— determining a technical basis for judging the accuracy of simulated wind fields. Current criteria for evaluating wind field modeling performance have not been applied in wind field generation for ozone modeling. Performance evaluation of the meteorological model, independent of the air-quality model, is necessary to ensure that compensating errors are not introduced into predicted ozone concentrations through unjustifiable modifications of the wind field.

There are several important meteorological variables other than wind field. Of particular importance is the treatment of photolysis rates and of the effects on these rates of clouds, urban aerosols, and ozone aloft. Clouds have traditionally been neglected in photochemical grid models because these models have focused on gas-phase pollutants, even though clouds can have a significant effect on the vertical distribution of pollutants (see Chapter 4) and on the attenuation (below cloud) or enhancement (near top of cloud) of photolysis rates. Objective or diagnostic techniques and prognostic modeling also can be used to calculate mixing heights. The key questions relate to the level of accuracy of the spatial and temporal variability in mixing heights. There are no currently accepted procedures for calculating mixing heights, and the mixing-height profile will strongly influence the predicted ozone concentrations in the modeling domain.

BOUNDARY AND INITIAL CONDITIONS

When a grid-based photochemical model is applied to simulate a past pollution episode, it is necessary to specify the concentration fields of all the species computed by the model at the beginning of the simulation. These concentration fields are called the initial conditions. Throughout the simulation it is necessary to specify the species concentrations—called the boundary conditions—in the air entering the three-dimensional geographic domain.

Three general approaches for specifying initial and boundary conditions for urban-scale applications can be identified: Use the output from a regional scale photochemical model; use objective or interpolative techniques with ambient observational data; or, for urban areas sufficiently isolated from significant upwind sources, use default regional background values and expand the area that is modeled and lengthen the simulation period to minimize uncertainties due to a lack of measurements.

In the ideal case, observed data would provide information about the concentrations at the model's boundaries. In practice, however, few useful data are generally available—a result of the difficulty in making measurements aloft and the fact that monitoring stations tend to be in places where air-quality standards are expected to be violated. An alternative approach is to use regional models to set boundary and initial conditions. This is, in fact, preferred when changes in these conditions are to be forecast. In any event, simulation studies should use boundaries that are far enough from the major source areas of the region that concentrations approaching regional values can be used for the upwind boundary conditions. Boundary conditions at the top of the area that is being modeled should use measurements taken from aloft whenever they are available. Regional background values are often used in lieu of measurements. An emerging technique for specifying boundary conditions is the use of a nested grid, in which concentrations from a larger, coarse grid are used as boundary conditions for a smaller, nested grid with finer resolution. This technique reduces computational requirements compared to those of a single-size, fine-resolution grid.

Simulations of a multiday episode, beginning at night, when concentrations of ozone precursors are the lowest, minimize the influence of initial conditions on ozone concentrations predicted 2 and 3 days hence. Initial conditions are determined mainly with ambient measurements, either from routinely collected data or from special studies. Where spatial coverage with data is sparse, interpolation can be used to distribute the surface ambient measurements. Because few measurements of air-quality data are made aloft, it is generally assumed that species concentrations are initially uniform in the mixed layer and above it. To ensure that the initial conditions do not dominate the performance statistics, model performance should not be assessed until the effects of the initial conditions have been swept out of the grid.

DEMONSTRATION OF ATTAINMENT

Chapter 3 introduced the SIP concept wherein the demonstration of attainment of the ozone NAAQS is based on future-year simulations. These demonstrations require projections of emissions and of initial and boundary conditions. The use of "background" boundary conditions would require an estimate of how these background conditions might change in response to regional changes in emissions. Boundary conditions can be based on future-year regional modeling. Initial conditions are typically reduced in proportion to emission reductions from the base to future year.

A major question is "What is an acceptable procedure for demonstrating attainment?" Previous model applications for control strategy development have, for the most part, avoided complex scientific issues by focusing on one to three worst-case episodes, interpreting model results in a deterministic form without regard to modeling uncertainties and the statistical form of the ozone NAAQS. There is a critical need to identify and investigate methodologies for transforming results from deterministic models into a probabilistic form, so that informed decisions can be made about the efficacy of the selected emissions control strategy in achieving compliance with the ozone NAAQS. Other issues have not been adequately addressed in the context of ozone attainment demonstration. These include the number of episodes that need to be modeled and the duration of each episode.

When the episodes to be modeled have been selected, it must be decided whether attainment should be demonstrated for all modeled episodes. For a particular episode, it is not a simple matter to decide which measure will be used to show attainment. For example, how should bias—under- or over-prediction of the peak ozone concentration—in the base year be addressed? One approach is to ignore any bias in the base year and consider attainment to be achieved if future-year predictions produce peak values below the NAAQS. Alternatively, the future-year peak ozone predictions could be normalized relative to the bias in the base-year simulation; if peak ozone is underpredicted by 10% in the base year, the future-year predicted peak ozone is increased by the same amount. Such questions related to how attainment is demonstrated have not been addressed adequately by the regulatory community.

An aerometric data base is a critical component of a modeling application. Such a data base is needed to provide input to the model and to serve as a tool for assessing model performance. The elements of an aerometric data base for photochemical air-quality modeling are presented in Table 10-2. Most aerometric data bases use routine surface meteorological and air-quality measurements, either no upper-air meteorological data or routine National Weather Service balloon soundings only, and no air-quality data from aloft. A data base with such limitations does not contain enough information to characterize the three-dimensional meteorological and air-quality fields in space and time required for the model. An issue that must be addressed in an attainment demonstration is whether special aerometric measurements should be made over a limited period to provide a more extensive data base than is routinely available or whether a larger number of episodes should be considered with the more routine data base.

Table 10-2   Aerometric Data Base Elements

	Extent of database	
	Extensive/ Intensive	Routine
Upper air meteorology (wind speed and direction, temperature, relative humidity, pressure)		
Number of soundings per day	4-8	None or limited to routine National Weather Service or military installation observations
Vertical resolution	100 meters	None
Air quality aloft		
Species	Ozone, NO$_x$, NO, speciated VOCs	None
Number of profiles per day per site	≥ 3	None

	Extent of database	
	Extensive/ Intensive	Routine
Surface meteorology		
Additional characteristics	Ultraviolet radiation	
Temporal resolution	Continuous and hourly averaged	Continuous and hourly averaged
Surface air quality		
Species and temporal resolution	Ozone, NO_x, NO (continuous) Speciated VOCs (2-4 times per day) Secondary species (continuous)	Ozone, NO_x, NO (hourly averaged)

Another major issue is what should be done when sufficient air-quality data do not exist to perform a definitive model performance evaluation, but a SIP is still required. For some areas in this category, a grid-based, three-dimensional model simulation is inappropriate. However, because government agencies must make billion-dollar decisions concerning emissions control strategies, they should require that sufficient data be collected to support a proper air-quality model analysis. Use of a model in a data-poor situation, with a large number of default inputs, can delude regulators into believing they have a viable plan when none exists, but model use in a data-poor situation can lead to better analysis in the future. A model can be used as a tool to guide the collection of data and identify the most important data needs for a given region, such as was done in planning the Southern California Air Quality Study (SCAQS) (Lawson, 1990) and the San Joaquin Valley Air Quality Study (SJVAQS)/Atmospheric Utility

Signatures, Predictions, and Experiments (AUSPEX) (Ranzieri and Thuillier, 1991).

REGIONAL GRID MODELS

Grid-based models have been developed to simulate oxidant production and acid deposition over the eastern United States. The grid size of these models ranges from 18.5 to 127 kilometers (km). Only one of them, EPA's ROM, has been used extensively to examine the effects of emission reductions on ozone concentrations in urban, suburban, and rural areas. Development of this model began in 1977, after EPA-sponsored field programs revealed the regional nature of the ozone problem in the northeastern United States. Grid-based models and their evaluation are described here in detail, as these models are relatively new compared to urban models.

In the 1980s, Liu et al. (1984) developed a three-layer regional transport model with a grid size of 80 km that covered the domain of the Sulfate Regional Experiment (SURE) sites. (See Chapter 4.) Winds and mixing heights were obtained by interpolating observations from the National Weather Service radiosonde network, and the model used a version of the carbon bond mechanism (CBM-II). Liu and co-workers evaluated the performance of the model by comparing the time-series of ozone concentrations at the SURE sites for July 16-23, 1978—the most intense episode of high concentrations of ozone that year. The model has been updated to incorporate a nested plume model that treats large point sources; it has been applied, with different grid sizes, to western Europe, to the eastern United States, and to central California (Morris et al., 1988). Performance in the East was evaluated by comparison with ozone data from the SURE sites for Aug. 15 to Sept. 15, 1978. The overall correlation coefficient between the model and observations for all the daylight hourly data was 0.64; the model tended to underpredict the higher ozone concentrations and overpredict the lower ones. The model also overpredicted nighttime concentrations at many sites, possibly because of the poor representation of surface removal by dry deposition.

Two comprehensive models, RADM (regional acid deposition model) and ADOM (acid deposition and oxidant model), have been applied principally to simulations of acid deposition, with only limited evaluation of results for ozone (Middleton et al., 1988; Venkatram et al., 1988). RADM is driven by meteorological input from a mesoscale model (Anthes and Warner, 1978). It uses a lumped chemical mechanism (Stockwell, 1986; Stockwell et al., 1990), and it has a grid size of 80 km and either 6 or 15 vertical layers (Chang et al., 1987). Middleton and Chang (1990) have made comparisons of daily maximum ozone concentrations for three periods, one of which had high ozone concentrations;

data from three to six stations were used.

ADOM is driven by meteorological input from the Canadian numerical weather prediction model and a high-resolution boundary layer model. It uses a condensed chemical mechanism adapted from Lurmann et al. (1986b) and has a grid size of 127 km and 12 vertical layers (Venkatram et al., 1988). Karamchandani et al. (1988) made comparisons with ozone time series at four rural stations, and showed ratios of observed to modeled ozone for June 9-17, 1983, for the northeast. Limited emissions reduction scenarios were examined, with results presented only for Ontario (Karamchandani et al., 1988).

The model of McKeen et al. (1991a) is similar in many respects to RADM. It is driven by the same mesoscale model, and it uses a version of the Lurmann et al. (1986b) chemical mechanism. It has a grid size of 60 km and 15 vertical layers and covers the eastern half of the United States. A preliminary evaluation was made for July 4-7, 1986. This period of high ozone concentrations was selected because it coincided with an intensive field measurement program at Scotia, Pennsylvania (Trainer et al., 1987). Model results for ozone (averages for 1:00 p.m. to 5:00 p.m.) were compared to data from the Aerometric Information Retrieval System (AIRS) (EPA, 1987) for the final two days of the simulation, and maps were shown of model results for NO_x and isoprene, averaged over 0900-1700 hr, as were average vertical profiles for odd-nitrogen species and various VOCs. The budgets of ozone, NO_x, and isoprene, averaged over the model domain, were discussed. Detailed comparisons with the Scotia data are not yet available (M. Trainer, NOAA, pers. comm., Nov. 1991).

The major shortcoming with the studies (Liu et al., 1984; Middleton and Chang, 1990; McKeen et al., 1991a) described above is that the model evaluations focus almost exclusively on surface ozone in rural air, usually for a single pollution episode. This is in part because of the lack of high-quality data for NO_x and VOCs in rural locations (see Chapter 7). Only the study by McKeen et al. (1991a) shows model predictions for NO_x, and none of the studies shows the spatial concentration fields predicted for any of the anthropogenic VOCs. None of the models has been evaluated systematically by rigorous comparison of predictions for ozone and its precursors for the range of meteorological conditions found in the eastern United States.

The Regional Oxidant Model (ROM) is a grid-based photochemical air-quality model with a grid size of ~ 18.5 km, designed to simulate ozone formation and transport over the eastern United States (Figures 10-1 and 10-2). The transport is governed by wind fields interpolated on an hour-by-hour basis from observations at surface and upper air stations. The model includes parameterizations for the effects of cumulus clouds on vertical transport; nighttime wind shear and turbulence episodes associated with the nocturnal jet stream; mesoscale vertical motions induced by the interaction of terrain and the large-scale flow; and the

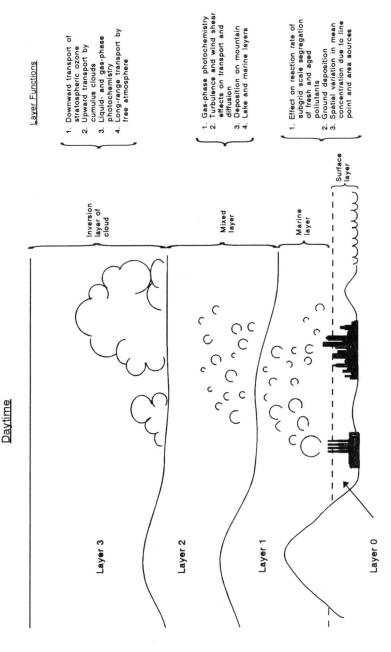

FIGURE 10-1. Regional oxidant model (ROM) vertical structure of the atmosphere during daytime conditions.

Source: Schere and Wayland, 1989.

effects of differences in terrain on advection, diffusion, and deposition. Lamb (1983, 1984) described the theoretical basis and design of ROM. The model incorporates the Carbon Bond 4.0 (CBM-IV) chemical mechanism (Whitten and Gery, 1986). Most studies with ROM2.0 have used the National Acid Precipitation Assessment Program (NAPAP) 1980 emissions inventory (Version 5.3) for anthropogenic species and the inventory of biogenic VOCs described by Novak and Reagan (1986), and have focused on the northeastern United States (Figure 10-2a), although a few studies have been conducted for the Southeast (Figure 10-2b).

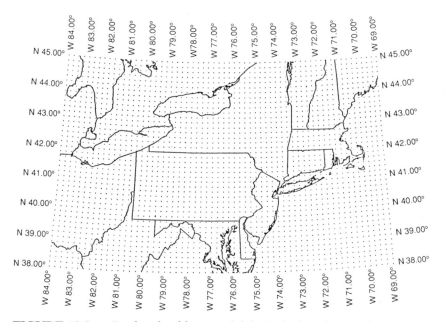

FIGURE 10-2a   Regional oxidant model (ROM) domain, Northeastern United States. Source: Schere and Wayland, 1989.

The advantage of the physical layered structure of ROM (Figure 10.1) over a finite-difference representation of the vertical is that it allows high horizontal resolution without excessive amounts of computer memory being consumed by many vertical layers. This is particularly important for a model that uses 28 chemical tracers as well as meteorological fields. There are problems with its approach, however, in mountainous regions, such as the Appalachians, Rocky Mountains, or along the West Coast of the United States, where the atmospheric boundary layer has more than one level. Pollutants thus could be emitted in any

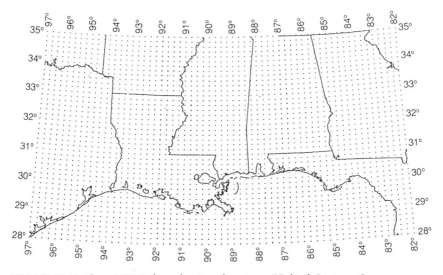

FIGURE 10-2b   ROM domain, Southeastern United States. Source:
Schere and Wayland, 1989.

of the layers shown in Figure 10-1, or even above the model layers, depending
on environmental stability and winds. Moreover, the transfer rates between
layers are affected by the presence of mountains whose heights are greater than
the height of the boundary layer. Multi-atmospheric boundary layers also can
form along coastal regions because of sea breeze circulations. In some cases
their effects can be characterized in a three-layer atmosphere; in other cases
such a simple parameterization will not work. The same holds true for
parameterizations of the effects of cumulus transport and wet deposition on
transfer rates. There often is not enough information in a three-layer model to
diagnose the predominant mode of convection (by ordinary cumulus clouds, by
deep convection, or by mesoscale convective systems). The deeper modes of
convection require information about the deeper tropospheric structure and its
modulation by convection.

A major limitation of ROM is that it is a "hardwired" model such that resolution cannot be readily expanded in the vertical direction as better computer
resources become available. There has been a revolution in supercomputers and
workstations in the decade since ROM was designed. Numerical prediction
models with 15 to 30 levels in the vertical direction and 15 km resolution later-
ally are now being run operationally in the United Kingdom and elsewhere. The

output of such models could be used to drive an improved ROM with vertical resolution that matches the larger-scale model grid and uses some of the physical routines, such as the convective parameterization schemes, available in the larger-scale model.

The domain of ROM is relatively small, a consequence of its high spatial resolution. The effect of initial and boundary conditions is therefore of concern. The model is initialized with "clean" tropospheric background conditions—ozone at 35 parts per billion (ppb), NO_x at 2 ppb, and nonmethane VOCs at 15 ppb carbon—and it is allowed to run chemically for 10 hr before the actual start of a run. This procedure reduces the value for NO_x to ~0.2 ppb (Schere and Wayland, 1989). ROM takes boundary conditions for ozone to be the average of values (6-hour mean) at three "relatively rural" sites in Ohio, West Virginia, and Virginia; the model is then re-equilibrated for 1.5 hours, starting with the equilibrated initial conditions and the new ozone value, to obtain daytime and nighttime boundary conditions. The equilibrated initial and boundary conditions for NO_x are exceedingly (and probably unrealistically) low. The effects of the boundary conditions are expected to be greatest in the west and south of the modeled area, and least in the Northeast corridor, the focus of the model.

ROM2.0 has been evaluated by making comparisons with measurements of ozone, NO_x, and nonmethane VOCs collected between July 12 and August 31, 1980 (Schere and Wayland, 1989). The model was run without re-initialization for this six-week period, which was chosen to coincide with three major field programs in the Northeast: the Northeast Regional Oxidant Study (NEROS) (Vaughan, 1985), the Persistent Elevated Pollutant Episode (PEPE) study, and the Northeast Corridor Regional Modeling Program (NECRMP). These programs were designed to provide data for input to and evaluation of regional and urban models applied to the Northeast, and each included a major component of aloft data obtained by aircraft. The frequency of upper air soundings was increased from two to four per day at the stations within the domain of ROM2.0 during intensive NEROS field study periods. The major shortcoming of the data available for this period is the relative dearth of rural data, particularly for NO_x. None of the NO_x monitors were sensitive to concentrations below 5 ppb, and concentrations below this level are often found at rural sites (see Chapter 7). The ozone data were obtained from EPA's SAROAD data base (now AIRS [EPA, 1987]) supplemented by data for southern Ontario (obtained from Environment Canada and the Ontario Ministry of the Environment). In all, 214 sites were studied, primarily in or near urban areas.

ROM2.0 results for ozone were evaluated (Schere and Wayland, 1989) by examining frequency distributions for groups of measurement sites that had common characteristics as discussed below. This method of comparison was chosen over a point-by-point comparison because of the inherent uncertainty in

TABLE 10-3   Mean of Normalized Frequency Distribution of Daylight (8:00 a.m. to 7:00 p.m. Local Standard Time) Observed Ozone Concentrations at Monitoring Sites in Six Groupings. Standard Deviation Given in Parentheses.

	Concentration range, ppb			
	5-20	5-40	40-80	>80
Group 1,	0.084	0.271	0.385	0.344
35 sites	(0.043)	(0.063)	(0.045)	(0.056)
Group 2,	0.149	0.378	0.406	0.216
39 sites	(0.058)	(0.067)	(0.072)	(0.033)
Group 3,	0.159	0.438	0.466	0.096
64 sites	(0.061)	(0.068)	(0.073)	(0.046)
Group 4,	0.218	0.590	0.355	0.054
54 sites	(0.063)	(0.060)	(0.043)	(0.040)
Group 5,	0.423	0.799	0.187	0.014
20 sites	(0.133	(0.093)	(0.081)	(0.024)
Group 6,	0.044	0.262	0.627	0.111
2 sites	--	--	--	--

the wind-driven transport component of the model. The sites were divided into groups based on observed frequency distributions, as shown in Table 10-3. Groups 1 and 2 (Figure 10-3) consisted of sites where concentrations above 80 ppb were found most often, and tended to be located along the Northeast corridor. Groups 3 and 4 had lower concentrations of ozone and were distributed throughout the model's domain. Group 5 had many sites where concentrations were below 20 ppb, suggesting contamination by a nearby source of NO_x, or problems with the monitor. Group 6, Whiteface Mountain, New York, and Long Point Park in southern Ontario, had few high or low values, characteristic of more remote or mountaintop sites. Comparisons of the frequency distributions for these groups for 8:00 a.m. to 7:00 p.m. LST (local standard time) with model results are shown in Figure 10-4. The figure illustrates the tendency of the model to overpredict ozone at lower concentrations (<50 ppb); some of the

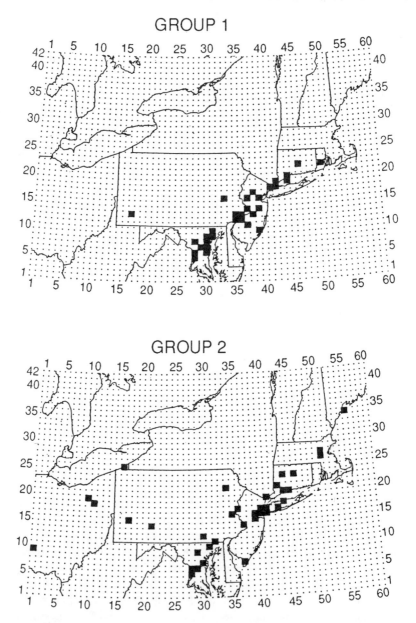

FIGURE 10-3   ROM grid cell locations (darkened) of monitoring sites within groups 1 through 6 (see Table 10-3). Source: Schere and Wayland, 1989.

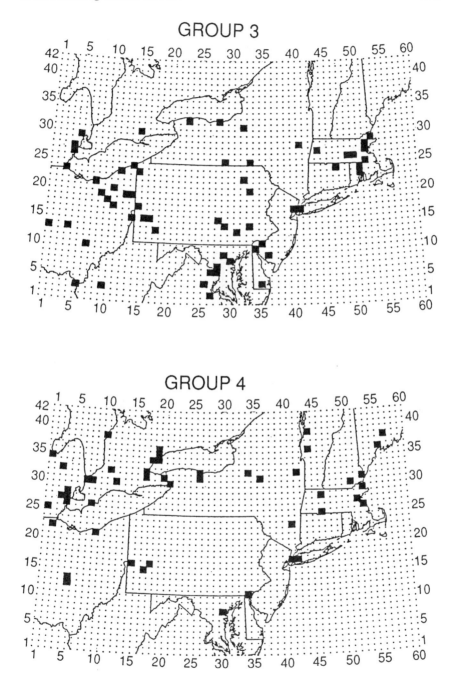

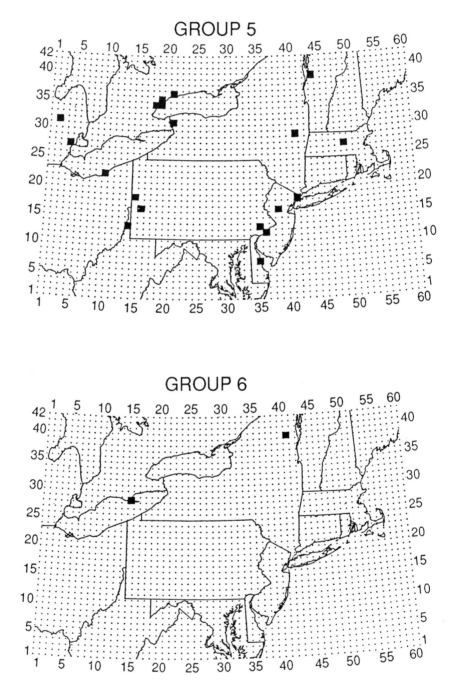

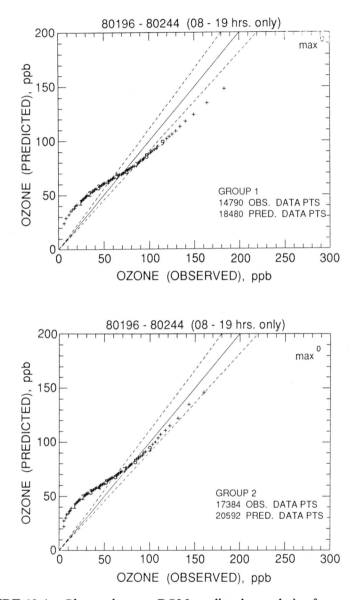

FIGURE 10-4   Observed versus ROM-predicted cumulative frequency
distributions of daytime (8:00 a.m. to 7:00 p.m., LST) hourly ozone concentrations at each of six groups of receptor locations over the period
from July 14 to August 31, 1980. Source: Schere and Wayland, 1989.

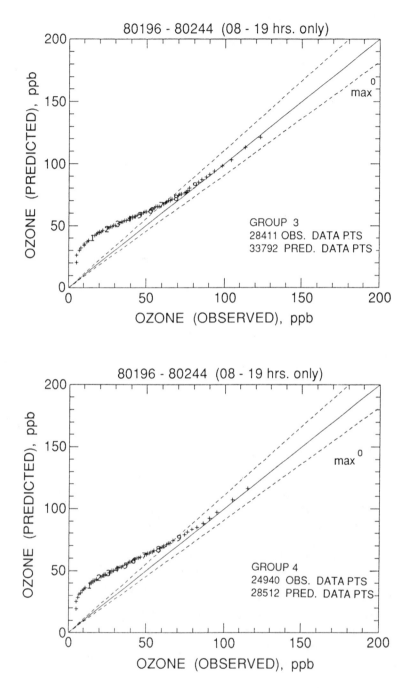

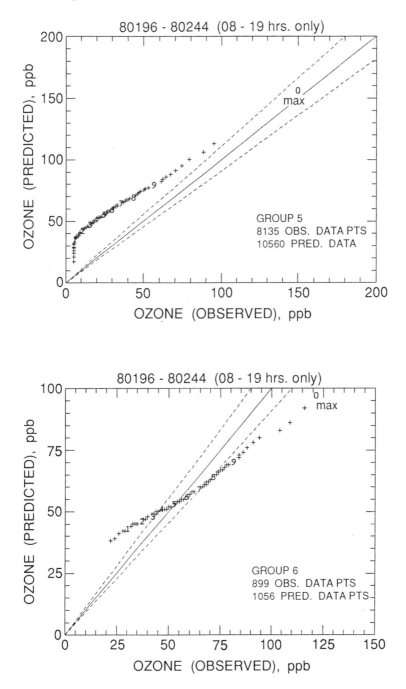

low values could result from local titration of ozone by NO_x at the predominantly urban monitoring sites, an effect that the model would not reproduce. The model underpredicts ozone concentrations above 100 ppb in group 1, which contains the sites where the highest concentrations of ozone are found, but reproduces values from 60 ppb to 120 ppb at sites in groups 3 and 4, most of which lie outside the Northeast corridor. The model also underpredicts concentrations above 75 ppb at the two most remote sites (group 6).

The NAAQS for ozone is based on maximum 1-hour daily concentrations, so these were evaluated separately, using the same groups of stations (Schere and Wayland, 1989). The bias was calculated as the difference between the daily maximum at a site and that in the model (the data were matched in space but not in time on a given day). The bias values are shown in Figure 10-5, as a function of observed ozone, sorted in 20-ppb ranges. The figure shows increasing underprediction of ozone with increasing concentration, and overprediction of low concentrations of ozone. The model performs best for maximum concentrations of 60-100 ppb. The mean underprediction is ~30 ppb for observed values of 120-140 ppb, and 50-70 ppb for observed values above 160 ppb.

Schere and Wayland (1989) examined the spatial pattern of ozone in the Northeast corridor for five episodes in the six-week period. They compared model results to the pattern of maximum concentrations (the maximum for each station during the episode), as shown in Figure 10-6. The pattern of high concentrations agrees well for July 20-22, less well for July 25-27. The tendency to underpredict the highest concentrations is obvious. Examination of all the episodes shows some characteristic problems. There is often a systematic underprediction of ozone concentrations downwind of Philadelphia, Pennsylvania, such that the urban plume is not apparent in the model results. The urban plumes from Washington, D.C.; Baltimore, Maryland; New York, New York; and Boston, Massachusetts were discernable in the predictions. There appears to be a significant bias in the wind field toward a westerly direction, such that the plumes were carried eastward in the model, whereas the observed data suggest transport to the northeast. Aircraft data show that the model tends to underpredict regional background concentrations of ozone by 20-30 ppb; observations were typically 40-90 ppb whereas the model predicted 40-60 ppb. This could result in part from the low upper boundary value for ozone and in part from underprediction of NO_x. The model also underpredicted ozone concentrations aloft in urban plumes.

Model results for NO_x were evaluated in the same manner as for ozone, by forming groups of sites. ROM2.0 significantly underpredicts NO_x and NO_2 (nitrogen dioxide) for urban sites at both the 50th and 90th percentiles of the frequency distributions, as shown in Table 10-4. This could result from the coarse grid of the model or from a systematic underestimate of NO_x in the

TABLE 10-4   Average Ratio (Observation/Prediction) over Station Groups at 50th and 90th percentiles of Cumulative Frequency Distributions

Daytime, 8:00 a.m. to 7:00 p.m.		All hours	
Percentile		Percentile	
50th	90th	50th	90th
NO$_x$	1.8	2.3	1.9	2.5
NO$_2$	2.2	2.2	2.1	1.9

emissions inventory. Rural measurements of NO$_x$ were not available for the evaluation.

Schere and Wayland (1989) evaluated data for VOCs (excluding methane) on a site-by-site basis. The ratios of observed to predicted VOC concentrations for 6:00 to 9:00 a.m. LDT (local daylight time) were typically around 6 for urban areas with a range of 4-10 for specific cities, and somewhat lower, 1.3-8, for data collected after 9:00 a.m. The ratio was lower, about 2, at sites outside large urban areas. The model gave a more accurate prediction of the distribution among the various VOC species than of the total amount. It underpredicted VOCs aloft by a factor of only 1.4 to 2.8, and there were some overpredictions; the more reactive hydrocarbons were underpredicted to a greater extent than were the less reactive ones. VOC emissions appear to be seriously underestimated by the NAPAP 1980 inventory (cf. Chapter 9), and the effect of inaccurate inventories on ROM predictions for ozone requires investigation.

Schere and Wayland (1989) concluded their evaluation of ROM2.0 with the caution that "a number of issues have arisen that prevent recommending use of the model in a simple unassisted manner for studies of violations of the ozone air-quality standard." They stressed that careful analysis of the simulated wind fields and background concentrations of ozone must be made before an episode is used in a model for regulatory application.

EVALUATION OF MODEL PERFORMANCE

Air-quality models are evaluated by comparing their predictions with ambient

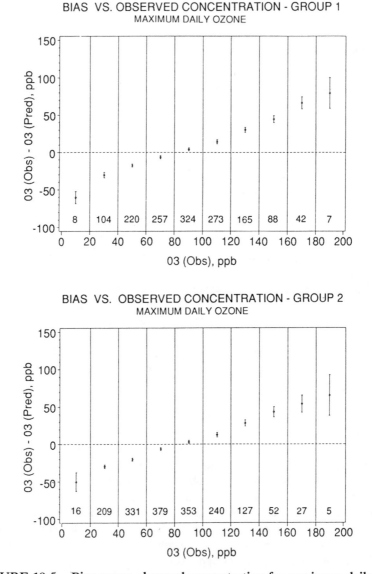

FIGURE 10-5   Bias versus observed concentration for maximum daily
ozone over the simulation period from July 14 to Aug. 31, 1980, for
groups 1 through 6 (see Table 10-3). The 95% confidence interval is
shown about the mean value for each concentration interval. Source:
Schere and Wayland, 1989.

BIAS VS. OBSERVED CONCENTRATION - GROUP 3
MAXIMUM DAILY OZONE

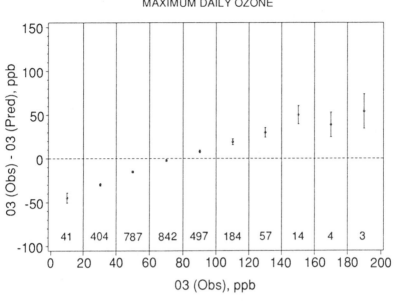

BIAS VS. OBSERVED CONCENTRATION - GROUP 4
MAXIMUM DAILY OZONE

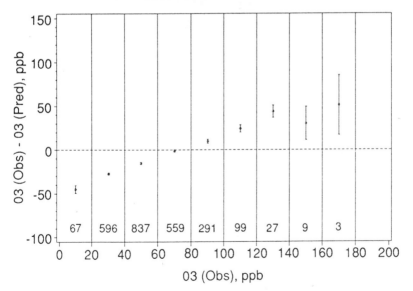

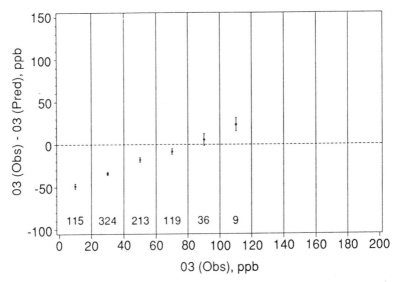

BIAS VS. OBSERVED CONCENTRATION - GROUP 5
MAXIMUM DAILY OZONE

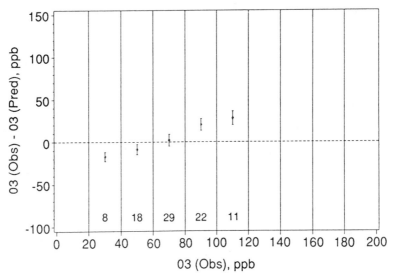

BIAS VS. OBSERVED CONCENTRATION - GROUP 6
MAXIMUM DAILY OZONE

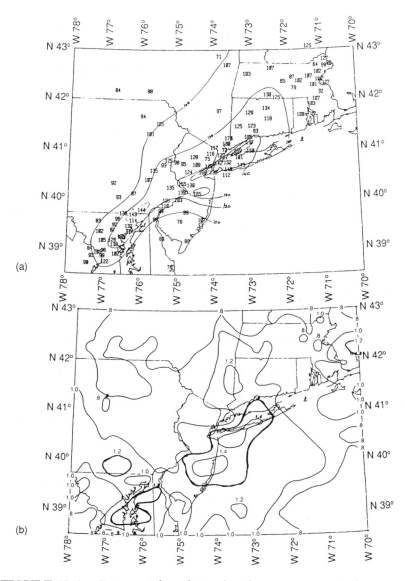

FIGURE 10-6   Contours of maximum hourly ozone concentrations over the period July 25-27, 1980, for (a) observed and (b) predicted data sets. Contours of observed data are in concentration units of parts per billion and contours of predicted data are in units of ppb/100. Source: Schere and Wayland, 1989.

observations. Because a model's demonstration of attainment of the ozone NAAQS is based on hypothetical reductions of emissions from a base-year-episode simulation, the accuracy of the base-year simulation is essential. Emissions control prescriptions estimated for an area to meet the NAAQS are sensitive to the base-year model performance. Performance requirements for the base-year simulation are instituted to ensure that the model is adequate to evaluate emissions control strategies. An adequate model should give accurate predictions of current peak ozone concentrations and temporal and spatial ozone patterns. It should also respond accurately to changes in VOC and NO_x emissions, to differences in VOC reactivity, to changes in temperature, and to spatial and temporal changes in emissions patterns for future years. Associated with model performance evaluation is the quantification of the uncertainty in predicted emissions control estimates; for example, determination of the relationship between uncertainties in the emissions inventory and uncertainties in model predictions.

A performance evaluation should answer several questions:

• If successfully evaluated for a base case, does the model hold for significantly reduced emissions?

• Does the model correctly reveal the effect of VOC and NO_x emissions controls and the influence of chemical reactivity in ozone reduction?

• Has the sensitivity of model performance to variability in input data been established? Have the evaluations been stringent enough to clearly reveal a flawed model?

• What are the component uncertainties in the modeling analysis, and what is the aggregate uncertainty in the control strategy requirement obtained from the model results?

• What is the ability of the model to correctly simulate the effects of individual sources or upwind source regions on local air quality?

• What is the probability that desired air quality will be achieved in the model, given the use of a deterministic model and a limited number of modeling episodes?

The terms "model validation" and "model verification" are frequently used to describe the process of comparing model predictions with observations. These terms can be confusing. If a model is valid, its predictions will agree with the appropriate observations, given a perfect specification of model inputs. A "model performance evaluation" is the process of comparing a model's predictions with observations. The term "verification" might be reserved to describe a successful, or positive, outcome of the model evaluation process; to determine whether a model is in fact valid.

Traditional photochemical model performance evaluations do not provide enough information to decision makers about the suitability of models for use in regulatory applications (Dennis and Downton, 1984; Seinfeld, 1988). Because evaluation methods have traditionally focused only on the degree to which predictions and observations agree, they are not specifically designed to reveal flaws in a model, in its data base, or in the procedures used to exercise the model. Instead, they are aimed at quantifying the correspondence between predicted and observed ground-level concentrations of pollutants, generally ozone. Although current operational evaluation procedures can indicate how well a model performs in an overall sense relative to similar applications in the past; little direct information is provided about a model's suitability for predicting the effects of emission reductions.

An important question in performance evaluation is whether specific performance goals are specified, or whether minimum criteria for rejection of the results should be set. This is a subtle issue. Key model inputs needed to simulate past episodes have important areas of uncertainty. These inputs include boundary and initial conditions, both on the ground and in the air, and emissions. Sometimes the uncertainties in these inputs are large enough that the temporal and spatial features of ozone behavior can be reproduced reasonably well by variation of the inputs within their ranges of uncertainty. At other times, such variation does not produce adequate agreement between predictions and observations.

Figure 10-7 shows an ozone time series for a UAM simulation of an episode of high concentration of ozone on Sept. 16-17, 1984, in the South Central Coast air basin of California. The solid line represents the hourly base-case model predictions for the Simi monitoring station. The boxes indicate the ozone concentrations observed for each hour. The vertical lines represent the estimated overall uncertainty and representativeness of the spatial distribution of ozone measurements. The thin solid lines enclose an ensemble of time series profiles obtained from more than a dozen model runs focusing on uncertainties in the predicted mixing heights. These uncertainties were derived from a numerical mixed-layer model (Tesche et al., 1988c). Ideally, the ensemble of ozone predictions (enclosed by the thin solid lines) would trace a path within the upper and lower uncertainty bounds of the hourly ozone measurements. If this were the case, the predictions would match the observations within experimental uncertainty. This is not the case in the example presented here; the model underestimates the peak ozone concentration, and the variability in afternoon ozone predictions that results from mixing-height uncertainties appears to be comparable to the estimated uncertainty in the ozone measurements.

Model performance evaluation procedures and tests must be designed to reveal flaws in assumed input information and model components in order to ensure that a model is producing the right answer for the right reason. In the past, ozone-modeling protocols accepted photochemical model results as ade-

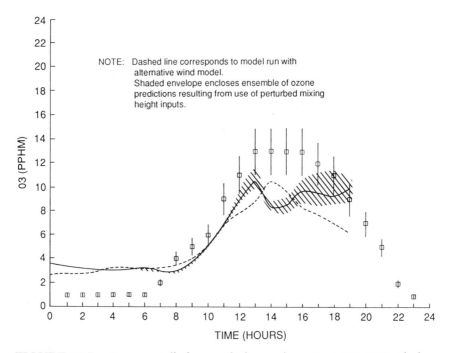

FIGURE 10-7   Ozone predictions and observations; Sept. 17, 1984, Simi
monitoring station. Source: Tesche et al., 1990.

quate if the bias and error statistics were comparable to those of previous, similar studies, but little attention was given to whether the previous simulations
adequately met policy-making needs.

The performance of photochemical models, judged by commonly reported
statistical procedures, appears to have reached a plateau (Seinfeld, 1988). Current grid-based photochemical models reproduce hourly averaged ozone concentrations to within 30-35% of measured values, and the peak 1-hr concentration
is often reproduced to within 15-20%. Tesche (1988) has compiled most of the
ozone simulations reported prior to 1988, and the results are summarized in
Figure 10-8. The statistics on which Figure 10-8 is based are pairs of hourly
averaged predictions and observations at the various monitoring stations in each
region. Sixty-seven percent of the single-day simulations exhibit underestimation of ozone, and 62% of the multiday simulations show underestimation of
ozone concentrations. This performance is consistent with an underestimation
of VOC emissions.

Despite more than a decade of accumulated experience in photochemical
model evaluation and refinement, there has been no focused attempt by regu-

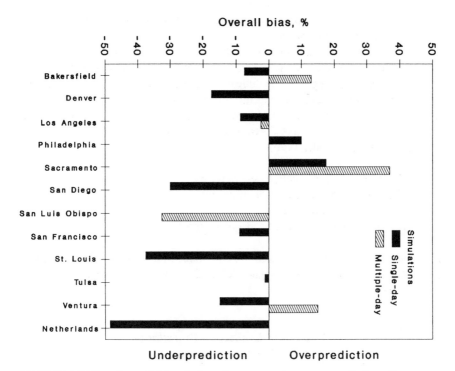

FIGURE 10-8   Overall bias in hourly averaged ozone predictions by urban area for single- and multiple-day simulations of episodes of high concentrations of ozone for model applications prior to 1988. Source: Tesche, 1988.

latory agencies, model developers, or others to derive a consistent set of evaluation procedures from among the techniques in common use—notwithstanding the widely recognized need for such methodology.

Model Performance Evaluation Procedures

Before a photochemical model is applied to simulate the effects of emissions controls, the model must be shown to reproduce the chemical and physical processes that govern ozone formation. This kind of assessment, referred to as model performance evaluation, involves the compilation of emissions, meteorological, air quality, and chemical data drawn from an actual episode. This information is used in the exercise of the model. Through various statistical and graphic means, the model predictions and air-quality observations are compared to determine the fidelity of the simulation.

Specific numerical and graphic procedures have been recommended for evaluation of the performance of grid-based photochemical models (Tesche et al., 1990). The methods suggested include the calculation of peak prediction accuracy; various statistics based on concentration residuals; and time series of predicted and observed hourly concentrations. Four numerical measures appear to be most helpful in making an initial assessment of the adequacy of a photochemical simulation (Tesche et al., 1990):

- The paired peak prediction accuracy.
- The unpaired peak prediction accuracy.
- The mean normalized bias.
- The mean absolute normalized gross error.

Graphic procedures are suggested to complement the statistical approaches. Finally, a minimum set of six diagnostic simulations is recommended as part of performance evaluations.

Diagnostic Evaluation Procedures of Model Performance

The diagnostic evaluation procedures for models discussed in this section are used to determine the causes of failure of a flawed model; to stress a model to ensure failure if the model is flawed; and to provide additional insight into model performance beyond that supplied through the operational evaluation procedures previously discussed. Frequently, operational model evaluation does not convey enough information about the model and data base to allow their use in emissions control strategy development and testing. Particularly in complex air basins, such as the South Coast air basin of California, the model evaluation process should be supplemented with additional diagnostic analyses that probe further into the comparisons between prediction and observation, attempting to ensure that the ozone response given by the model is correct. This section discusses a series of tests and comparisons that are useful in diagnostic analysis of photochemical models. The general categories of these analyses are: diagnostic simulations, testing of species other than ozone (when adequate data permit), examination of model-predicted fluxes and pollutant budgets, and sensitivity-uncertainty testing. In some cases the tests must be supported by high-quality aerometric data bases, only now available in some areas due to the high cost of their acquisition.

Five diagnostic simulations are suggested to accompany the numerical and graphic comparisons of predictions and observations (Tesche et al., 1990): zero emissions, zero initial conditions, zero boundary conditions, zero surface deposition, and mixing-height variations.

Zero Emissions

The purpose of the zero-emissions simulation is to ensure that the base-case simulation results are influenced appropriately by the emissions inputs or, conversely, are not over-influenced by initial and boundary conditions. Removing all emissions should lead to much lower concentrations of reactive species. The zero-emissions simulation is performed by running the base-case scenario with all emission values reduced to zero. All other model input files remain unchanged from the base case.

The diagnostic run should produce significantly reduced concentrations—close to background or to the concentrations representing the inflow boundary conditions. If not, there is reason to question the accuracy of the simulation. Lack of sensitivity to emissions can indicate inappropriately high initial conditions, improper boundary conditions, or some flaw in the model itself. Quite apart from these concerns, insensitivity to emissions raises serious questions about the usefulness of the simulated episode for control strategy development and assessment.

Zero Initial Conditions

The zero-initial-conditions simulation reveals how many of the predictions from the second (or third) day result from the initial field used to start the base-case simulation. This test simulation is performed by setting all initial concentrations as close to zero as possible, while avoiding numerical instabilities. Deficit enhancement and time series graphs can be used to display the results of this simulation. If the effect of the initial field is completely gone by the second or third day, the deficit enhancement graphs will indicate essentially no differences between the diagnostic and base-case runs on the following days. For stagnation episodes, some residual effects of initial conditions could be seen even on the third day of a multiday simulation (Tesche and McNally, 1989).

Zero Boundary Conditions

The zero-boundary-condition simulation examines the influence of boundary values on second (or third) day concentrations, particularly in regions where the base-case predictions are highest. This simulation helps identify situations in which the base-case results are greatly affected by the boundary conditions. The simulation is performed by setting all inflow and outflow boundary values to zero, including those for the top surface of the modeling region.

Zero Surface Deposition

The zero-surface-deposition simulation addresses the influence of dry surface deposition removal on concentrations of primary and secondary species. The diagnostic run is performed by setting deposition velocities for all species to zero in the base-case simulation. Although deposition tests have not been reported in previous model evaluation studies, some general guidelines can be suggested. For primary species, such as NO_x and VOCs, when deposition is neglected the downwind concentration fields should increase relative to the base case in a manner consistent with the deposition velocities for each primary species. For secondary species, such as ozone, the effect of a change in the deposition of primary species will depend on how that change propagates through the nonlinear chemistry.

Mixing-Height Variations

Mixing heights have a direct and often significant influence on ozone concentrations. The objective of the mixing-height diagnostic simulation is to reveal the degree to which ozone concentrations are influenced by the height of the mixed layer. At a minimum, one diagnostic run is suggested in which the hourly mixing-height values are uniformly increased by 50% above the base-case account. This increase is somewhat larger than the expected uncertainty in estimates of mixing heights typically encountered. Therefore, this simulation should provide a bound on the change in ozone predictions resulting from uncertainties in this input. Increased mixing heights typically reduce ozone concentrations, although the reduction is usually less than a one-to-one change. This effect can be seen by moving toward the origin along a line of constant VOC/NO_x ratio on an ozone isopleth plot (see Chapter 6). One might choose, instead, to reduce the hourly mixing heights by 50%. The resultant changes in ozone concentrations under this scenario will typically be comparable in magnitude but of opposite sign to those for the mixing height increase.

Dependence on Initial and Boundary Conditions

Given the dependence of model results on initial and boundary conditions, it is of interest to quantify that dependence. Russell et al. (1989) performed a series of diagnostic simulations to identify how grid-based airshed model predictions change with changes in boundary and initial conditions. Initial conditions alone (in a simulation without emissions) led to the development of an air mass with elevated concentrations of ozone and VOCs that would last for several days before being slowly depleted by ground-level deposition and atmospheric chemistry. Predictions on the first and second day of a three-day simulation were dependent on initial conditions of VOC and NO_x. By the third day, most of the initial-condition-dominated air mass had left the modeling region, although part of the region was still influenced. Boundary conditions primarily influence the regions near the boundary, with a small influence in the central modeling region. Reducing both boundary and initial conditions to background concentrations led to less than a 4% reduction in peak ozone and exposure predictions from the base-case simulation that used more representative values.

However, if emissions are reduced by 50% or more, the role of initial and boundary conditions increases and can become significant. For example, diagnostic simulations show that there are enough VOCs in the boundary conditions to form significant amounts of ozone when there are no VOC emissions from any source. Trajectory model studies show similar findings, and this is particularly important because trajectory model simulations are usually for less than 1 day. Simulations of less than 1 day are very sensitive to the choice of initial and upper level boundary conditions. Studies of boundary and initial condition effects suggest that large modeling domains and multiday (preferably, 3 or more) simulations are necessary for testing the effects of control strategies, and it is preferable to have model domain boundaries in relatively clean, rural regions.

Assessing Simulation Results

Decision makers and regulatory agencies seek quantitative performance standards by which to judge new models as acceptable. Each photochemical modeling episode exhibits distinctive aerometric and emissions features. Each model's available data base also is unique in the amount and quality of observations available to support model evaluation and testing. In addition, the particular set of modeling procedures and codes makes each application distinctive. Therefore, automatic use of standards for acceptance or rejection raises the risk of accepting a model evaluation that gives seemingly "good" performance statistics but for the wrong or misleading reasons. It also could lead to rejection of a model evaluation that violates criteria for reasons related to input inaccuracies rather than to fundamental flaws.

Instead of prescribing fixed performance standards, Tesche et al. (1990) suggest the following approach. From more than 15 years of photochemical model development and testing, photochemical grid model simulations generally produce peak (unpaired) prediction accuracy, overall bias, and gross error statistics in the approximate ranges of 15-20%, 5-15%, and 30-35%, respectively (Figure 10.8). A study that follows an approved ozone-modeling protocol but falls outside all these ranges would not be rejected unless evidence from the model's diagnostic simulations and the other numerical measures and diagnostic tests suggested unusual or aberrant behavior. For model simulations falling within these ranges, some additional diagnostic analyses could be appropriate to lend further support to the contention that the simulation is acceptable. For model results outside the ranges given for any one of these areas, it should be incumbent on the modeler to explain why the performance is poorer than that commonly achieved in similar applications and whether the causes of poorer performance will adversely affect the use of the model in control strategy evaluations. This method provides reviewing agencies with a general model performance target, but still guards against the inappropriate rejection of less accurate model simulations when appropriate explanations can be provided.

Multispecies Comparisons

The development of evaluation procedures that test photochemical model performance for species other than ozone can provide a basis for accepting or rejecting a model (or a model simulation); they significantly improve the chances that a flawed model will be identified. Adequate model performance for several reactive species increases the assurance that correct ozone predictions are not a result of chance or fortuitous cancellation of errors introduced by various assumptions. Multispecies comparisons could be the key in discriminating among alternative modeling approaches that provide similar predictions of ozone concentrations.

To date, most model evaluation studies present results only for ozone (Tesche, 1988), although there are a few limited tabulations of NO_2 predictions (Wagner and Ranzieri, 1984). Studies reported by Roth et al. (1983), Tesche (1983), Russell and Cass (1986), Wagner and Croes (1986), and Russell et al. (1988a,b) are among the few that present performance evaluation statistics for associated pollutants. Lack of ambient measurements for such pollutants is the major reason for the limited number of past studies. The data bases for the

Southern California Air Quality Study and the San Joaquin Valley Air Quality Study (SJVAQS)/Atmospheric Utility Signatures, Predictions, and Experiments (AUSPEX) now allow for several comparisons with species other than ozone. The SCAQS data in particular afford a level of testing of photochemical models and modules, such as the chemistry mechanism, not previously possible. The availability of ambient air measurements for speciated organics and for key species such as formaldehyde (HCHO), peroxyacetyl nitrate (PAN), nitrogen dioxide (NO_2), hydrogen peroxide (H_2O_2), nitric acid (HNO_3), and organic acids will allow not only more extensive operational model testing but also diagnostic and comparative evaluations. Finally, the SCAQS data base, or similar data bases that will be assembled in the future, such as one from the SJVAQS/AUSPEX (Ranzieri and Thuillier, 1991), also offer the potential for evaluations of alternative chemical kinetic mechanisms.

Evaluation of model performance for precursor and intermediate species as well as for product species other than ozone is recommended when ambient concentration data for these species are available. Comparisons of observed and predicted concentrations for all important precursors, intermediates, and products involved in photochemical air pollution—such as individual VOCs, nitric oxide (NO), nitrogen dioxide (NO_2), PAN, ozone, H_2O_2, nitrous acid (HONO), and HNO_3—are useful in model evaluation, especially with respect to the chemistry component of the model (Dodge, 1989, 1990). Accurate matching of ozone alone may not be sufficient to indicate that a chemical mechanism is correct. Comparisons of predictions and observations for total organic nitrates (mainly PAN) and inorganic nitrates (HNO_3 and nitrate aerosol) can be used to test qualitatively whether the emissions inventory has the correct relative amounts of VOCs and NO_x. However, HNO_3 and nitrate aerosol cannot be included in the data set for model comparisons if the model does not include an adequate description of the HNO_3 depletion process associated with aerosol formation.

Depending on the availability of other measurements and the incorporation of aerosol dynamics and thermodynamic processes in the model, the above comparisons can be supplemented with others. For example, because HNO_3 is a sink for the OH radical and because H_2O_2 is a sink for the HO_2 radical, the ratio of HNO_3 to H_2O_2 is an indicator of the extent to which OH and HO_2 radicals are adequately simulated (if nighttime HNO_3 formation processes are properly accounted for).

There are practical limitations in evaluating a model's performance in identifying speciated VOCs. Conceptually, comparisons can be made between observed and predicted total VOC concentrations as well as between observed and predicted concentrations of classes of VOCs. The second type of comparison requires aggregations of ambient VOCs into the classes used in the particular chemical mechanism of the model under consideration. However, emissions of nonreactive organic compounds, which can constitute 5 to 30% of actual VOC emissions, usually are not included in the simulation. Thus, predictions of VOCs could have an inherent bias toward underestimation relative to observed VOCs unless the difference is accounted for by explicitly excluding the nonreactive compounds from the observed concentrations. Also, approximations must be made in the schemes used to group the VOCs (Middleton et al., 1990). In some cases, the assignment of individual compounds to a class is based more on the similarity of their ozone formation potential to that of the model species than on the rate at which they react. Hence, perfect agreement is not expected and, in fact, agreement to within 20% for VOC classes is probably the best that can be expected.

Mass Fluxes and Budgets

Only recently have attempts been made to derive mass balances and carry out flux calculations for photochemical grid model simulations. This has occurred more routinely for regional grid-based models. Four mass balance and flux calculation procedures are suggested to accompany detailed performance evaluations. The first involves computing the mass fluxes into and out of the domain boundaries. The second procedure involves the mass fluxes into and out of the mixed layer. Third, the surface deposition fluxes should be estimated; hourly and daily average surface deposition rates should be calculated and reported for each species removed at the ground. In the final procedure, emissions, transport, transformation, and removal terms are reconciled in a simplified, closed mass budget over the whole modeling domain. The various flux terms described above, when combined with the hourly emissions rates, can be used in a simple mass budget to apportion the total mass in the modeling domain into emission, transport, and removal components. The transformation term is obtained by taking the difference between masses flowing into and out of the model domain, assuming a closed budget.

Another test that should be performed to ensure mass consistency in meteorological and air-quality models is a simulation with an inert tracer version of the model. This simulation should be initialized with a homogeneous boundary and three-dimensional concentration field and should cover a period of at least 12 hours. The concentration field at the end of the simulation should be the same as the initial field.

Sensitivity-Uncertainty Analysis

Sensitivity analysis consists of systematically studying the behavior of a model over ranges in variation of inputs and parameters. This process can extend to studying the behavior of the model for changes in its basic structure—for different assumptions in its formulation. When model inputs and parameters are varied over their ranges of uncertainty to provide estimates of the range of uncertainty in predicted concentrations due to these input uncertainties, the process can be called sensitivity-uncertainty analysis. The diagnostic simulations discussed earlier fall within the general category of sensitivity analysis.

Sensitivity analysis can be used to determine whether the predictive behavior of a model is consistent with what is expected on the basis of its underlying chemistry and physics—whether the model responds "properly" when its inputs and parameters are varied. Sensitivity-uncertainty analysis is just a sensitivity analysis in which the variations in inputs and parameters correspond to their estimated uncertainties, and it is used to estimate the uncertainty in a model prediction. Sensitivity analysis of air-quality models meets two objectives: to determine qualitatively whether a model responds to changes in a manner consistent with what is understood about the basic physics and chemistry of the system, and to estimate quantitatively the uncertainty in model predictions that arise from uncertainties in the inputs and parameters.

Various methods applicable to sensitivity-uncertainty analysis of photochemical air-quality models are available (Dunker, 1980, 1984; Seigneur et al., 1981; Tesche et al., 1981; Tilden et al., 1981; McRae et al., 1982; Brost, 1988; and Derwent and Hov, 1988). An overview and synopsis of major results of sensitivity testing and analyses of photochemical air-quality models can be found in Seinfeld (1988). There are several parameters of interest in the sensitivity analysis of photochemical air-quality models (Tesche et al., 1990):

• Structure and design parameters of the model, including the horizontal and vertical dimensions of the computational grid cell, the number of cell layers in the vertical direction, and the size of integration time steps. In sensitivity testing, changes in these areas are deliberate and are related to model use. The objective is to identify values for each element that will lead to an optimal combination of computational efficiency and accuracy of prediction.

• Constitutive parameters of the model, including chemical reaction rate constants and deposition velocities. Sensitivity analysis usually focuses on the effects on model predictions of uncertainty in these values.

• Input parameters. These are calculated from the input data and, as discussed below, carry the uncertainties inherent in these data.

When comparing model predictions and observations, one must remember that observations contain uncertainties due to measurement errors and the naturally random character of the atmosphere. Futhermore, models predict volume average concentrations, whereas observations reflect point measurements. Fox (1981) gives a good introduction to the concept of uncertainty in air-quality modeling. Beck (1987) offers further discussion of the concept of uncertainty in environmental models and data.

The results of a sensitivity-uncertainty analysis can be presented graphically as indicated in Figure 10-7. The solid bold line in the figure represents base-case ozone predictions for particular air-monitoring stations. The boxes indicate the observed ozone concentrations at each hour, and the vertical lines associated with each box represent the estimated uncertainty of the ozone measurement. The magnitude of these measurement uncertainties has been estimated to include a component related to the spatial representativeness of the monitoring station. The solid lines enclose an ensemble of time series profiles obtained from several sensitivity runs involving different increases and decreases in the base-case mixing heights. In the example shown, these mixing-height uncertainties were derived from more than a dozen simulations of a numerical mixing-height model (Tesche et al., 1988c). Ideally, the ensemble of photochemical model predictions (enclosed by the thin solid lines) would trace a path within the upper and lower uncertainty bounds of the hourly ozone measurements.

TESTING THE ADEQUACY OF
MODEL RESPONSE TO CHANGES IN EMISSIONS

It is important to assess the ability of models to correctly simulate the effects of emissions changes because of the direct connection between changes in emissions and the intended regulatory application of photochemical models. Traditionally, photochemical models are evaluated for a variety of meteorological conditions over periods of time too brief to involve major changes in emissions. Then the critical assumption is implicitly made that the models will be applicable under conditions of drastically altered emissions. The work of Dennis and co-workers (Dennis et al., 1983; Dennis and Downton, 1984; Downton and Dennis, 1985; Dennis, 1986) has shown that grid-based photochemical models that perform adequately for a range of meteorological conditions do not necessarily work well when the evaluation involves a large change in emissions. They have found that different versions of the UAM that give similar performance results under conditions of changing meteorology perform very differently when tested for emissions changes. (These versions of the UAM represent progressive improvements in chemistry, numerical methods, and the treatment of meteorology.) It is imperative, therefore, to evaluate photochemical models intended for use in the development of air pollution control strategies to determine their ability to simulate the effects of emissions changes.

The problems raised in this kind of evaluation are serious. To allow a meaningful model performance evaluation, detailed emissions inventories of comparable accuracy are required for base years long enough apart (at least 10 years) that major emissions changes have taken place. It also is necessary to identify episodes that occur in periods of similar meteorology for the two representative years. Ideally, this kind of evaluation, which uses historical emissions and air-quality records, is the preferable one, but generally the lack of detailed inventories and historical aerometric data prohibits this approach. Even if the required data are available, the evaluation would have to account for all the changes that have occurred over the years in the procedures for developing emissions inventories, monitoring air quality, and so on.

The use of weekday versus weekend emission rates has been suggested as an alternative to retrospective modeling. Even assuming that inventories are accurately estimated, it is not likely that the difference in emissions would be sufficient for a meaningful evaluation of model performance, although such studies would be valuable. An exception might be to conduct such an evaluation after the implementation of a major regulation such as RVP (Reid vapor pressure) reduction of motor vehicle fuel. A third promising approach to evaluating a model's ability to correctly predict the effects of major changes in emissions would be to thoroughly test the model for different urban areas, using input data sets of similar levels of detail.

Because evaluation using historical inventories and aerometric data does not appear feasible now, this third option appears to be a good alternative. A fixed version of the photochemical model (same horizontal and vertical resolution; identical input data preprocessors, chemistry, and removal modules; and so on) could be applied to all regions selected for the evaluation. The evaluation could span a wide range of meteorological conditions for the urban areas under consideration, corresponding to high-, moderate-, and low-ozone days. Such an evaluation would not test uniform changes in emission; instead, it would evaluate overall model performance for different spatial and temporal distributions, source strengths, and speciation of emissions.

Even if the problems of availability and quality of input data (emissions and aerometric) are solved, allowing one to evaluate a model's ability to simulate significant emissions changes, one must still account for the fact that the sensitivity of a photochemical modeling simulation to emissions changes will vary according to meteorology. Wagner and Wheeler (1989), reporting on sets of simulations performed by Tesche et al. (1988a,b) and Wagner (1988), concluded that "the location and amount of maximum sensitivity to emissions changes vary with the meteorology. This may mean that more than one episode should be used in evaluating the effects of emission changes upon peak ozone concentrations." Indeed, the selection of particular ozone episodes on which to design emissions controls can have a substantial effect on the projected control levels. It is therefore important to examine several episodes to determine the sensitivity of controls to meteorology.

SUMMARY

There are several classes of photochemical air-quality models (Table 10-5), and although grid-based models are the best for simulating atmospheric chemistry and transport processes, they require relatively large data bases, and many areas of the country do not have the resources to support their use. These areas will continue to rely on trajectory models like EKMA (empirical kinetic modeling approach) to determine ozone abatement strategies. Two questions attend the use of EKMA-type models and reduce the confidence that can be placed in guidance derived from them: First, what inaccuracies result if vertical heterogeneity is not included in the trajectory model? Second, to what extent is a trajectory model simulation an adequate representation of three dimensional airshed behavior? Some results suggest that EKMA-type trajectory models are too limited in their formulation to account for multiple-day episodes of high ozone concentrations. For example, application of EKMA-type models to Boston (Chang et al., 1989) and Philadelphia (Whitten et al., 1986) showed either little change or an increase in ozone concentrations in response to NO_x reductions. However, multiple-day ROM simulations (Possiel et al., 1990) found that NO_x reductions led to decreased ozone concentrations in both locations.

The model inputs needed to simulate historical ozone episodes—boundary and initial conditions, both on the ground and aloft, and emissions inventories—have associated uncertainties, often of a magnitude difficult to estimate. Compilation of simulations over many episodes and many regions indicates a general underprediction of peak ozone concentrations. The most consistent explanation for this behavior is a general underestimation of volatile organic compound (VOC) emissions. In some cases, however, the uncertainties in model inputs are large enough that the temporal and spatial features of ozone behavior can be reproduced by selection of the inputs within their ranges of uncertainty. Indeed, the "play" in inputs that has been used to improve ozone predictions might have compensated for an underestimated VOC emissions inventory.

TABLE 10-5   Classes of Photochemical Models

Model Type	Strengths	Limitations
EKMA (Empirical kinetic modeling approach)	Easy to apply, detailed chemistry, computationally rapid.	Lacks physical detail, short model simulations. Does not accurately simulate multiday events or long-range transport.
Urban, grid-based	Physically detailed. Suited for multiday modeling of urban areas (~ 400 km).	Computationally demanding. Sensitive to boundary conditions when long-range transport is important.
Regional	Physically detailed. Suited for studying regional areas (~ 1000 km).	Computationally demanding, limited spatial resolution. (ROM also has limited vertical resolution.) Not well suited to studying pollutant dynamics in cities.
Nested[a]	Advantages of both urban and regional models.	Computationally awkward and demanding. Information travels in one direction.
Multiscale[b]	Advantages of nested, urban, and regional models. Computationally straightforward.	Computationally demanding. In development.

[a]Nested models are models in which finer grid scales are embedded.
[b]Multiscale models are in effect nested models whose character on different scales may be different.

If so, this is a cause for concern when models are then used to determine degrees of VOC and NO_x control needed to attain the ozone National Ambient

Air Quality Standard (NAAQS). The critical question is "What is the effect of uncertainties in base-year inputs on projections of control levels for future years?" Once this question is answered, these uncertainties need to be incorporated in the State Implementation Plan. For example, it is important to know whether such uncertainties affect the choice of control strategies—for example, control of VOCs versus control of oxides of nitrogen (NO_x) versus both. Model performance evaluation procedures must be designed to reveal flaws in a base-case simulation to ensure that a model gives the right answer for the right reason.

Computational constraints historically have limited the use of advanced three-dimensional, photochemical air-quality models. Instead, less-comprehensive, less computationally intensive, and more limited models, such as EKMA have been used. Such models are not capable of fully characterizing ozone dynamics in urban and regional areas over multiple days, nor the response of ozone to emission changes. Rapid increases in computational power and algorithmic efficiencies now allow for more widespread use of advanced models, and are playing a significant role in the ability to understand atmospheric pollutant dynamics. The continued evolution of computational capability will allow for in-depth studies using more chemically and physically comprehensive models.

11

VOC Versus NO$_x$ Controls

INTRODUCTION

Although ozone concentrations in many areas of the United States violate the National Ambient Air Quality Standards (NAAQS), the circumstances of the violations can be quite varied. A few areas, such as Los Angeles, California, are isolated from regional influences, although most are not. Some have relatively high concentrations of VOCs (volatile organic compounds) compared with NO$_x$ (the oxides of nitrogen), whereas others do not. One goal of this report is to assess current understanding of the relative effectiveness of VOC versus NO$_x$ controls in ozone abatement in the United States. Knowledge of the atmospheric chemistry leading to ozone formation, together with the use of ozone isopleth diagrams (Chapter 6), provides a qualitative understanding of the relationship between ozone concentrations and VOC and NO$_x$ emissions. To actually evaluate the effectiveness of potential control strategies requires the use of photochemical air quality models (Chapter 10) that incorporate the best possible information about an area's initial and boundary conditions, emissions, and meteorology. In this chapter, we synthesize and assess much of the information from air quality models about the relative effectiveness of VOC and NO$_x$ controls in various regions of the country.

The most widely used method for determining ozone control requirements for urban areas has been the U.S. Environmental Protection Agency's Empirical Kinetic Modeling Approach (EKMA). The limitations of EKMA are discussed in Chapter 6: In practice, only periods of less than one day are simulated, and, as a result, the method cannot capture the multiday nature of

351

episodes of high concentrations of ozone. EKMA also simulates ozone formation only along a single trajectory, not providing any regionwide information
about the effects of controls.

The number of analyses carried out using grid-based air quality models is
limited but growing (see Chapter 10). Urban scale models, such as the urban
airshed model (UAM) and the CIT model, have been applied to a number of
cities in the United States and elsewhere. The Regional Oxidant Model
(ROM) has been applied recently to the northeastern United States and to
urban areas in that region.

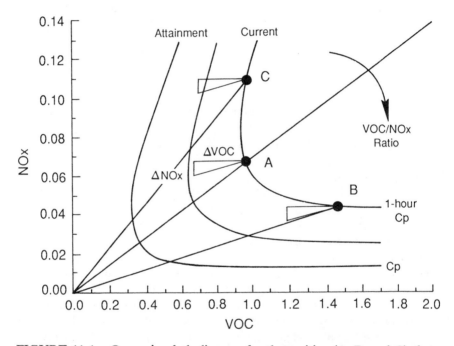

FIGURE 11-1   Ozone isopleth diagram for three cities (A, B, and C) that
have the same peak 1-hour ozone concentrations (Cp). The VOC/NO$_x$
ratios differ: a low ratio (C), a high ratio (B), and a medium ratio (A).
Isopleths = lines of constant 1-hour peak ozone.

EKMA-BASED STUDIES

EKMA is used to generate ozone isopleth diagrams for cities, and EPA and other agencies have used it to determine the fractional VOC and NO$_x$ reductions needed to meet the ozone NAAQS from particular base-year conditions. The ozone isopleth diagram, introduced in Chapter 6, shows the peak 1-hr ozone concentration in terms of the initial VOC and NO$_x$ concentrations. Figure 11-1 shows a hypothetical diagram for three cities that have the same peak 1-hr ozone concentration but for which the ambient ratios of VOCs to NO$_x$ differ: a low ratio (city C), a high ratio (city B), and a medium ratio (city A). As illustrated in Figure 11-1, at low VOC/NO$_x$ (city C), reductions in NO$_x$ can have little effect or actually can cause increases in ozone; for city B, reductions in NO$_x$ can lead to substantial decreases in ozone. At moderate VOC/NO$_x$ (city A), reductions in NO$_x$ can lead to small or moderate decreases in ozone, depending on the shape of the isopleth and the amount of NO$_x$ reduction. City A is located along what is often called the ridge line. If the molar ratio of carbon to NO$_x$ is greater than about 20 ppbC/ppb, NO$_x$ control is clearly more effective than VOC control, whereas at a ratio of about 10 or less, VOC control is more effective. At ratios between 10 and 20, control of either VOC or NO$_x$ or both might be preferred; specific situations must be carefully evaluated to determine the relative effectiveness of alternative abatement strategies (Blanchard et al., 1991).

Although the VOC/NO$_x$ ratio is a useful measure of the overall nature of the VOC-NO$_x$-ozone system, it is at best a qualitative measure of the reactivity of a given city's air because, as noted in Chapter 6, VOC/NO$_x$ ratios vary both spatially and diurnally in a given city and from one episode to the next for the same city. Variation among proximate cities is observed as one travels from west to east in the Los Angeles basin; the VOC/NO$_x$ ratio in the atmosphere varies from that of city C to that of city A and, as one goes sufficiently far east, to that of city B. In areas where the VOC/NO$_x$ ratio is between roughly 10 and 20, control of NO$_x$ may reduce the effectiveness of VOC controls. At a fixed level of VOC emissions, NO$_x$ control in such cases may cause ozone concentrations to decrease in downwind areas and increase in near-source areas. In some downwind areas, ozone concentrations may decrease less than they would have decreased if VOC emissions alone had been reduced (Blanchard et al., 1991).

Results of sample State Implementation Plans (SIPs) for various cities generated by EKMA are given in Table 11-1. In each case, only VOC control was considered. In general, the higher the original ozone concentration, the greater the VOC control predicted. (Biogenic VOC emissions are not accounted for in the calculations in Table 11-1.)

Chang et al. (1989) used EKMA to study the effect of conventional and methanol-fueled vehicles on air quality in 20 cities. In that study they calculated the effect that removing light-duty VOC emissions (primarily emissions

TABLE 11-1 Ozone Design Values, VOC Concentrations, VOC/NO$_x$, Mobile Source Emissions, and Estimated VOC control requirements[a,b]

Area	Ozone design value, ppb	Median VOC, ppbC	Median VOC/NO$_x$	On-road mobile source percent of emissions[c]		Required VOC % control to meet standard[d]
				VOC	NO$_x$	
Akron, Ohio	125	600	12.8	39	--[e]	--
Atlanta, Georgia	166	600	10.4	52	43	25 - 50
Boston, Massachusetts	165	380	7.6	51	47	35
Charlotte, North Carolina	149	390	10.4	52	--	25 - 50
Cincinnati, Ohio	157	740	9.1	42	40	> 50

355

City						
Cleveland, Ohio	145	780	7.5	49	36	25 - 50
Dallas, Texas	160	730	11.8	52	43	25 - 50
El Paso, Texas	160	670	11.9	66	--	25 - 50
Fort Worth, Texas	160	630	11.8	52	--	--
Houston, Texas	200	740	12.9	36	--	> 50
Indianapolis, Indiana	130	690	10.9	49	--	--
Kansas City, Missouri	130	410	8.5	50	--	--
Los Angeles, California	360	--	7.8	46	60	85
Memphis, Tennessee	146	127	13.9	48	--	25 - 50
Miami, Florida	130	103	13.3	61	--	25 - 50
New York, New York	217	--	9.6	47	40	--
Philadelphia, Pennsylvania	180	570	8.0	41	41	25 - 50
Portland, Maine	140	430	11.6	38	--	0 - 15
Richmond, Virginia	125	450	11.2	44	--	25 - 50
St. Louis, Missouri	160	570	9.6	43	23	25 - 50
Washington, D.C.	140	600	8.7	55	56	25 - 50

Wilkes Barre, Pennsylvania	125	430	14.3	43	--	--
Average of nonattainment areas	--	--	--	48	50	--

[a] Based on Chang et al. (1989), SCAQMD (1989), OTA (1989), EPA (1983), Systems Applications Incorporated (1990), E.H. Pechan (1990).

[b] "VOC" refers to volatile organic compounds excluding methane.

[c] Mobile-source percentage of total NO_x emissions primarily derived from EPA (1983) and might not correspond to same period used to derive mobile-source VOC percentage from Chang et al. (1989).

[d] Based on EKMA. A range indicates uncertainty in the amount of control needed.

[e] --No data were available or no calculation was made.

from automobiles and pickup trucks) would have on ozone concentrations, and how varying NO$_x$ emissions would affect these results. The response—the percentage of ozone reduction achieved in relation to a given percentage VOC reduction—is highly variable. The ratio of those two can be construed as the sensitivity of ozone formation to VOC emissions, and is a measure of the effectiveness of VOC control.

The sensitivity of ozone to VOCs for 20 cities is given in Table 11-2. The sensitivity is seen to correlate with the ambient VOC/NO$_x$ ratio in that higher sensitivities are associated with lower ratios. Biogenic emissions were not included in the calculation. At VOC/NO$_x$ ratios less than 8, reductions in VOCs were found to be particularly effective, and at higher ratios, the sensitivity slowly declines as the ratio increases. Carter and Atkinson (1989b) obtained similar results for air parcels in urban areas.

Most assessments of control strategies using EKMA have not considered the effects of biogenic VOCs. Chameides et al. (1988) argued that biogenic VOCs must be considered, particularly in southern cities where warm temperatures lead to significant emissions of isoprene. They showed that for Atlanta, anthropogenic VOCs need to be reduced by 30% to meet the NAAQS according to the standard EKMA calculation with no biogenic VOCs. Inclusion of isoprene increases the necessary reduction in anthropogenic VOCs to 70%. With inclusion of other biogenic hydrocarbons, ozone concentrations were predicted to exceed the NAAQS with no anthropogenic VOC emissions. Once isoprene is included, the percent reduction in NO$_x$ emissions needed to meet the NAAQS is less than the required reduction in VOC emissions.

Intercomparison of model results indicates that EKMA and other single- and double-layer trajectory models are too limited by their mathematical formulation and lack of physical detail to assess ozone control strategies accurately. A major shortcoming is that high-ozone episodes are multiday events, and EKMA simulations are generally less than one day long. NO$_x$ is removed from the photochemical system faster than the bulk of the VOCs, leading to more NO$_x$-limited conditions on subsequent days of an episode. Grid-based airshed models provide a much stronger foundation on which to build ozone control strategies. Because of EKMA's inherent limitations, our assessment of the relative effectiveness of VOC and NO$_x$ controls will focus on applications of grid-based models.

TABLE 11-2   Sensitivity of Ozone Formation to VOC Emissions

Area	Median VOC/ NO_x	Sensitivity to light-duty vehicle VOC emissions, $\Delta O_3 / \Delta VOC$[a]
Akron, Ohio	12.8	0.44
Atlanta, Georgia	10.4	0.56
Boston, Massachusetts	7.6	1.08
Charlotte, North Carolina	10.4	0.54
Cincinnati, Ohio	9.1	0.52
Cleveland, Ohio	7.5	0.92
Dallas, Texas	11.8	0.53
El Paso, Texas	11.9	0.54
Fort Worth, Texas	11.8	0.51
Houston, Texas	12.9	0.59
Indianapolis, Indiana	10.9	0.51
Kansas City, Missouri	8.5	0.67
Memphis, Tennessee	13.9	0.45
Miami, Florida	13.3	0.55
Philadelphia, Pennsylvania	8.0	1.45
Portland, Maine	11.6	0.43
Richmond, Virginia	11.2	0.49
St. Louis, Missouri	9.6	0.58
Washington, D.C.	8.7	0.64
Wilkes Barre, Pennsylvania	14.3	0.44
Average		0.62

[a] $\Delta O_3 / \Delta VOC$, ratio of percent reduction in ozone concentration to percent reduction in VOC emissions.
Source: Chang et al., 1989.

GRID-BASED MODELING STUDIES

Two areas of the United States—the Northeast corridor, which extends from the Washington, D.C. area to beyond Boston, and the Los Angeles basin—have received a large share of attention in the evaluation of ozone abatement strategies. Los Angeles has little influence from upwind sources, whereas each city in the Northeast corridor is affected by transport of ozone and precursors from upwind regions. In essence, the Northeast corridor acts as a system, and the effectiveness of ozone controls in one urban location will depend on controls throughout the region. Limited studies are available for other areas of the country, such as the Southeast and the Midwest.

Los Angeles Basin

The effects of controlling VOC and NO$_x$ emissions in the Los Angeles basin have been explored in a variety of studies, for example those of the South Coast Air Quality Management District (1989) and Milford et al. (1989). Basinwide control of VOC emissions was predicted in these reports to reduce ozone concentrations everywhere. Controlling NO$_x$ emissions was predicted to lead to increased ozone concentrations in the downtown and midbasin areas but decreased ozone concentrations in the far eastern portion of the region. Since the emissions inventories used in those studies apparently underpredicted VOCs, it is likely that a larger portion of the basin would respond favorably to NO$_x$ reduction than was predicted.

Studies of ozone abatement strategies in Los Angeles have used both the urban airshed model (UAM) and the CIT model (see Table 10-1). The UAM was used in developing the air quality management plan for the South Coast air basin (SCAQMD, 1989). The CIT model was used to determine the effects of VOC and NO$_x$ controls on ozone, nitric acid (HNO$_3$), nitrogen dioxide (NO$_2$), peroxyacetyl nitrate (PAN), and aerosol nitrate for an episode that occurred Aug. 30-31, 1982 (Russell et al., 1988a,b, 1989; Milford et al., 1989).

Milford et al. (1989) used the CIT Model to develop ozone isopleth diagrams across the Los Angeles Basin, showing how the effectiveness of NO$_x$ and VOC controls varies spatially (Figure 11-2). Likewise, they developed the isopleth diagram for peak ozone in the basin (Figure 6-4). In those figures, the base level of emissions corresponds to the upper right hand corner, and increasing levels of VOC and NO$_x$ control are plotted along the horizontal and vertical axes, respectively. Milford et al. (1989) found that NO$_x$ controls were most effective in the downwind regions, e.g., around San Bernardino.

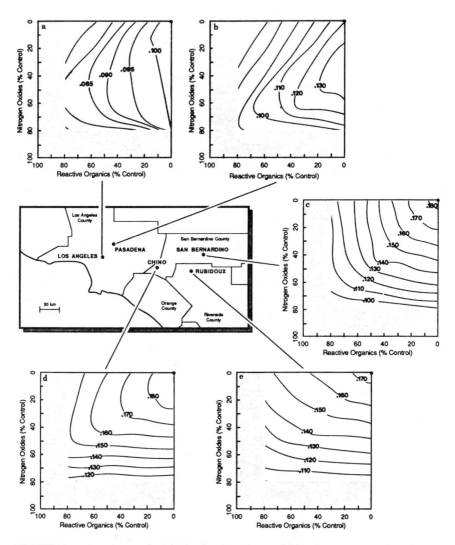

FIGURE 11-2   Ozone isopleths for locations within the Los Angeles air
basin from an airshed model for spatially uniform reductions of VOC
and NO$_x$. Source: Milford et al., 1989.

These regions also had the highest ozone concentrations. In the central regions, such as downtown Los Angeles and Pasadena, where peak ozone concentrations were lower, VOC controls were most effective, and NO$_x$ reduc-

tions could inhibit ozone reduction. Control of both VOCs and NO$_x$ led to basinwide reductions. Correction of the likely underestimation of VOC and CO emissions from motor vehicles would enhance the effectiveness of NO$_x$ controls.

The spatial variation in ozone response has been explained by Milford et al. (1989) based on the emission patterns. In Los Angeles and Pasadena, the VOC/NO$_x$ ratio is estimated to be between 5 and 10 at 9:00 a.m. Especially in Pasadena, this ratio does not increase dramatically by noon, apparently because of high local NO$_x$ emissions compared to downwind locations. Thus the situation corresponds to the region to the left of the ridge line in Figure 11-1, and NO$_x$ controls can lead to increased ozone. At the downwind locations, however, there are lower local emissions, and much of the NO$_x$ has been lost due to deposition and chemical transformations. The resulting VOC/NO$_x$ ratio is much larger (≥ 25), corresponding to the region to the right of the ridge line in Figure 11-1, where the chemistry is NO$_x$ limited, and hence NO$_x$ controls are most effective.

As noted in Chapter 6, a significant advantage of using grid-based airshed models to generate ozone isopleths is that these models show the effect of precursor controls on peak concentrations of ozone, regardless of where the peak occurs in the air basin. For example, in the studies of Milford et al. (1989), reducing NO$_x$ or VOCs by 25%, or each by 15%, shifts the location of the ozone peak westward from San Bernardino to Chino. Figure 11-3 shows the effect of VOC and NO$_x$ controls on peak O$_3$ in the Los Angeles air basin as a whole, i.e. irrespective of the location of peak ozone. A more L-shaped isopleth results. The isopleths in Figure 11-3 show that when the Los Angeles air basin as a whole is considered, up to 80% control of VOCs alone will not result in attainment of the NAAQS.

Tesche and McNally (1990) applied the UAM to the South Central Coast air basin, in the Santa Barbara area of California, for Sept. 5-7, and Sept. 16-17, 1984. They predicted ozone isopleths for this air basin which, like those of Milford et al. (1989) for Los Angeles, are more L-shaped than are the EKMA-type isopleths shown in Figure 11-1. As Milford et al. (1989) and Tesche and McNally (1990) pointed out, although the calculations are specific to the southern California area, the approach and issues involved (e.g. downwind areas) have general validity and applicability.

Nonlinearities in the response of ozone concentrations to emissions changes generally result in smaller ozone reductions than might be expected or desired from reducing emissions. For example, by the year 2000, mobile sources in Los Angeles are expected to account for about 30% of total VOC emissions. Airshed model calculations indicate that removing this fraction of VOCs would decrease peak ozone 16% from 270 to 230 ppb for the particular set of

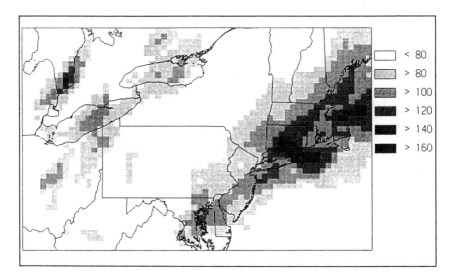

FIGURE 11-3a   Maximum predicted ozone concentration (ppb) over the
six-day simulation period for the model run with anthropogenic emissions only. Source: Roselle and Schere, 1990.

episode conditions studied (Russell et al., 1989). Exposure would decrease by
20%. Chang and Rudy (1989), using a trajectory model, found that eliminating mobile-source VOC emissions would result in a 10-15% reduction in peak
ozone concentrations—seldom to below 120 ppb. This has an important rami-
fication: even though mobile sources are the single largest source, their control alone will not solve the smog problem in urban areas.

Northeastern United States

High concentrations of ozone occur in the eastern United States concurrently across urban and rural areas that can span more than 1000 km. Concentrations often exceed 90 ppb in rural areas, and the greatest concentrations
(sometimes exceeding 200 ppb) are found downwind of the largest urban and
industrial centers. Episodes of such high concentrations of ozone are associated with the slow-moving high-pressure systems that provide weather conducive
to ozone production (see Chapter 4). These episodes occur several times each
year, usually between May and September, and ozone concentrations can stay
high from late morning into early evening for several consecutive days during

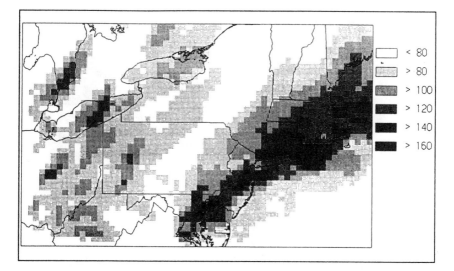

FIGURE 11-3b   Maximum predicted ozone (ppb) over the six-day simulation period, for the AB run, which contains both anthropogenic and BEIS biogenic emissions. Source: Roselle and Schere, 1990.

these episodes (e.g. Logan, 1989).

Pioneering studies by Hov et al. (1978) and Isaksen et al. (1978) showed that ozone can build up to concentrations exceeding 100 ppb in a few days in air subjected to anthropogenic emissions of NO_x and VOCs. The ozone can persist for several days, permitting long-range transport. More recently, Liu et al. (1987) examined the relationship between ozone and NO_x in detail, focusing on data from Niwot Ridge, Colorado, a remote site affected by urban plumes. They showed that ozone production per unit NO_x is greater for NO_x <1ppb than for NO_x >1ppb, and that the relationship between ozone and NO_x is nonlinear in the range of concentrations found in nonurban air, <0.3-10 ppb. The higher production rates at lower concentrations of NO_x imply that ozone is generated more efficiently in rural areas than in urban areas, and may explain why rural ozone concentrations are often similar to those in cities (Linn et al., 1988).

Several recent studies have shown that ozone in rural areas of the eastern United States is limited by the availability of NO_x rather than hydrocarbons, and that reductions in NO_x probably will be necessary to reduce rural ozone values (Trainer et al., 1987; Possiel et al., 1990; Sillman et al., 1990b; McKeen et al., 1991b). Trainer et al. (1987) examined the mechanisms responsible for high concentrations of ozone (110 ppb) observed at a rural site in Pennsylva-

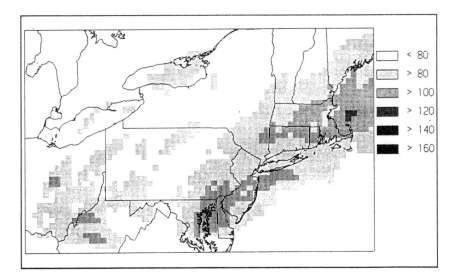

FIGURE 11-3c   The six-day maximum predicted ozone concentration
(ppb) for the run with Biogenic Emissions Inventory System (BEIS)
biogenic emissions and no anthropogenic VOC emissions ("A(NO$_x$)B").
Source: Roselle and Schere, 1990.

nia, where local NO$_x$ concentrations were only 1 ppb and isoprene was the
dominant VOC. Model simulations showed that oxidation of isoprene in the
presence of anthropogenic NO$_x$ could result in ozone concentrations of more
than 100 ppb. The addition of anthropogenic VOCs to the model made little
difference, because the chemical system is in the NO$_x$-limited regime. Trainer
et al. argued that reduction of NO$_x$ would be needed to reduce rural ozone
concentrations. Sillman et al. (1990b) examined the sensitivity of rural ozone
to NO$_x$ and VOCs for the range of values found in the eastern United States.
They found that rural ozone increases as NO$_x$ increases, when NO$_x$ < 2 ppb,
but is almost insensitive to anthropogenic VOCs. These conclusions rely only
on observations and the chemical mechanisms in the models and are independent of emissions inventories. Sensitivity studies with regional models using
the 1980 and 1985 National Acid Precipitation Assessment Program (NAPAP)
emissions inventories also demonstrated that NO$_x$ reductions will probably be
necessary to reduce rural ozone concentrations (Sillman et al., 1990a,b; Possiel
et al., 1990; Possiel and Cox, 1990; McKeen et al., 1991b). We focus below
on studies using the Regional Oxidant Model (ROM).

    ROM is the only regional model available for assessment of control strategies for urban and rural ozone in the eastern United States. It has the smallest grid size of the regional models discussed in Chapter 10 and has been evaluated in the greatest detail. ROM predicts the occurrence of high ozone concentrations and the spatial distribution of ozone reasonably well, but it systematically underpredicts the highest concentrations of ozone (Schere and Wayland, 1989).

ROM has been used in a variety of regulatory applications, primarily using ROM2.0 with the 1980 NAPAP inventory. These applications are summarized in Tables 11-3 and 11-4, which describe, respectively, the individual emission control strategies and the particular ROM simulations. Many of these simulations are discussed only in unpublished EPA reports. Here we describe the most recent applications to the Northeast, which demonstrate the role of biogenic VOCs in generating high ozone concentrations (Roselle et al., 1991) and the greater efficacy of NO$_x$ control compared with VOC control in most of the region (Possiel et al., 1990; Possiel and Cox, 1990). These applications use ROM2.0, discussed in Chapter 10, and ROM2.1. The more recent version of the model includes some changes in the way the wind data are processed, so that more reliance is placed on surface wind data than on upper air data in the lowest layer of the model. This leads to improvement in the westerly bias of the model, which was discussed in Chapter 10 (Schere, pers. comm., NOAA, Research Triangle Park, N.C., 1990).

Roselle et al. (1991) examined the sensitivity of ozone to biogenic emissions for the period July 12-18, 1980 (see Figure 11-3a,b,c). Anthropogenic emissions were taken from the 1980 NAPAP version 5.3 inventory, and biogenic emissions came from the latest inventory developed for EPA (Pierce et al., 1990). Total VOC emissions came about equally from anthropogenic and biogenic sources. Figure 11-4 shows predicted ozone concentrations for the Northeast for three scenarios: A, with anthropogenic VOCs, but without biogenics; AB, with both kinds of VOCs; and A(NO$_x$)B, with only anthropogenic NO$_x$ and biogenic VOCs. The charts show the maximum ozone concentrations for the six-day period. The combination of biogenic VOCs and anthropogenic NO$_x$ gives rise to ozone concentrations greater than 80 ppb in the entire Northeast corridor, the Ohio Valley, and most of the southern half of the region. There are large areas with ozone in the range of 100 ppb to 120 ppb. Biogenic emissions are generally higher in the south, and large point sources of NO$_x$ in the Ohio Valley are predicted to interact with the biogenic VOCs there to produce significant amounts of ozone. These results concur with the earlier analysis by Trainer et al. (1987) of data from rural Pennsylvania. The addition of anthropogenic VOCs leads to much higher ozone concentrations (120-180 ppb) downwind of major urban areas such as Detroit, Michigan; Pittsburgh, Pennsylvania; Cleveland, Ohio; and the Northeast corridor cities.

TABLE 11-3　　Emission Control Scenarios used with ROM

VOC0 Base emissions projected to 1987 including the effects of existing controls. Largest changes in VOC emissions; some smaller changes to NO_x emissions.

VOC1 Base emissions from 1980 reduced to include the effects of existing VOC controls in 1982 SIP. Only VOC emissions are affected.

VOC2 VOC1 plus additional reductions expected by 1995 due to the Federal Motor Vehicle Control Program (FMVCP).

VOC3 VOC2 plus additional reductions in the Northeast corridor[a], Pittsburgh, Cleveland, and Detroit.

VOC4 VOC3 including a higher level of reductions, up to 90% control in the Northeast corridor.

NO_x1 NO_x controls on utility boiler emissions from 1980 resulting in a 39% cut in utility emissions (11% cut in regionwide NO_x)

NO_x2 Utility-industrial boiler plus FMVCP NO_x controls only in Detroit (27% resulting cut in NO_x emissions) and the Northeast corridor (22% resulting cut). Overall 10% cut in regionwide NO_x.

NO_x3 Utility-industrial boiler plus FMVCP NO_x controls applied regionally. (This resulted in a 22% cut in NO_x emissions for the Northeast corridor and a 27% cut in regionwide NO_x for the northeastern United States domain.)

NO_x4 NO_x3 except $\approx$ 70% cut in point-source NO_x emissions in the Northeast corridor.

---

[a]Northeast corridor extends from Washington, D.C. to Boston and beyond.

TABLE 11-4   ROM simulations

Episode	Base case or control strategy	Model	Purpose[a]	Seasonal extrapolation[b]
Northeastern U.S. Domain				
Aug. 3-4, 1979	Base case	ROM1.0	R,E	
	VOC0	ROM1.0	R	
April 14-29, 1980	Base case	ROM2.0	R	✓
	VOC1	ROM2.0	R	✓
	VOC2	ROM2.0	R	✓
	VOC3 + No_x3	ROM2.0	R	✓
July 12-17, 1980	Without biogenic VOC emissions	ROM2.0	S	
July 12-26, 1980	Base case	ROM2.0	R	✓
	VOC1	ROM2.0	R	✓
	VOC2	ROM2.0	R	✓
	NO_X1	ROM2.0	R	
	NO_X2	ROM2.0	R	
	VOC2 + NO_X3	ROM2.0	R	✓
	VOC3 + NO_X3	ROM2.0	R	
	VOC4 + NO_X3	ROM2.0	R	
	VOC4 + NO_X4	ROM2.0	R,S	
	With TSDF[c] emissions			
July 12-Aug. 31, 1980	Base case	ROM2.0	E	
July 22-31, 1980	Base case	ROM1.0	R	
	VOC0	ROM1.0	R	
July 26-28, 1980	Base case	ROM1.0	S	
	with other grid sizes	ROM1.0	S	

Episode	Base case or control strategy	Model	Purposeª	Seasonal extrapolationᵇ
August 22-31, 1980	Base case	ROM2.0	R	✓
	VOC1	ROM2.0	R	✓
	VOC2	ROM2.0	R	✓
	NO$_x$1	ROM2.0	R	
	NO$_x$2	ROM2.0	R	
	VOC2 +	ROM2.0	R	
	NO$_x$3	ROM2.0	R	✓
	VOC3 +	ROM2.0	R,S	
	NO$_x$3 With TSDF emissions			
Southeastern U.S. Domain:				
April 14-29, 1980	Base case	ROM2.0	R	✓
	VOC1	ROM2.0	R	✓
	VOC2	ROM2.0	R	✓
	VOC2 + NO$_x$3	ROM2.0	R	✓
June 29-July 14, 1980	Base case	ROM2.0	R	✓
	VOC1	ROM2.0	R	✓
	VOC2	ROM2.0	R	✓
	VOC2 + NO$_x$3	ROM2.0	R	✓
August 10-September 1, 1980	Base case	ROM2.0	R	✓
	VOC1	ROM2.0	R	✓
	VOC2	ROM2.0	R	✓
	VOC2 +	ROM2.0	R	✓
	NO$_x$3 With TSDF emissions		R,S	

ªE, Model evaluation study, S, model sensitivity analysis, R., regulatory analysis.

ᵇ✓, This simulation was used, along with others, to extrapolate episodic model results to a full season.

ᶜTSDF, treatment, storage, and disposal facility.

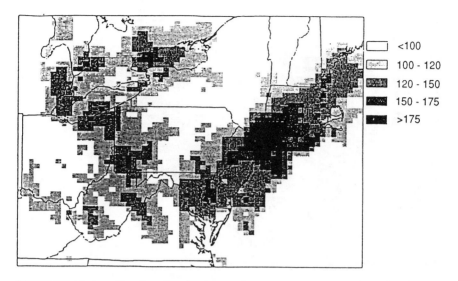

FIGURE 11-4a   Predicted episode maximum ozone concentrations (ppb) for the 1985 base case (July 2-17, 1988). Source: Possiel et al., 1990.

Possiel et al. (1990) examined the effects of proposed regional control strategies on ozone in the Northeast for July 2-17, 1988, the most severe episode in this region between 1980 and 1988. The model was run for a base case of 1985. Its emissions data were taken from the 1985 NAPAP inventory, adapted for the above-average temperatures that prevailed during the episode, and from Pierce et al. (1990) for biogenic emissions. The target year was 2005, with projected emissions that accounted for existing federal and state controls. With these controls, anthropogenic VOC emissions were 20% lower than in 1985, carbon monoxide emissions were 43% lower, and total NO_x emissions were the same. Several other control scenarios were also applied to the 2005 calculation.

Results for the 1985 base case and for the 2005 case with existing controls are shown in Figure 11-4a,b in terms of maximum ozone concentrations for the episode. Predicted reductions in peak concentrations ranged from 5-10% in and downwind of most major source areas, to as much as 20% in New York City. For both simulations, ozone concentrations exceeded 120 ppb r and downwind of all major source regions. Changes in ozone relative to the 2005 case with existing controls are shown in Figure 11-5 for two scenarios: In one scenario (Figure 11-5a), VOC emissions were reduced throughout the United States, leading to a 45% reduction; other emissions were at levels

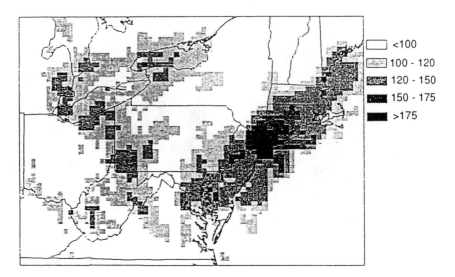

FIGURE 11-4b   Predicted episode maximum ozone concentrations (ppb)
   for the 2005 case with existing controls (July 2-17, 1988). Source: Possiel et al., 1990.

assuming existing controls. In the other scenario (Figure 11-5b), anthropogenic VOCs were reduced by 49% in the Northeast corridor and by 26%
elsewhere, with corresponding reductions in NO$_x$ of 26% and 34%, respectively. The reduction in VOCs alone produced the greatest effect north of
Philadelphia, with reductions in peak ozone of as much as 25-50% in the
immediate area of New York City. There was little change elsewhere, including most of New England and the southern part of the corridor. The predictions for the combined NO$_x$-VOC strategy were quite different. Peak ozone
was reduced by 10-15% across much of the domain, and the reductions were
generally greater than with the VOC-only strategy. In the New York City
area, however, the combined controls were less effective in reducing ozone.

The frequency distribution of maximum ozone concentrations was also
examined for the two scenarios. The combined strategy was more effective
in the Washington-Baltimore area, Philadelphia, and Boston; the VOC-only
strategy was more effective in New York City. There was little difference
between the two strategies in Connecticut. The combined strategy led to
decreased population exposure in all regions of the corridor except the New
York City area, but 43% of the corridor's population lives there. Even with
the control strategies, ozone concentrations were predicted to exceed 120 ppb

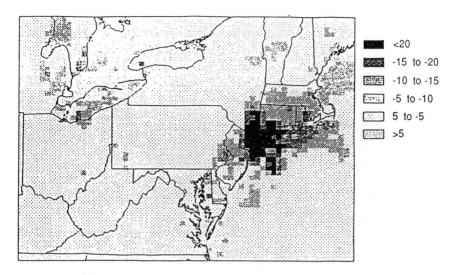

FIGURE 11-5a   Percentage change in episode maximum ozone concentrations, 2005 base case versus a VOC-alone reduction strategy (July 2-17, 1988).

along the Northeast corridor.

Possiel and Cox (1990) examined another set of control scenarios for the period July 2-17, 1988; they showed results for NO$_x$ control alone, VOC control alone, and simultaneous NO$_x$ and VOC control, all relative to a somewhat different base case for 2005 emissions (see Table 11-4). These scenarios, based on application of "maximum technology," resulted in reduction of NO$_x$ by 58% and anthropogenic VOCs by 63% compared with the 2005 base case. Total VOC emissions were reduced by 40% within the Northeast corridor but by only 20% outside the corridor, because of the preponderance of natural emissions there. Control of NO$_x$ alone caused large reductions in ozone throughout most of the U.S. portion of the model domain, including the Northeast corridor. The exception was New York City, where ozone increased (see Table 11-5). Outside the Northeast corridor, control of VOCs alone led to ozone concentrations 15-25 ppb higher than did control of NO$_x$ alone. However, the VOC-only strategy was more effective in lowering peak ozone in New York City. VOC control resulted in a much larger area with ozone above 120 ppb than did NO$_x$ control. Combined control of NO$_x$ and VOCs reduced ozone outside the Northeast corridor only slightly more (<5%) than did NO$_x$ control alone. Within the corridor, the combined controls were

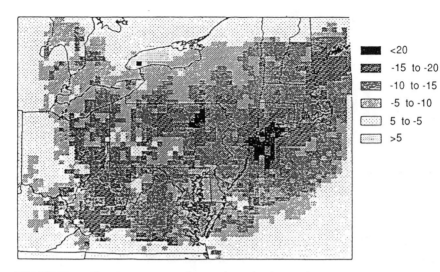

FIGURE 11-5b   Percentage change in episode maximum ozone concentrations, 2005 base case versus a combined NO_x-VOC reduction strategy (July 2-17, 1988). Source: Possiel et al., 1990.

more effective than either alone, except in New York City, where VOC-only control was most effective.

McKeen et al. (1991b) also found that control of NO_x was more effective than control of VOCs in reducing peak ozone values across most of the eastern United States and that control of NO_x alone led to increases in ozone in a few areas of high NO_x emissions. Their model had a grid size of 60 km and simulated a different meteorological period. Nevertheless, this model gave results similar to those of ROM for similar scenarios.

The results from ROM for the effect of a NO_x-VOC versus VOC-only strategy agree with analyses based on much simpler models (Sillman et al., 1990a,b), as do the studies of the role of biogenic VOCs (Trainer et al., 1987; Chameides et al., 1988; McKeen et al., 1991b). The results from ROM emphasize that a combined NO_x-VOC strategy should be more effective than a VOC-only strategy in reducing ozone over a large geographic area in the Northeast. A VOC-only strategy would be more effective in some areas of high population density (New York City) but less beneficial downwind. Although the general nature of these results is likely to be correct, the details of the predictions should be viewed with caution, because of known deficiencies in the base-case emissions inventories (see Chapter 9) and because of the possible deficiencies in the model itself, as implied by evaluation studies (Schere and Wayland, 1989).

New York Metropolitan Area

Rao and co-workers (Rao et al., 1989; Rao and Sistla, 1990) have applied the UAM to study the control of ozone in the New York City area. Rao and Sistla (1990) studied imposition of 75% control on NO$_x$, VOCs, or both (Table 11-6). Control of NO$_x$ alone decreased ozone concentrations in Connecticut and New Jersey but increased them in New York. Ozone exposure above 120 ppb increased. VOC-only control led to substantial reductions in ozone in all three regions and to a concomitant reduction in exposure, and combined controls were less effective in reducing ozone than were VOC controls alone. These results indicate that VOC controls are necessary to reduce ozone concentrations in the New York area, although downwind areas can benefit from NO$_x$ reductions. It also was found that biogenic VOC emissions alone, in concert with emissions of anthropogenic NO$_x$, would lead to ozone concentrations above 120 ppb. The sensitivity of the peak ozone and exposures can be estimated from these results and are given in Table 11-6. The sensitivities to VOCs are for the anthropogenic portion only and are generally less than those found by Chang et al. (1989) for other cities using EKMA. The results of Chang et al. (1989) also showed that the effects of NO$_x$ and VOC controls are not additive.

In accordance with the UAM study by Rao and Sistla (1990), peak ozone concentrations as predicted by ROM fell in New York in response to VOC controls but not NO$_x$ controls (Possiel and Cox, 1990). Virtually all other cities in the ROM domain responded favorably to control of VOCs or NO$_x$ or both. ROM results indicate that eight-hour exposure to ozone would decrease in all cities, including New York, when NO$_x$ controls are imposed in addition to VOC controls. This is contradictory to the UAM results discussed above. NO$_x$ reductions led to a regionwide decrease in ozone exposure.

A shortcoming of current regional air quality models is that they do not have the spatial resolution required to accurately assess the chemical transformation and transport within urban areas. A solution to this problem is to embed, or "nest" a model with finer spatial resolution. For example, an urban model can be embedded in a regional model. The regional model then prescribes transport into and boundary conditions for the urban domain. A UAM-ROM interface has been developed to serve this purpose (Rao et al., 1989). In this case, the nesting is one way; information flows from ROM to UAM, but not back.

The results of the nested-grid study are interesting in that they compare

TABLE 11-5 Ozone Response in Northeast to VOC and NO$_x$ Controls Found Using ROM[a,b]

Area	Base[c,d]	63% VOC control[c]	58% NO$_x$ control[c]	Combined VOC, NO$_x$ controls[c]	Sensitivity of peak ozone to VOC	Sensitivity to NO$_x$
Washington,D.C.	140/87	132/74	122/77	113/74	0.09	0.22
Philadelphia, Pennsylvania	143/92	129/85	116/74	111/73	0.16	0.32
New York, New York	234/107	140/87	257/107	163/83	0.63	-0.17
Rhode Island	138/85	120/78	112/68	105/66	0.20	0.32
Boston, Massachusetts	137/84	121/80	103/67	100/67	0.18	0.43
Pittsburgh, Pennsylvania	136/80	126/77	103/66	101/66	0.12	0.41
Detroit, Michigan	120/76	112/74	104/70	102/69	0.11	0.22

[a] In parts per billion.
[b] From Possiel and Cox (1991).
[c] First value is 90th percentile 1-hr ozone, and second value is peak 8-hr average.
[d] Base case refers to the year 2005.

TABLE 11-6   Effect of Controls on Ozone in New York[a]

	Peak Ozone (second day), ppb		Exposure to ozone > 120 ppb (1000 population hours)
	New York	Connecticut	
Base Case	167	173	28,127
75% VOC control	144(0.18)[b]	136(0.28)[b]	10,418(0.84)
75% NO$_x$ control	185(-0.14)[b]	154(0.14)[b]	33,120(-0.23)
75% VOC and NO$_x$ control	149	138	20,598

[a]Derived from Rao and Sistla (1990).
[b]Sensitivity of peak ozone and ozone exposure shown in parentheses.

model calculations of a regular (nonnested) simulation with various nesting procedures (Table 11-7). Predicted ozone concentrations vary considerably depending on the kind of nesting employed. Although Rao et al. (1989) did not test how these variations affected estimates of necessary control levels, the scatter in the predictions indicates that the calculated effect of controls would differ greatly depending on which nesting procedure is used.

SUMMARY

Application of grid-based air quality models to various cities and regions in the United States shows that the relative effectiveness of controls of volatile organic compounds (VOCs) and oxides of nitrogen (NO$_x$) in ozone abatement varies widely. Most major cities experience ozone concentrations that exceed the National Ambient Air Quality Standard (NAAQS) one-hour concentration of 120 ppb—a result of the density of precursor emissions in those areas. The predominant sources of emissions are mobile, although other sources contribute significantly. These cities share an ozone problem, but differ widely in the

TABLE 11-7   Comparison of Nesting Techniques for Peak Ozone Predictions[a], ppb

	New Jersey	New York	Connecticut
Observed	145	240	303
Regular UAM[b]	180	199	202
Nested, ROM[c] ICs, BCs, winds	155	138	148
Nested, ROM ICs, BCs	156	143	167
Nested, ROM BCs	126	259	233

[a]From Rao et al. (1989), for July 21, 1980.
[b]UAM, urban airshed model, without nesting.
[c]ROM, Regional Oxidant Model, and refers to the ROM supplying initial conditions (ICs), boundary conditions (BCs), windfields.

magnitude of the problem and in the relative contributions of anthropogenic VOCs and NO_x and biogenic emissions. As a result, the optimal set of controls relying on VOCs, NO_x, or, most likely, reductions of both, will vary from one place to the next. In cities where the VOC/NO_x ratio is high, VOC control provides less ozone reduction per unit of VOC reduction than in cities with a low VOC/NO_x ratio. Cities with a high VOC/NO_x ratio benefit from NO_x control, but less so if the ratio is low. Studies have predicted that in some areas—downtown Los Angeles and New York City, for example—ozone will increase in certain locations (not necessarily those where the peak occurs), if NO_x emissions are lowered.

Few urban areas in the United States can be treated as isolated cities unaffected by regional sources of ozone. The regional nature of the ozone problem east of the Mississippi, as demonstrated by ozone observations (see Chapter 4) and model simulations (e.g., Roselle et al., 1990); McKeen et al., 1991a) requires the use of regional models for assessment of control strategies. The Regional Oxidant Model (ROM) has been the major tool used for regulatory studies of areas affected by regional ozone. A significantly greater effort needs to be devoted both to understanding the reason's for the model's failures and to further developing the model itself. The present regional models do not have sufficient spatial resolution for detailed studies of major urban areas, and a nested model approach is likely to be necessary. A fully interactive, two-way nested multiscale model is desirable for studying intercity and regional pollutant transport. Ozone air quality is intimately related to other air-quality issues, such as acid deposition and visibility, and a comprehensive modeling system with high spatial resolution is ultimately necessary.

Biogenic VOCs, in combination with anthropogenic NO$_x$, are capable of generating ozone concentrations above 80 ppb in favorable meteorological conditions across much of the eastern United States, with values of more than 100 ppb downwind of a number of major cities. Future assessments of control strategies must include biogenic emissions, given their potential for generating ozone concentrations close to the (NAAQS) concentration.

Many simulations conducted to date have relied on emissions inventories that are suspected of significantly underestimating anthropogenic VOC emissions (see Chapter 9) and that have not included biogenic emissions. The result is an overestimate of the effectiveness of VOC controls and an underestimate of the efficacy of NO$_x$ controls (Chameides et al., 1988; McKeen et al., 1991b). Underestimates in the VOC inventories might be partly responsible for the underprediction of ozone concentrations in central urban areas (SCAQMD, 1989; Rao et al., 1989). The consequences of an underestimate in the VOC inventories on predicted concentrations of ozone and its precursors and on control strategies must be investigated.

Even with the limitations of present models and emissions inventories, certain robust conclusions emerge when the modeling studies are synthesized. Production of ozone is limited by the availability of NO$_x$ and is much less sensitive to anthropogenic VOCs in most rural environments in the eastern United States, where NO$_x$ concentrations are less than ~ 2 ppb and the VOC/NO$_x$ ratio is high. Control of NO$_x$ is also effective in lowering peak ozone concentrations in many urban areas, although it is predicted to lead to an increase in ozone in some places, such as downtown Los Angeles and New York City. The ozone increases in these urban cores, however, are predicted to be accompanied by decreases in ozone downwind, in the Los Angeles basin and Connecticut, respectively. While control of VOCs never leads to a significant increase in ozone, there are many areas where control of VOCs is either ineffective or does not bring an area into compliance with the NAAQS. Hence NO$_x$ control will probably be necessary in addition to or instead of VOC control to alleviate the ozone problem in many cities and regions. The optimal set of controls of NO$_x$, VOCs, or both will vary from one region to another, as discussed above.

12

Alternative Fuels

INTRODUCTION

Current estimates show that automobiles and trucks account for about 45% of the anthropogenic VOCs (volatile organic compounds), 50% of the NO_x (oxides of nitrogen) (Figures 12-1, 12-2, 12-3, 12-4), and 90% of the CO (carbon monoxide) in cities where the ozone NAAQS is not met (OTA, 1989). This is by far the largest category of emissions, and recent studies indicate that these estimates could be low for VOCs and CO (Pierson et al., 1990; Lawson et al., 1990b). So, although strategies to improve air quality should consider all sources, it is clear that reductions in automotive emissions are necessary. Some reductions will occur as older, dirtier vehicles are taken off the road and replaced with new, well-controlled vehicles. However, these reductions will be small, and by the year 2004, the increase in vehicle use is expected to lead to an increase in VOC emissions (OTA, 1989). A longer term solution is necessary. One possibility is using alternative fuels. However, the use of alternative fuels alone will not solve the ozone problems of the most severely affected areas. Moreover, it will not necessarily alleviate the most critical problem associated with motor vehicle emissions, which is the increase in emissions that occurs over time with in-use vehicles (see Chapter 9).

The central role of VOCs in the formation of tropospheric ozone suggests that changing fuels could be effective by reducing the reactivity of the VOC emissions, the total mass emitted, or both. Gasoline and the exhaust from conventionally fueled vehicles are highly reactive in the atmosphere because

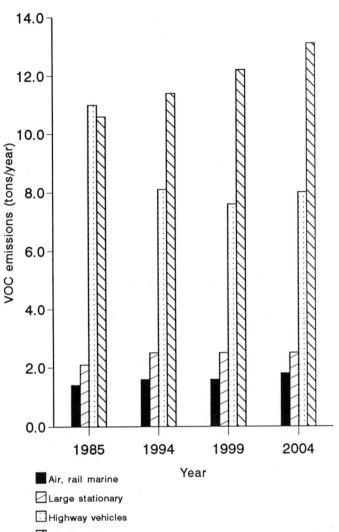

FIGURE 12-1   Estimated nationwide VOC emissions by source category, by
year. Assumes no regulations other than those in place in 1987. The estimates are representative of the emissions on a typical nonattainment day,
multiplied by 365 days per year, rather than estimates of true annual emissions. The baseline does not include reductions due to the limit on gasoline
volatility of 10.5 psi Reid vapor pressure (RVP). Stationary sources that
emit more than 50 tons per year of VOC are included in the "Large" category. Source: OTA, 1989.

they are rich in aromatics and alkenes. Alternatively fueled vehicles, such as those that run on natural gas or methanol, would have less reactive emissions and hence reduce ozone. There is also the possibility of reducing the mass emission rates of VOCs, CO, and NO_x when these fuels are used (Austin et al., 1989; Williams et al., 1989). Electric vehicles would run virtually without emissions. As discussed elsewhere in this report, the relative benefits of VOC and NO_x controls vary by location, and this phenomenon is critical to determining the benefits (or lack thereof) of alternative fuel use. For example, reducing the reactivity of VOCs can reduce ozone in NO_x-rich urban centers, such as downtown Los Angeles and New York, but would likely provide little help in regions with high VOC-to-NO_x ratios, such as Houston and Atlanta. The degree to which the different alternative fuels could improve air quality is discussed below. It should be noted that the choice of alternative fuel cannot be made on environmental considerations alone, nor are different locations going to be affected similarly. Economics, politics, energy security and diversity issues, consumer acceptance, technological advances, and the relative prices of fuels will enter into the choice, but such considerations are beyond the scope of this report. Concern over global warming and the need to improve air quality suggest that the use of alternative fuels, and the choice of those fuels, will need to be considered thoroughly and carefully.

FUEL CHOICES

A variety of alternative fuels and technologies are or could become available for automotive use, including natural gas; methanol (and methanol blends); ethanol (and ethanol blends), liquid petroleum gas (LPG), including propane; hydrogen; electricity; and reformulated gasoline. Although much interest has focused on use in light-duty vehicles (cars and light trucks), most of the fuels also can be used in heavy-duty vehicles. Different technologies also exist to use these fuels: the traditional internal combustion engine (using organic fuels and hydrogen) and electric motors coupled with battery storage or fuel cells (running on hydrogen or methanol). Each of the fuels and the associated technologies will have various environmental effects and could be viable at different times in the future.

In the near term (0-5 years) the only alternative fuel that can effect ozone concentrations is reformulated gasoline, whose potential is uncertain. No new technology or distribution system is required, although refining capacity for these fuels is limited. Most major oil companies have environmentally improved, reformulated gasolines for sale in limited markets, targeting those urban areas with the most severe air-quality problems. These fuels are formu-

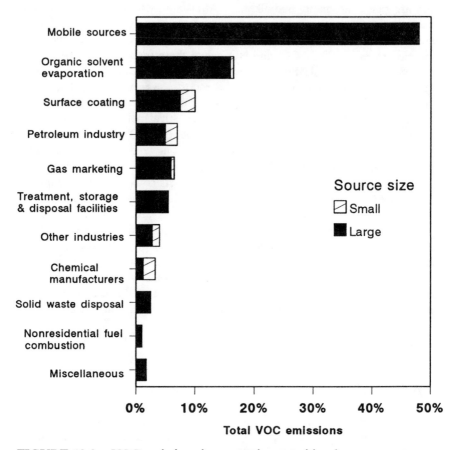

FIGURE 12-2   VOC emissions in nonattainment cities, by source category, 1985. Stationary sources that emit more than 50 tons per year of
VOCs are included in the "Large" category. Total emissions, 11 million
tons/year. Source: OTA, 1989.

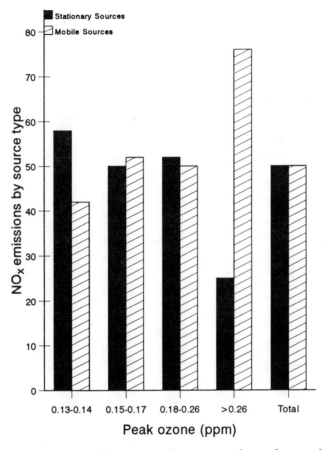

FIGURE 12-3   NO$_x$ emissions an peak concentrations of ozone in non-attainment cities, 1985. Source: Adapted from OTA, 1989.

lated to have lower emissions of highly reactive or toxic organic compounds. These fuels are new, and their ability to improve air quality is not clear. Reformulated gasoline is obtained from refined petroleum. As such, it is not considered a true alternative, but it is discussed here because it is viewed as a candidate means for improving air quality.

In the middle term (5-20 years), likely candidates to replace gasoline- and diesel-powered vehicles are similar vehicles powered by internal combustion engines that run on methanol, natural gas, or reformulated gasoline. Electric vehicles are another middle-term possibility.

It is unclear which fuels and technologies have the greatest potential or will

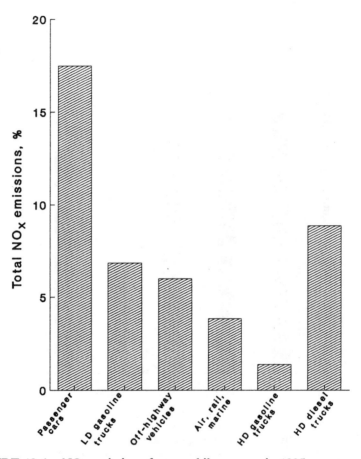

FIGURE 12-4 NO_x emissions from mobile sources in 1985 as a percentage of total (mobile plus stationary) emissions. LD refers to light duty;
HD refuers to heavy duty. Source: OTA, 1989.

even be viable for the long term.   Reformulated gasoline, methanol, and
natural gas would still be available.   It is clear that there are ample supplies
of both petroleum and natural gas worldwide (Sperling, 1991).   Other fuels,
in particular hydrogen and electricity, are more attractive from an environmental standpoint.   Advances in battery technology could make electric vehicles an attractive alternative to vehicles with internal combustion engines.

      Because many variables affect which fuel is best, it is important to consider the attributes of each. Below, the environmental attributes of each fuel are discussed, followed by a summary of the studies that have looked at their effects on air quality. Results of the air-quality studies are then used to estimate the relative effectiveness of each fuel in lowering ozone. For each of the fuels, there are significant uncertainties in the likely environmental benefits. Still, some general conclusions can be developed, and these are discussed at the end of the chapter.

ATTRIBUTES OF ALTERNATIVE FUELS

Reformulated Gasoline

As the name implies, reformulated gasoline is a refined petroleum product whose composition is similar to gasoline. It can be used in conventional engines with no modification, although achieving the greatest air-quality benefits could require modification of the automotive control system. The composition of the gasoline is altered to make exhaust products less photochemically reactive and toxic and to lower total emissions. This is accomplished by modifying the refining process and by adding oxygenates. The resulting fuel is lower in olefins and aromatics, and has a lower Reid vapor pressure (RVP). (RVP is the constrained vapor pressure of the fuel at 100°F.) Oxygenates serve two purposes: First, they enhance the fuel's octane rating, which is lowered when the aromatic and olefin content is reduced. Second, they can improve the efficiency of combustion, and the presence of fuel oxygen tends to decrease CO emissions.

Commonly used oxygenates include ethers, particularly methyl t-butyl ether (MTBE) and ethyl t-butyl ether (ETBE), and alcohols (methanol and ethanol). The trend is to use the ethers, although ethanol is used extensively in the Midwest, especially to reduce CO emissions in the winter. The addition of alcohols can raise the vapor pressure of the gasoline, increasing evaporative emissions and decreasing the fuels' positive benefits during the summer. Oxygenates also can increase the rate of formaldehyde (HCHO) emissions (Anderson et al., 1989). Oxygenates may comprise a few to more than 15% of the reformulated fuel.

Reformulated gasolines have been introduced in limited markets—those that have severe photochemical smog problems. The potential for using reformulated gasoline to improve air quality is uncertain. First, reformulated gasolines are new, and their compositions could evolve further. Second, it needs to be determined to what degree automobile control systems and fuels can be matched to lower emissions; this would likely require standardization of fuel composition. Still to be fully explored is the possibility of reducing total mass emissions of VOCs and NO_x by using different blends of reformulated gasolines. A major effort, the Auto/Oil Air Quality Improvement Research Program, is under way to help accomplish these goals (Burns et al., 1991).

Natural Gas

Natural gas is primarily methane (>90%), with other light hydrocarbons, including ethane, ethene, propane, propene, and butane, as impurities. For use as an automotive fuel, it is compressed and stored at pressures up to 30 megapascals (4500 psi), or liquified. Because liquified natural gas (LNG) requires cryogenic cooling and storage, it has not been used as much as compressed natural gas (CNG). (In this section, reference is made to natural gas vehicles [NGVs.]. Unless stated otherwise, these are vehicles that use either CNG or LNG. For the most part, the advantages and disadvantages of the two fuels are similar.)

VOC emissions from NGVs mimic the fuel, and are largely methane (Table 12-1). Given its very low atmospheric reactivity, methane has great potential for reducing ozone formation. However, impurities in natural gas can greatly reduce its benefit. Ethane and propane contribute up to 25 times as much ozone on a mass basis (grams ozone/grams VOC), as does the less reactive methane, and the presence of alkenes would lead to even less benefit (Carter, 1990b). Assuring a low alkene content of the natural gas is crucial to achieving the maximum benefit of NGVs. Also, the products of incomplete combustion (aldehydes) are highly reactive, although the mass emission rates of these species are likely to be small (Alson, 1988; Austin et al., 1989; CARB, 1989a,b).

Emissions of CO from NGVs, which are usually operated under lean-burn conditions (i.e., more air is used than is required for complete combustion of the fuel), are generally much less than from gasoline-powered vehicles (Alson, 1988; Austin et al., 1989). Tests consistently show reductions on the order of 90%. In addition to lowering ambient CO, the decrease in CO emissions lowers, slightly, the ozone formation resulting from the vehicles. This is because CO oxidation produces hydroperoxyl (HO_2), resulting in increased nitric oxide (NO) oxidation. It is difficult to state how NO_x emissions will compare in an optimized NGV engine, and tests show results in both directions (Alson, 1988; Austin et al., 1989). Meeting the NO_x emission standard limits the ability to use lean-burn engines in NGVs or methanol-fueled vehicles. Both fuels can be burned very lean, leading to increased fuel economy and lower engine-out VOC, CO, and NO_x emissions. (Engine-out emissions are those

TABLE 12-1 VOC Composition of Exhaust and Evaporative Emissions from Gasoline (Indolene) and Alternative Fuels[a,b]

VOC	Indoline	Methanol[c]		Ethanol	Liquid petroleum gas	Compressed natural gas[d]
		M85	M100			
Alkanes	0.632	0.224	0.023	0.077	0.797	0.170
Alkenes	0.040	0.007	0.001	0.002	0.062	0.031
Formaldehyde	0.021	0.067	0.050	0.010	0.041	0.023
Aldehydes	0.004	0.004	0.001	0.050	0.005	0.005
Ethene	0.031	0.005	0.001	0.034	0.082	0.017
Toluene	0.199	0.032	0.009	0.023	0.007	0.007
Aromatics	0.059	0.023	0.005	0.010	0.003	0.014
Methyl ethyl ketone	0.015	0.005	0.001	0.002	0.003	0.009
Methanol	0	0.633	0.911	0	0	0
Ethanol	0	0	0	0.791	0	0

[a]Compositions given in decimal form.

[b]The tests relied on very few vehicles, and the individual tests are not likely to be representative of the actual fleet of vehicles operating on each fuel.

[c]M85, 85% methanol by volume; M100, 100% methanol.

[d]The composition of exhaust is primarily methane, and the composition shown reflects that about 70% of the VOC is nonreactive. Methane comprises a much smaller fraction of the exhaust when other fuels are used.

Source: CARB, 1989a.

released prior to catalyst reduction by a catalytic converter.) However, the exhaust is so oxygen-rich and CO-lean that the performance of the NO_x reduction catalyst is very poor, leading to higher exhaust NO_x emissions. Although it could be possible to reduce the rate of emissions to 0.4 g NO_x/mile, estimates indicate that California's standard of 0.2g NO_x/mile would not be met using a lean-burn engine (Austin et al., 1989). Given the recent findings that further NO_x reductions are effective for reducing ozone, using a lean-burn engine fueled by natural gas (or methanol) could be counterproductive in many environments.

One advantage of NGVs, from an air-quality perspective, is that the emissions from a "super-emitting" NGV are not as likely to increase to the same levels as those from a super-emitting conventional vehicle. Engine-out emissions of VOCs, CO, and NO_x from NGVs are substantially lower than from conventional vehicles; studies indicate that malfunctioning vehicles account for 10% of the conventional fleet and are responsible for 60% of fleet CO emissions (Lawson et al., 1990a). A second inherent advantage is that evaporative emissions from NGVs would be very small (if any) and only slightly reactive. This could be a substantial advantage if it turns out that evaporative emissions are the reason for the discrepancy between the inventories and ambient measurements discussed in Chapter 8. Refueling emissions also would be small.

Methanol

Methanol and methanol blends have recently been the subject of great interest. Methanol fuel is primarily methanol, although significant amounts (up to 15%) of other compounds are added for safety and performance reasons.

Pure methanol is a colorless, toxic liquid with low vapor pressure and high heat of vaporization. Although the latter two attributes can lower emissions when an engine is warm, they also make it difficult to start a vehicle running on pure methanol when cold, and high cold start emissions ensue. Also, the flame from a pure-methanol-based fire is nearly invisible in daylight, which may be a safety problem. Additives raise the vapor pressure, help cold start, and add color to the flame. The most common additive is unleaded gasoline, from 5 to 15% by volume. The resulting blends are designated by an "M" followed by the volume fraction of the methanol. Thus, M85, the most common blend, is 85% methanol by volume, but there is no standard blend. Dedicated methanol-fueled vehicles (MFVs) and flexibly fueled vehicles have been developed. The flexibly fueled vehicles are designed to operate on any combination of gasoline and M85 fuel, and in this way they ease the transition during periods when methanol is not widely available.

On a mass basis, unburned methanol is the largest component of MFV exhaust whether M85 or M100 fuel is used (Table 12-1). The second most abundant species of methanol exhaust is formaldehyde (HCHO). Gasoline-like hydrocarbon components comprise 35-50% of the exhaust of M85-fueled vehicles (Horn and Hoekman, 1989; Snow et al., 1989; Gabele, 1990). HCHO is often cited as the most deleterious species emitted by MFVs because of its toxicity and high photochemical reactivity. Engine-out HCHO emissions are substantial, up to 600 mg/mile, and to derive any air-quality benefits the catalyst system must remove most of that. Tests of current vehicles show that after catalytic reduction, M85 vehicles emit 30-40 mg HCHO/mile—three to six times as much as is emitted by a conventional vehicle (Horn and Hoek-man, 1989; Gabele, 1990), although amounts as high as 180 mg/mile have been reported for older (pre-1989) cars and trucks (Snow et al., 1989). Recent tests show that M100-fueled vehicles have 25% higher HCHO emissions than do vehicles with similar control systems fueled by M85 (Gabele, 1990). This is attributable to the poor cold start when using M100. During cold start, the catalyst takes about 30 seconds to light off (become hot enough to cause VOC oxidation), and engine-out HCHO is readily emitted during the transient warming. The high HCHO emissions, if not properly controlled, could negate any potential advantage of MFVs.

Cold start is responsible for a bulk of the emissions from MFVs, and various technologies are being studied to help solve this problem. One promising direction is a more closely coupled catalyst that is electrically heated before or during start-up. Tailpipe emissions of HCHO of as little as 4 mg/mile have been measured from an M100 vehicle with an electrically heated catalyst (Hellman, et al., 1989). This is a 55% decrease compared to emissions using an unheated catalyst, and this amount of HCHO emissions is comparable to that from a gasoline-fueled vehicle, although it has not been established that these results can be sustained in vehicles with higher mileage (more than 20,000 miles, for example). The resistively heated catalyst also decreases methanol, although NO_x emissions increase. Tests consistently show that CO emissions from MFVs are lower than from their conventionally fueled counterparts (Williams et al., 1989; Gabele, 1990), because of the higher oxygen content of the fuel. Reductions of 50% could be realized. This is, however, a direct function of the control system used, and might not be realized in practice. This amount of CO reduction would lead to substantial decreases in ambient CO and slight decreases (0-2%) in ozone.

It is not clear whether methanol use would lead to lower NO_x emissions. Limited tests on flexibly fueled vehicles show some NO_x reductions, but a dedicated MFV would use a higher compression ratio, which could lead to higher NO_x emissions. It is likely that the NO_x emissions would be catalytically controlled to meet emission standards, and there might be no substantial change in net NO_x emissions.

Evaporative emissions from MFVs would be less reactive than those from a conventional vehicle. It is also possible that the mass emissions rate would be lower. To minimize evaporative emissions, the RVP of gasoline in California and in the northeastern United States is mandated to be less than 9.0 psi. Pure methanol has an RVP of 4.6 psi, and M85 has an RVP of 7.5-8.0 psi (Horn and Hoekman, 1989; Gabele, 1990). The lower RVP results in lower emissions. However, as the gasoline content of a methanol blend increases, so does the RVP, until the mixture is almost pure gasoline at which point RVP decreases rapidly (Figure 12-5). Thus, although using M100 or M85 could decrease evaporative emissions, a vehicle using mostly gasoline and small amounts of methanol would have higher evaporative emissions than would a vehicle running on pure gasoline. This suggests the need for some method to ensure that the fuel composition in an average commuter's tank is a high methanol blend, or that the evaporative control system is built to handle 11 RVP fuel. Dedicated vehicles would not have this problem.

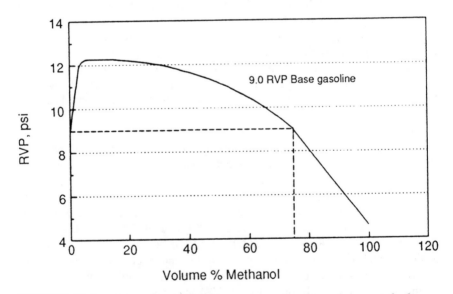

FIGURE 12-5   Approximate Reid vapor pressure dependence on fuel composition. From Black, 1991.

Ethanol

Ethanol's use would be similar to methanol's, and flexibly fueled vehicles built to use methanol also can run on ethanol. The differences are in the sources of ethanol and in its smaller air-quality benefits. Ethanol is slightly more reactive than methanol, and it leads to peroxyacetyl nitrate (PAN) formation. Carter (1991) reports that ethanol produces about 15% more ozone than methanol does on a carbon-atom basis. There is considerably less information on the emission characteristics of well-controlled ethanol vehicles, and the limited tests show high emissions (CARB, 1989b). It should be noted that using ethanol as a blending agent in gasoline and simultaneously allowing for an increase in RVP would not achieve significant air-quality benefits, and in fact would likely be detrimental. The increased evaporative emissions and increased concentrations of ethanol and acetaldehyde in the atmosphere would lead to increases in such pollutants as PAN (Tanner et al., 1988). Ethanol is more reactive than methanol and other blending agents such as MTBE (Carter, 1990b, 1991; Japar et al., 1990)

Hydrogen

Hydrogen is the cleanest fuel that can be used in an internal combustion engine. In the longer term, the possibility of producing hydrogen from nonfossil electricity (solar, nuclear, or hydropower) is seen as a nearly emission-free source of transportation fuel. Hydrogen produced from solar or hydropower would be renewable as well. Hydrogen-powered internal combustion vehicles would not be entirely clean. They emit NO_x (DeLuchi, 1989) and tiny quantities of carbon dioxide, CO, and VOCs (from burning oil).

LPG and Propane

LPG, which is primarily propane, shares many of the attributes of compressed natural gas, with several disadvantages. The supply of LPG is limited, and it is a petroleum-refining byproduct (Sperling, 1988). Also, it would not provide as great a reduction in exhaust reactivity as CNG would (Carter, 1990b; Chang and Rudy, 1990). An advantage of LPG over CNG is its higher energy per unit volume, so a smaller fuel tank is required. LPG is not expected to be a major alternative transportation fuel outside of the current limited applications.

Electricity

Electricity-powered vehicles are the cleanest of the alternatives. Some of the benefits are obvious: Electric vehicles produce virtually no on-road emissions, and there is a wide variety of potential energy sources, including fossil fuels, nuclear and solar energy, and hydropower.

Use of electric vehicles would not eliminate all smog-producing emissions. In the near term, production of electricity would come largely from fossil fuel power plants that emit NO_x, oxides of sulfur, and small quantities of VOCs and CO. These plants would contribute to urban or regional ozone formation. Depending on the source (coal, gas, nonfossil), the net change in NO_x emissions could be a small increase (coal) or large decrease (nonfossil). Emissions of oxides of sulfur would likely increase. Direct emissions of particulate matter would increase, although the total attributable to increased electricity production has not been determined. Reductions in NO_x would lower particulate nitrate, and increases in oxides of sulfur would increase particulate sulfate. Net CO and VOC emissions would be reduced by 95-99% (Krupnick et al., 1990; Sperling and DeLuchi, 1989).

Alternative Fuels
For Heavy-Duty Vehicles

Most of the current discussion of alternative vehicles centers on light-duty vehicles (cars and small trucks), even though use in heavy-duty vehicles could be of more immediate benefit. There are demonstration fleets of alternatively fueled heavy-duty vehicles. Diesel vehicles are notorious for their particulate emissions and also have high emissions of HCHO, polynuclear aromatic hydrocarbons (PAHs), and NO_x. Use of either methanol or CNG would virtually eliminate the particulate and PAH emissions. Evidence also suggests lower NO_x emissions by a factor of about two. Heavy-duty vehicles, especially buses, are more commonly centrally fueled than are light-duty vehicles, making transition to an alternative fuel easier.

ALTERNATIVE FUELS AND AIR QUALITY

Knowledge of the emissions characteristics of the different alternative fuels can be used to compare air-quality benefits. Very few detailed, extensive studies of any of the various fuels have been completed so far because interest in these fuels has only recently been renewed and the emissions data are extremely sparse (especially for vehicles under typical driving conditions and with high mileage). To date, the most extensively studied fuel is methanol, followed by electricity. There is some recent information on the potential air-quality effects of using reformulated gasolines, and extremely limited information on natural gas, LPG, and hydrogen.

Each of the fuels is targeted for its ability to reduce ozone, although all of them will affect concentrations of other pollutants. Accordingly, ozone has been the focus of most studies, and accompanying effects are sometimes given. Because ozone formation is the result of a complex interaction of photochemistry, transport, and emissions, the primary tools for testing the effects of alternative fuels are photochemical air-quality models; box, trajectory, and grid-based models have been used. In addition, smog chamber experiments have been used to test mathematical models. The results of air-quality studies are discussed below starting with those for methanol, for which the most information is available.

Methanol

Interest in using methanol as either a fuel additive or a base fuel arose in the late 1970s in response to the energy crisis. Since then, several studies, using mathematical models or smog chambers, have focused on the response of ozone to use of methanol (Bechtold and Pullman, 1980; O'Toole et al., 1983; Whitten and Hogo, 1983; Pefley et al., 1984; Balentine et al., 1985; Nichols and Norbeck, 1985; Jeffries et al., 1985; Carter et al., 1986b; Whitten et al., 1986; Chang et al., 1989; Chang and Rudy, 1989, 1990; Russell, 1989, 1990; Russell et al., 1989, 1990; Dunker, 1990). Although these studies generally agree that methanol use would decrease ozone levels, the amount of reduction predicted by the studies varied.

Experimental Studies

Carter et al. (1986b) performed a series of methanol-related smog chamber experiments in a 6400-liter indoor Teflon chamber (ITC) and in a 50,000-liter outdoor Teflon chamber (OTC). Compounds similar to those in a very dirty urban atmosphere were introduced into a chamber, and the pollutant concentrations were followed for periods of up to three days. This was done for pollutant mixes corresponding to current emissions and also for mixes corresponding to methanol-fueled engine emissions replacing one-third of the base mixture. On the first day, peak ozone levels were significantly lower for the methanol emissions mix than for the conventional emissions mix. However, this difference decreased over time, and by the third day ozone concentrations were similar. This result raised a question about the effects of methanol during multiday smog episodes. It was thought that methanol could build up, negating any benefits, and that a large fraction would react within three days. This study also showed that the ozone decreases were sensitive to the amount of formaldehyde emissions. In an earlier study, researchers at the University of North Carolina (Jeffries et al., 1985) conducted 29 smog chamber experiments that used methanol with varying fractions of formaldehyde. One-third of the base organic mixture was replaced with methanol or methanol-formaldehyde mixture, either a 90:10 $CH_3OH/HCHO$ ratio or an 80:20 ratio. Ozone reductions varied widely, from zero to 80%, depending on the HCHO content and amounts of VOC emissions.

Differences between smog chambers and the atmosphere make it difficult to use smog chamber results directly for predicting urban and regional air-quality changes that would result from methanol use. One difference is that a smog chamber does not replicate atmospheric diffusion and transport of chemical species. A second difference is that for the experiments conducted, all pollutants were present at the beginning of the smog chamber experiments. Fresh pollutants, however, are emitted throughout the day into the atmosphere. Consequently, the smog chambers had very low NO_x concentrations on days two and three of the simulations, making ozone concentrations relatively insensitive to changes in VOCs. (This is also pertinent to understanding the benefits of reducing VOC in NO_x-limited regions, such as rural and downwind areas.) A third difference is that even a very large smog chamber has a ratio of surface area to volume that is many orders of magnitude greater than the atmosphere's, and surface reactions have a much larger effect on pollution formation in the smog chamber than they do in the atmosphere. These experiments did, however, highlight issues that needed to be addressed further—especially the need for multiday simulations—and they have been crucial to the understanding of the chemical system.

Mathematical Modeling Studies

Atmospheric simulations of methanol have been accomplished by adapting existing photochemical models to include methanol chemistry. The chemistry of methanol is relatively simple, and it has been treated explicitly by addition of the following reaction,

$$CH_3OH + OH \xrightarrow[k_{meth}]{O_2} HCHO + HO_2 + H_2O \qquad k_{meth} \approx 1200/ppm-min$$

(1)

to the existing chemical mechanism. The input emissions are then adjusted accordingly to approximate the change to methanol-fueled vehicles. Results of these studies are summarized in Table 12-2. Most of the early studies used single-day trajectory model simulations. Early studies that concentrated on Los Angeles showed relatively large reductions in ozone when methanol is substituted for conventional fuels (O'Toole et al., 1983; Whitten and Hogo, 1983). Later studies (e.g., Russell et al., 1990) showed less benefit from using methanol. This difference is due in part to the use in the earlier studies of trajectory or box models, which were very sensitive to initial conditions and had limitations in their formulation (see Chapter 10). Russell et al. (1989) found significant differences in the results of a trajectory model and grid model applied to the same period for methanol use. Much of the model response, or lack thereof, is the result of treatment of initial conditions. As noted above, the Carter et al. (1986b) study that used a smog chamber found that the relative ozone reductions on the second and third days of multiday experiments were less than on the first day. High ozone episodes are multiday events, and carryover of pollutants from one day to the next is critical. This too explains, in part, the higher ozone reductions found in early studies.

Several studies by Russell and co-workers (Russell, 1989; Russell et al., 1989, 1990) use a grid-based airshed model to look at the air-quality effects methanol could have in the Los Angeles basin in two future years: 2000 and 2010. Russell et al. (1990) found peak ozone reductions of 9-17% in 2000, depending on the fuel type simulated, and of about 4% in 2010. In these simulations all light- and medium-duty vehicles, and some heavy-duty vehicles, were assumed to be converted. The 2010 ozone peak occurred far east of central Los Angeles and was found to be NO_x-limited and relatively insensitive to methanol emissions. Central basin ozone was lowered by about 15%. Ozone exposures were predicted to be reduced by 12-20% in both years. When coupled with reductions in stationary-source emissions, a later study by Russell (1990) found that a 50% penetration of vehicles fueled with M85 would lead to a 9% reduction in peak ozone and a 19% reduction in population exposure. The reason for the larger reduction in exposure in both studies is that the population is more concentrated in regions with higher NO_x concentrations, and ozone formation in those areas is more sensitive to VOC reactivity than is the ozone peak. Despite the limitations of trajectory models,

TABLE 12-2 Summary of Peak Ozone Reductions from Modeling Studies

Application area	Model type[a]	Modeling period	Vehicles substituted	Base year[b]	Composition of exhaust[c]	Peak ozone reduction in percent	Reference
Los Angeles	EKMA	1 day	All gasoline	1987	100/0/0 90/10/0 90/20/0	31 22 13	Whitten and Hogo, 1983
Los Angeles	Multilevel trajectory	1 day	All gasoline	2000	57/17/26	14	O'Toole et al., 1983
Los Angeles	Box	1 day	All gasoline	1987	90/10/0	18	Pefley et al., 1984
20 Cities	EKMA	1 day	Light-duty[f] gasoline	1982	100/0/0 90/10/0	13[d] 3.5[d]	Nichols and Norbeck, 1985
Philadelphia	Trajectory and others[h]	1 day	Mobile sources	2000	100/0/0 110/6/0	10 7	Whitten et al., 1986

20 Cities	EKMA	1 day	Light-duty gasoline	2000	90/10/0 M85[f] M100[g] 100/0/0	1.9[e] 1.3[e] 2.6[e] 2.9[e]	Chang et al., 1989
4 Cities	EKMA	1 day	Light-duty gasoline	2000	M85 M100	2 3	Chang et al., 1989
Los Angeles	Airshed	3 days	Post-1990 mobile sources	2000 2010	M85[i] M100[i] M85[i] M100[i]	9.3/13[j] 17/20[j] 4/12[j] 4.3/19[j]	Russell et al., 1990
Los Angeles	Airshed	3 days	Clean Air Act[k]	2010	M85	9/19	

Specifics of each study can be found in the cited reference. This tabulation is a summary of those findings and follows the format of Chang et al. (1989).

[a]Mathematical models used include box, trajectory, and grid-based photochemical models. Some studies used more than one, in which case the most representative results are used.
[b]Base year is the year of emissions inventory used.
[c]Composition of exhaust is shown as percentage of nonmethane VOCs substituted by methanol, formaldehyde, or hydrocarbons.
[d]Average over 20 cities. Range was 1-36%.
[e]Average of four cities.
[f]A complex mixture simulating M85 exhaust measurements used.
[g]Mixture simulating M100 exhaust used.
[h]Results based on trajectory modeling.
[i]Complex mixture of methanol and other organics used. Synergistic effects on evaporative emissions

calculated. Formaldehyde emissions were 15-23 mg/mile from light-duty vehicles. Only post-1990 vehicles assumed to use methanol.

^{j}The two values reported are the reductions in peak ozone followed by the reduction in exposure to ozone concentrations over 120 ppb.

^{k}In this case, the 2010 base inventory included proposed emission reductions from stationary sources. The methanol case looked at the effect of introducing 300,000 vehicles per year (about half the fleet), as was proposed as part of the Clean Air Act amendments. The resulting act did not include specific targets for methanol fuel use.

^{l}Refers primarily to passenger vehicles weighing less than 3,750 lb.

the results of the other studies are not terribly inconsistent with the grid-based modeling studies considering the different treatment of emissions.

Ozone reduction from methanol use in most other areas is not expected to be as great as it is for Los Angeles. This is attributable to higher VOC-to-NO_x ratios in other cities and to lower traffic-related emissions. The Ford studies (Nichols and Norbeck, 1985; Chang and Rudy, 1989) show average reductions of 1-3% (depending on fuel) in the year 2000, although a few cities (Pittsburgh, Pennsylvania, for example) are predicted to see more substantial ozone reductions. Total removal of light-duty VOC emissions led to only a 7% ozone reduction in the cities studied. However, those calculations assumed that mobile sources in the year 2000 will account for only 13% of the total VOC, on average, or about one-third of the current mobile-source contribution. This is probably a low estimate. These results, along with those of Russell et al. (1990) and Dunker (1990), indicate that in NO_x-rich areas, vehicles fueled with M100 would contribute 45-75% as much ozone as would equal mass emissions from conventional vehicles, and M85 vehicles would contribute 70-80% as much. In VOC-rich areas, little effect would be expected.

Russell et al. (1989) also looked at the effect that methanol substitution would have on other species. It was speculated that the increased direct emissions of (HCHO) from methanol fueled vehicles could lead to unacceptably high ambient concentrations of HCHO. It was found, however, that ambient HCHO concentrations change very little, and in some cases decrease, when methanol use is simulated. This is because most atmospheric formaldehyde is formed photochemically as the product of VOC oxidation (Grosjean, 1982; Rogozen and Ziskind, 1984). Methanol's low reactivity slows atmospheric production, offsetting the increase in direct emissions. Direct emissions of HCHO from methanol-burning vehicles could constitute a problem in closed areas (parking garages and tunnels) under extreme circumstances; CO buildup would be a similar problem in those cases (Machiele, 1987; Gold and Moulis, 1988). Chang and Rudy (1989) found that eye irritation in some individuals may result in the most severe tunnel exposures. A similar study is under way for parking garages. Ambient exposure to HCHO from methanol-fueled vehicles would be a small fraction of the total individual exposure (Gold and Moulis, 1988). Unlike many compounds found in gasoline, methanol is not a precursor to PAN or to higher organic nitrates, and those compound concentrations were predicted to decrease by 25% (Russell et al., 1990). Predictions indicate that the lower reactivity of methanol would slow oxidation of nitrogen compounds, leading to reductions in nitrogen dioxide (NO_2), nitric acid (HNO_3), and particulate nitrate of 20-40%. Other particulate components, such as carbon and sulfate, would decrease when methanol is used in place of diesel fuels. In the study by Russell et al. (1989), the use of methanol in place of diesel fuel was found to be particularly attractive because of the reduction in particulate matter and NO_x emissions. It is expected that benzene concentrations, which are due predominantly to gasoline, also would decrease when methanol fuel is used (Gabele, 1990). Methanol-fueled vehicles will lead to increased atmospheric concentrations of methanol, although the predicted concentrations are likely to be less than the level of concern (e.g., the threshold limit value) (Chang and Rudy, 1990; Russell et al., 1990). Also, these vehicles will likely emit formic acid (Smith, 1982), which would contribute to acid deposition.

Methanol's low reactivity led to concern that it could be transported long distances and cause high ozone or formaldehyde concentrations downwind. Calculations by Russell (1990) and Sillman and Samson (1989) show that this is not the case. Reasons for this finding include the high VOC-to-NOx ratio during long-range transport and the slow atmospheric production of HCHO from methanol compared with other VOCs.

Both the modeling and experimental results indicate that methanol use can improve air quality by lowering nitrogen dioxide, aerosols, and some toxics (e.g., benzene), as well as ozone, although the expected ozone reductions are modest, especially from M85 fuels. If flexibly fueled vehicles are used, air-quality benefits likely will be achieved only if M85 or purer fuel is used consistently. If most vehicles were running on a more dilute blend (say M50 or M25), increased evaporative and other organic emissions could lead to increases in ozone. Use of methanol in heavy-duty applications has promise for reducing particulate matter and also could reduce ozone. Increased research on technologies to reduce methanol and HCHO emissions effectively is necessary to obtain these potential benefits.

Reformulated Gasoline

One outcome of the consideration of alternative transportation fuels has been an interest in how reformulating the composition of gasoline may improve air quality. This effort has been spearheaded by some individual companies, as well as the Auto/Oil Air Quality Improvement Research Program (AQIRP) (Burns et al., 1991). Ultimately the AQIRP provide extensive information about the air-quality impacts of reformulated gasoline and methanol blends, but only limited information is available at this time. In particular, only the predicted impacts of a limited variation in gasoline composition are available as used in 1989-model-year vehicles (Auto/Oil Air Quality Improvement Research Program, 1991; Hochhauser et al., 1991).

The AQIRP involves testing the effect of varying fuel compositions on emissions composition and quantity, and modeling the changes in air quality that would accompany use of the various fuels. In particular, the AQIRP is investigating changes in aromatic, olefin, sulfur, and oxygenate (MTBE) content and in the fuel vapor pressure. A suite of late model (1989) light-duty vehicles are run on these reformulated fuels, and the resulting exhaust and evaporative emissions are measured for composition and mass. These test results are then used to specify input data for air-quality models. An EKMA-type model is being used for screening purposes, and a grid model is used for more in-depth analysis. The grid model is being used to simulate air-quality impacts in Los Angeles, New York, and Dallas.

Airshed model results to date show that changing fuel composition can have an effect on air quality, due to changing the mass of emissions of VOC, NO_x and CO, as well as the reactivity of the VOC. In future year simulations (2005 for New York and Dallas, 2010 for Los Angeles), peak ozone concentrations were predicted to be reduced only from about 1% (Dallas) to 3% (Los Angeles) in response to gasoline changes. These small changes, however, are about one-fourth of the predicted peak ozone reduction due to the complete removal of all light-duty vehicle emissions. Increasing vapor pressure and olefinic content were predicted to lead to increased ozone peaks and exposure. The aromatic and MTBE content did not have as a marked effect on predicted ozone.

Studies of more extensively reformulated gasolines than those tested in the AQIRP have found greater potential ozone decreases (Boekhaus et al., 1991; DeJovine et al., 1991). For one advanced reformulated gasoline reductions in VOC mass emissions of 31% were found and the reactivity of the organics also decreased leading to an effective "reactivity-adjusted" VOC reduction of 39%. NO_x and CO also decreased by 26% each.

Air-Quality Benefits of Other Fuels

There have been few detailed studies of the likely air-quality benefits of alternative fuels other than methanol and reformulated gasoline. Carter's studies (1990b, 1991) provide the ability to estimate the likely benefits of using ethanol-based fuels, natural gas, and LPG. Krupnick et al. (1990) and Hempel et al. (1989) have conducted detailed studies of electric vehicle use in Los Angeles.

Carter's relative reactivity measure indicates that switching to ethanol-based fuel (both E85 and E95) would provide small decreases in reactivity, as compared with gasoline or diesel fuel. However, the ethanol-fueled vehicles tested had greater mass emission rates. (The CARB [1989a,b] vehicle tests relied on very few vehicles, and the individual tests are not likely to be representative of in-use vehicles optimized to run on ethanol.) The mass-weighted emissions are predicted to increase ozone compared with emissions from conventional vehicles. LPG emissions would lead to 47% as much ozone, and CNG about 25% as much, after accounting for reactively differences and reductions in mass emissions. Evaporative emissions from LPG and CNG vehicles are expected to be small and unreactive. When weighted by the mass emission rates measured, LPG vehicles would contribute about half as much ozone as conventional vehicles, and NGVs would contribute about one-fourth as much.

It has been proposed that gasoline using ethanol as a blending agent (gasohol) would be allowed to have an increased vapor pressure, and hence increased evaporative emissions. Given the minor reactivity reduction of ethanol compared with gasoline, the increased evaporative emissions would negate much of the ozone reduction, and actually could lead to increased ozone production. This is similar to the effect resulting from use of low methanol blends in flexibly fueled vehicles (Figure 12-3).

Although Carter and co-workers have not specifically considered the effects on production of pollutants other than ozone, the atmospheric chemistry of the emitted compounds provides much of the information needed to assess the likely response of other compounds. Acetaldehyde is the atmospheric oxidation product of ethanol, and it is also relatively abundant in ethanol-fueled vehicle emissions. Acetaldehyde is a precursor to peroxyacetyl radical and PAN formation. Increased PAN levels can be expected, as has been noticed in Brazil (Tanner et al., 1988; Grosjean et al., 1990b).

Natural Gas and LPG

NGVs and LPG vehicles emit relatively small amounts of formaldehyde. Given the low reactivity of their emissions, it is likely their use would reduce atmospheric HCHO and PAN concentrations. It also is reasonable to expect slower NO oxidation to NO_2 and HNO_3, lowering those species' concentrations. Emissions of toxics (such as benzene) also would be substantially less from NGVs and LPG vehicles than from conventional vehicles.

Electricity

Electric vehicles are the cleanest of the alternative vehicles. Their emissions are essentially displaced to centralized electricity-generating stations.

Utility boilers emit, in comparison to internal combustion engines, very little VOCs and CO. NO_x emissions depend on the type of fuel and control technology used. Krupnick et al. (1990) used an airshed model to find the benefits of using 500,000 to 1.5 million electric vehicles in Los Angeles (1.5 million vehicles would be about 17% of the light-duty fleet). Peak ozone reductions of as much as 4.1% were found. This study also showed that using electric vehicles would lead to almost three times the reduction as that from using M85 vehicles. A study by Hempel et al. (1989) found similar results, assuming a much larger portion of electric vehicles in fleet.

Of the alternatives, electric vehicles are the only vehicles that, unequivocally, will lead to large NO_x reductions in urban areas. Regional effects are less certain and will depend on how electricity is produced. Increased NO_x emissions from electricity-generating stations could lead to increased regional ozone that can be transported into cities. The expected change in NO_x emissions is an important consideration, given that mobile sources are the dominant source of NO_x, and recent studies show that ozone formation in some areas is NO_x-limited (Chameides et al., 1988; Milford et al., 1989). Also, concentrations of particulate matter and organic nitrates can be effectively reduced by lowering NO_x emissions (Russell et al., 1988b; Milford et al., 1989).

Incremental VOC Reactivity

A recognized limitation of the studies conducted so far is the uncertainty about the emission composition of exhaust from conventional and methanol-fueled vehicles in the future. The studies discussed above use forecast emissions data or data from limited measurements of prototype methanol-fueled and flexibly fueled vehicles. The limited availability of emissions data has led to an alternative approach to estimating air quality that is applicable not only to methanol but to the other fuels as well. Instead of using a simulated exhaust mixture, the sensitivity of the ozone-forming potential of increased emissions of individual compounds is predicted (Carter and Atkinson, 1989b; Carter, 1991; Russell, 1990). Then the contributions of each emission species, multiplied by the fraction of that species in the exhaust, can be added to find the net emissions reactivity. This allows the flexibility of rapidly estimating impacts of future vehicles as the data become available. This also can guide manufacturers and regulators toward what constitutes a cleaner fuel and how fuels compare. A shortcoming of this approach is that it does not fully account for the nonlinearities in the chemistry of ozone formation.

Carter (1991) and Carter and Atkinson (1989b) used an EKMA-type model

with a chemically explicit mechanism to find the incremental reactivity (IR)
of individual organic compounds. They define IR as

$$IR = \lim_{\Delta ORG \to 0} \frac{R(Base + \Delta ORG) - R(Base)}{\Delta ORG} \approx \frac{\partial R}{\partial ORG} \quad (2)$$

R is a measure of the ozone formation and ΔORG is the change in the organic gas input. Carter and Atkinson (1989b) defined R as the maximum of the
difference between the ozone concentration and NO concentration during the
simulation (see also Chapter 5). In essence, this is the local sensitivity of
ozone formation to emissions of specific organic gases, and it is given as the
maximum number of moles of ozone formed by a one-mole increase of carbon in VOCs. A summary of their calculation results is given in Table 12-3
for some species typically emitted from alternate and conventional vehicles.
Note that the incremental reactivities of many compounds are sensitive to
VOC/NO_x. Some, such as toluene, have a negative sensitivity at high $VOC/
NO_x$ because of NO_x and radical scavenging by oxidation products. Chang
and Rudy (1990) found similar results from a similar study with a less explicit
mechanism. An important finding of both studies was that the maximum
incremental reactivity of the organics was found at a VOC-to-NO_x ratio of
about 6.

Two aspects of the above studies must be noted. First, they used predominantly single-day simulations, whose results depend on how initial conditions
and emissions scheduling are treated. Second, the emissions and other inputs
do not correspond to a particular air basin, and ozone responses to changes
in VOCs are location specific. Russell et al. (1991; 1992) conducted a similar
analysis using multiday airshed calculations for Los Angeles. However, the
chemical mechanism employed was more lumped (see Chapter 5) than that
used by Carter and Atkinson (1989b). Russell et al. (1991; 1992) calculated
local sensitivities to lumped organic classes, methanol, and formaldehyde
(Table 12-4), relative to ozone sensitivity to CO emissions. Individual compound sensitivities multiplied by the exhaust compositions given in Table 12-1
yield the relative reactivity of the exhaust. If the exhaust emissions rates also
are used, the relative reactivities per mile can be calculated. These measures,
for methanol and other fuels, are given in Table 12-5.

Comparison of the incremental reactivity method with the model results
that look specifically at switching to methanol show general agreement. VOC
emissions from M100 would be a little more than half as reactive in terms of
ozone formation as would emissions from conventional fuels, and M85 would

TABLE 12-3   Incremental Reactivities of CO and Selected VOCs in Alternative Fuels as a Function of the VOC/NO$_x$ Ratio for an Eight-Component VOC Mix and Low-Dilution Conditions, Moles Ozone/Mole
Carbon

Compound	VOC/NO$_x$, ppbC/ppb		
	4	10	20
Formaldehyde	2.42	0.77	0.24
Ethane	0.024	0.031	0.015
n-Butane	0.10	0.031	0.052
Ethene	0.85	0.64	0.30
Propene	1.28	0.61	0.25
Toluene	0.26	0.04	-0.058
m-Xylene	0.98	0.32	0.012
Methanol	0.12	0.12	0.055
Ethanol	0.18	0.14	0.038
Carbon monox-ide	0.011	0.018	0.010

Source: Carter and Atkinson, 1989b

be about 70% as reactive. This assumes that the total emissions of VOCs are
about the same, that HCHO emissions are not excessive (<30 mg/mile), and
that emissions are limited to areas that are not NO$_x$-limited (VOC/NO$_x$ <10).
Further air-quality benefits depend on lowering HCHO emissions.

REGULATORY IMPLEMENTATION
OF ALTERNATIVE FUEL USE

Because of the considerable contribution of mobile source emissions to

TABLE 12-4   Ozone Peak and Exposure Reactivities of Compounds Relative to Carbon Monoxide.

	Airshed (Russell, 1990)[a]		Carter (1989)
	Peak	Exposure	
CO	1	1	1
Aldehydes $>C_2$	64	64	93[b]
Alkanes	16	11	9.5[c]
Alkenes	67	83	51[d]
Aromatics	51	82	53[e]
Ethene	71	67	78
Formaldehyde	119	148	180
Toluene	15	24	25
Methanol	17	14	17

[a]Airshed model results were combined with the population distribution to find the sensitivities of both the peak ozone and ozone exposure as VOC emission rates are increased. Exposure, in this case, is defined as the ozone concentration multiplied by the population density for ozone concentrations in excess of the NAAQS concentration of 120 ppb.
[b]Used incremental reactivity for acetaldehyde.
[c]Used 50% C4-C5 alkanes, and 50% C6+ alkanes from Carter (1989)
[d]Used equal portions of C4-C5 and C6 alkenes from Carter (1989)
[e]Used equal portions of di- and tri-alkylbenzenes from Carter (1989)

California's air-quality problems, controlling these emissions has been a key aspect of the overall air pollution strategy of the California Air Resources Board (CARB). In recent years, CARB has adopted various measures to reduce mobile source emissions through better control of in-use emissions and more stringent emission standards. This process was furthered in September, 1990 with CARB's approval of the "low-emission vehicles and clean fuels" regulations (Resolution 90-58), which were approved by the California Office of Administrative Law on August 30, 1991.

The low-emission vehicles and clean fuels regulations, an integral part of

TABLE 12-5   Relative Reactivities of Emissions from Gasoline and Alternative Fuels

Fuel	Carter[a] (1991)		Dunker[c] (1990)	Russell (1990)/ Williams et al. (1989)[b]
	Reactivity/gram (exhaust + evap) nonmethane VOC[c]	Reactivity/mile Total emissions[c]		Reactivity/gram (exhaust + evap)[c]
Gasoline (indolene)	1	1	1[d]	1[d]
E95[f]	0.84	2.0	--	--
E85[g]	0.81	2.1	--	--
M85[h]	0.73	0.63	0.93	0.73 (0.55)[e]
Methanol	0.54	0.97	0.64	0.58 (0.48)[e]
Liquid propane gas	0.83	0.47	--	--
Compressed natural gas	0.44	0.24	--	--

[a]Carter (1991) used measurements conducted by CARB (1989a,b) to derive fuel-based reactivities;   [b]The measurements by Williams et al. (1989) are used for a flexibly fueled vehicle fueled on M85 and M100. The relative reactivities of the emissions components are taken from Russell. Dunker (1990) used the same emissions profiles and modeled the exact composition, but did not use relative reactivities.; [c]Reactivities are relative to gasoline (or indolene) being 1. Indolene is commonly used as a test fuel; [d]Composition profile developed by Sigsby et al. (1987); [e]The first value is for exhaust formaldehyde emissions as measured; the second, in parentheses, is for a low-formaldehyde (5 mg/mile)-emitting vehicle; [f]E95, 95% ethanol and 5% gasoline; [g]E85, 85% ethanol and 15% gasoline; [h]M85, 85% methanol and 15% gasoline.

CARB's Long-Range Motor Vehicle Plan, established stringent, reactivity-based exhaust emission standards for new passenger cars, light-duty trucks and medium-duty vehicles and required that any clean alternative fuels needed by these vehicles be made available to the public.

Under the low-emission vehicle regulations, four categories of low-emission vehicles, each certified to a particular set of exhaust emission standards, will be phased in during the mid-1990s. In order of increasing stringency, the vehicle categories are

- Transitional Low-Emission Vehicle ("TLEV")
- Low-Emission Vehicle ("LEV")
- Ultra-Low-Emission Vehicle ("ULEV")
- Zero-Emission Vehicle ("ZEV")

The standards for all four categories of low-emission vehicles represent significant reductions in emissions compared to previous standards. Table 12-6 summarizes the 50,000-mile certification standards for passenger cars and small light-duty trucks. The regulations also promulgated emission standards for light-duty trucks above 3750 pounds loaded vehicle weight (LVW) and for medium-duty vehicles and engines.

A regulatory problem that has plagued alternative fuels and reformulated gasoline has been how to treat all fuels equally, without apparent bias that is unwarranted on an air-quality basis. One strategy to account for the lower ozone-forming potential of alternative fuels, yet allow for all fuels and technologies to compete on an equal regulatory basis, is to use a "reactivity-adjusted" emission standard. CARB (1990) has adopted regulations that use the incremental reactivity of each compound in automobile exhaust (see Table 5-5) to calculate a Reactivity Adjustment Factor (RAF). The RAF is the reactivity of the alternative fuel exhaust compared to that of the baseline fuel. If the RAF is less than 1, proportionally more mass can be emitted, such that the total ozone-forming potential is equivalent to that from the baseline fuel. For example, if the RAF of the alternative fuel is 0.5, then 2 g of alternative fuel emissions would have the same ozone-forming potential as 1 g of baseline fuel emissions. At present, the method for calculating the reactivities of individual compounds involves using an EKMA-type box model, exercised for about 75 trajectories representing different days in different cities (Carter, 1991). The simulations are, in effect, less than one day long, which may cause the results not to reflect actual airshed behavior. As expressed elsewhere in this report, it is strongly suggested that multiday simulations using grid-based models be used for control strategy analysis. Application of a grid-based model over a three-day period found good agreement with the box model results, but the

TABLE 12-6   California's 50,000 Mile Certification Standards for Passenger Cars and Light-Duty Trucks $\leq$ 3750 lb. Loaded Vechicle Weight (g/mi).

Category	NMOG[a]	CO[b]	NO$_x$[c]	HCHO[d]
Conventional	0.25[e]	3.4	0.4	0.015[f]
Transitional Low-Emission Vehicle	0.125	3.4	0.4	0.015
Low- Emission Vehicle	0.075	3.4	0.2	0.015
Ultra-Low-Emission Vehicle	0.040	1.7	0.2	0.008
Zero-Emission Vehicle	zero	zero	zero	zero

[a]NMOG: Non-methane organic gases
[b]CO: Carbon monoxide
[c]NO$_x$: Oxides of nitrogen
[d]HCHO: Formaldehyde
[e]Standard is for non-methane hydrocarbons
[f]Applies to methanol vehicles only

less reactive compounds were relatively more reactive over the three days in the grid-based model than in the box model (Russell et al., 1991a, b). The box model's simulations of less than a day do not fully account for the multiday buildup and carryover of the less reactive compounds. While there are differences in the results of the two models, the studies to date support the use of reactivity scaling. However, there are important areas for further investigation, such as the more widespread use of advanced models and mechanisms, studies of the impact of different meteorological conditions and locations, and the development of uncertainty estimates.

SUMMARY

Alternative fuels have the potential to improve air quality, especially in some urban areas, by reducing concentrations of the precursors to ozone and other photochemical oxidants. However, alternative fuels alone will not solve the air quality problems experienced by major cities. They could be an effective addition to ozone abatement strategies, but they must be considered in combination with other possible controls. Also, there are significant uncertainties left to be resolved, many of which stem from the inability to predict what the emissions will be from alternatively fueled vehicles in the future. Also, it must be stressed that the use of alternative fuels other than electricity and natural gas may have little effect on the increase in emissions that occurs over time with in-use vehicles, particularly its "super-emitters." It is this emissions deterioration that is the most central aspect of the motor vehicle emissions problem (Chapter 9).

Alternatively fueled vehicles could reduce mass emissions of volatile organic compounds (VOCs) and oxides of nitrogen (NO_x) and could decrease the atmospheric reactivity of the emissions (by using methanol or natural gas). There are some regions where alternative fuels could work effectively and others where little benefit would result. Furthermore, the effectiveness of these fuels could vary within a single airshed, depending on the VOC/NO_x ratio. Their use must be considered for each location separately.

The expected air quality benefits from each of the choices are not equal. Estimates of the potential benefits can be derived from emissions composition and from recent studies on compound reactivities (or ozone-forming potential). The studies to date have been limited to a few cities under a few conditions, so increased study is warranted.

Electric vehicles would give the greatest improvement in air quality by virtually eliminating VOCs, CO, and NO_x (net NO_x emission changes would depend on the source). They would lead to ozone reductions in virtually any region.

The fossil fuel alternatives and reformulated gasolines should lead to ozone reductions in areas with low VOC/NO_x, such as downtown Los Angeles and New York City. However, their effects in areas with high VOC/NO_x (such as Houston, Atlanta, or regions downwind from urban centers) would be minimal, unless NO_x emissions are also reduced. The benefits in intermediate regions must be explored further.

Natural gas is the cleanest of the fossil fuels, and natural gas vehicles appear to have the lowest reactivity VOC emissions. Vehicles in areas that have carbon monoxide (CO) problems would especially benefit from use of natural gas. However, those vehicles are usually designed to burn fuel-lean, and might not be able to meet proposed NO_x standards.

If cold-start problems are overcome, burning essentially pure methanol gives substantial reductions in exhaust and evaporative emission reactivity, although the air quality benefits are degraded if M85 (or lower percent methanol) fuel is used or if formaldehyde emissions are not effectively controlled.

Flexibly fueled vehicles could provide an easy transition mechanism to extensive fleets of methanol-fueled vehicles, but they will not offer the air quality benefits of dedicated vehicles. Increased mass emissions from flexibly fueled vehicles using low-methanol-blend fuels could counter the decreased reactivity of those emissions. Also, formaldehyde emissions must be controlled over the life of the vehicle.

Reformulated gasolines offer the easiest transition to a cleaner fuel and studies in progress indicate that properly reformulated gasolines can meet or surpass reductions in the emission reactivity of methanol-gasoline blends (e.g. M85). Ethanol would provide less benefit and might even be detrimental (Table 12-5).

Recent studies indicate that mobile-source emissions have been seriously underestimated (Chapter 9). Also, if the excess emissions are from "super-emitters," this could affect the choice of alternative fuels used. Not enough is known to assess this issue in detail, although these findings could support the use of naturally cleaner fuels, such as electricity and natural gas. On the other hand, if the inventories have underestimated the VOC emissions from stationary and biogenic sources, the benefits of using alternative fuels would be less than currently predicted.

Although the degree to which alternatively fueled vehicles can improve air quality is not known, the use of alternative fuels can become part of an effective ozone control strategy. Certain applications now exist where these fuels can be used effectively. Heavy-duty-vehicle use of methanol or natural gas appears promising. It is not clear which choice is the best for light-duty vehicles, and fthat decision depends on location and goals.

13

| Tropospheric Ozone |
| and Global Change |

INTRODUCTION

This chapter addresses the scientific evidence that relates global change[1] in atmospheric gases and climate to tropospheric ozone. Such a consideration of global changes is important because they are likely to continue and might hamper local efforts to meet the ozone National Ambient Air Quality Standards (NAAQS). Discussions in this chapter must be mostly qualitative, because few examples of research couple global changes with air quality. This lack of information points to the need for a long-term coordinated research program (see Chapter 14).

GLOBAL CHANGE: OBSERVATIONS

Two recent reports on global change (WMO, 1990; IPCC, 1990) have presented detailed reviews of the observations and likely causes of the increases found in most long-lived atmospheric trace gases. Concentrations of gases such as carbon dioxide (CO_2), trichlorofluoromethane ($CFCl_3$), dichlorodifluoromethane (CF_2Cl_2), methane (CH_4), and nitrous oxide (N_2O) are increasing at typical observation sites (see Table 13-1).

---

[1]Global change refers to changes in climate and changes in atmospheric chemistry.

TABLE 13-1   Changing Atmospheric Composition

Species	Mean global concentration		Annual rate of increase during 1980s
	Pre-industrial	Circa 1987	
CO_2	~280 ppm	348 ppm	0.5%
CH_4	~600 ppb	1680 ppb	0.8%
N_2O	~285 ppb	307 ppb	0.2%
$CFCl_3$	0	240 ppt	4%
CF_2Cl_2	0	415 ppt	4%
CCl_4	0	140 ppt	1.5%
CH_3CCl_3	0	150 ppt	4%
CH_3Cl	600 ppt?	600 ppt	~0%
CO	?	90 ppb	~1% (northern hemisphere) <1% (southern hemisphere)

Source: WMO (1990)

These gases act as greenhouse gases that contribute to the radiative forcing of the atmosphere, increasing the radiative forcing at the tropopause by about 0.5 watts/meter^2 over the past decade. The record of global mean surface temperature exhibits fluctuations, but with an apparent increasing trend (Figure 13-1). Global mean surface air temperatures have increased by as much as 0.3°C to 0.5°C this century (Hansen and Lebedeff, 1988; Jones, 1988). The temperature trend for the United States is more ambiguous because of the smaller sampling area, but also shows a temperature increase, albeit smaller, about 0.1°C to 0.3°C (Hansen et al., 1989).

Column ozone has been decreasing over the past 2 decades in both hemispheres (see UNEP/WMO, 1990).[2] The decrease in stratospheric ozone

---

[2]Column ozone is the abundance of ozone, predominantly stratospheric, that is obtained by integrating the amount of atmospheric ozone in the vertical direction.

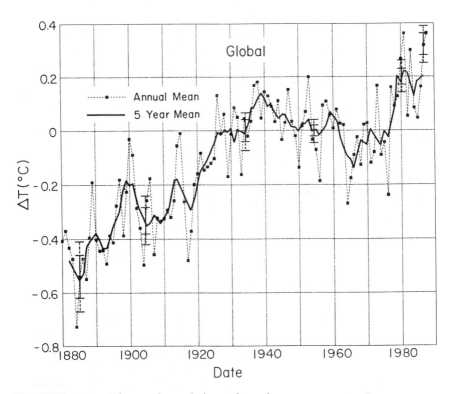

FIGURE 13-1   Observed trends in surface air temperatures. Source:
Hansen and Lebedoff, 1988.

over Antarctica is clearly one of the largest anthropogenic perturbations to
our planet. Ozone losses in the northern midlatitudes have been less dramatic
but still important—as much as 8% over the past decade in winter, with smaller but significant losses in summer (Stolarski et al., 1991).

Increases in CH_4 and CO, along with other photochemically active trace
gases such as odd-nitrogen compounds and volatile organic compounds
(VOCs), increase the potential for production of ozone throughout the troposphere.

Global tropospheric ozone is important as the primary source of tropospheric oxidation (mainly through OH) and as a greenhouse gas in the upper
troposphere (but not in the atmospheric boundary layer, where it radiates at
the same temperature as the surface). Tropospheric ozone is highly variable,
but exhibits systematic patterns: generally increasing from the equator to the

midlatitudes and from the surface to the tropopause (Chatfield and Harrison, 1977; Logan, 1985). Near the surface, average concentrations in the nonurban atmosphere (30-50 ppb) are greatest in spring and summer over northern midlatitudes. Interpretation of ozone trends is difficult because of the high variability in its concentrations. Reported trends (at the surface and from ozone sondes below 8 kilometers altitude) consistently show increases of about 1% per year over the past decade or two at northern midlatitudes (Angell and Korshover, 1983; Logan, 1985; Oltmans and Komhyr, 1986; Tiao et al., 1986; Feister and Warmbt, 1987; Bojkov, 1988; Crutzen, 1988)—a trend consistent with nineteenth-century measurements (Bojkov, 1986; Volz and Kley, 1988). Data for the Southern Hemisphere are scarce, but they indicate a small decline in two locations (Oltmans et al., 1989). The observed trends in tropospheric ozone should be considered as continental or possibly hemispheric in extent, but not global. As discussed below, it is not possible to make quantitative predictions regarding future trends in tropospheric ozone or its greenhouse effects.

GLOBAL CHANGE: EXPECTATIONS AND RESPONSE

It is difficult to make quantitative predictions of global change. Moreover, it would be impossible to make accurate predictions about the response of regional tropospheric ozone to global-scale changes in climate or atmospheric composition, even if we knew what the conditions in the future global atmosphere were going to be (see general discussions: Bachmann, 1988; Bernabo, 1989). Accordingly, qualitative estimates of some of the trends in temperature and trace gases that might be expected over the next two decades are presented. Mechanisms by which these global changes are likely to perturb tropospheric oxidants and local air quality are discussed, but in some cases even the direction (increase or decrease) of the effect is uncertain. Table 13.2 summarizes the links between human activities, global chemical and climate changes, and regional ozone.

Greenhouse gases are expected to increase substantially in the next century: CO_2 and CH_4 could double, N_2O could increase by 25%, and CFCs could more than double if not controlled under the Montreal Protocol (Prather and Watson, 1990). The sources of the CH_4 and N_2O increases are not yet fully understood, but the estimates are based on currently observed trends. Accurate prediction of CO_2 is hampered by a lack of understanding of the net biospheric source and oceanic uptake (IPCC, 1990).

T, the mean tropospheric temperature (Table 13-2), is predicted by climate models to continue to rise globally in response to the increased greenhouse forcing (Hansen et al., 1988; Raval and Ramanathan, 1989), with greater increases in large urban areas than in rural ones (Oke, 1973; Viterito, 1989). Average global increases of 0.5°C to 1.0°C are predicted over the next two to three decades (IPCC, 1990), but there is considerable uncertainty in these calculations. Late in the twenty-first century, the increase could be as large as 5°C (NRC, 1991).

There is an empirical relationship between worsened air quality and higher temperatures. As discussed in Chapter 2, a high temperature is generally a necessary but not sufficient condition for the occurrence of high ozone concentrations. This relationship is complex and cannot readily be extrapolated to a warmer climate because higher temperatures are often correlated empirically with sunlight and meteorology. Temperature increases will tend to destabilize peroxyacetyl nitrate (PAN) and related compounds, releasing more NO_x into the urban environment. There is much uncertainty about the effect of temperature on anthropogenic VOC emissions as well as on the possible enhancement of biogenic emissions of VOCs. Temperature has a direct calculable effect on the photolysis of ozone and other kinetic rates, but the effect on ozone concentration is expected to be small.

Water-vapor concentrations will increase concurrently with temperature, probably maintaining the same relative humidity, and hence increase by 6% per degree Celsius rise in temperature. Water-vapor increases will increase tropospheric ozone loss and OH production through the reactions involving water and photolysis products of ozone.

On average, global tropospheric OH concentrations would increase about 10% for a 25% increase in H_2O. The effect on aerosol chemistry and photochemical fogs is not known because the change in relative humidity is not known. The increase in water vapor would have a substantial effect on clean air chemistry, decreasing ozone, but should not significantly affect urban chemistry.

Stratospheric ozone is expected to decrease in response to the rise in atmospheric chlorine that is inevitable through the end of this century. The depletion in column ozone could be limited (no more than 5-10% at northern midlatitudes in spring) if an enhanced Montreal Protocol leads to a phaseout of CFCs (Prather and Watson, 1990; WMO, 1990). Depletion of stratospheric ozone leads to a direct increase in the penetration of solar ultraviolet light. The effect of increased ultraviolet light on tropospheric ozone formation depends on NO_x concentrations and other conditions that vary with latitude (Liu and Trainer, 1988; Gery et al., 1988b; Derwent, 1989). Enhanced photochemical activity in the troposphere will lead to increased production of ozone over most continental regions of the northern hemisphere, but can lead to reductions of ozone in clean maritime conditions where photochemistry is a net sink for ozone (Liu and Trainer, 1988; Schnell et al., 1991).

TABLE 13-2   Links Between Human Activities, Atmospheric Changes, and
Tropospheric Ozone

Results of human activities	Expected atmospheric changes	Effect on regional ozone
Increased greenhouse gases (CO_2, CH_4, N_2O, CFCs)	Warmer tropospheric temperature	Increase
	Increased tropospheric H_2O	Local effects small (global decrease small)
	Altered global circulation; possible enhancement of stagnation episodes	Unknown; possible increase
	Possible increase in stratospheric turnover rate, with greater injection of O_3 and NO_x into troposphere	Possible increase
Increased CFCs and halons	Less stratospheric O_3; more solar ultraviolet radiation reaching troposphere	Increase in polluted regions; decrease in remote global areas
Increased urbanization	Warmer local temperature	Increase
	Possible enhancement of stagnation episodes	Unknown; possible increase
Increased regional emissions of VOC, CO, and NO_x	Increased regional production of O_3	Increase (local and global)
Possible increased regional emissions of sulfur and aerosols	Enhanced cloud chemistry; possible increase in cloud cover	Unknown; possible decrease

Global tropospheric ozone is expected to respond to many aspects of climate change as outlined here. Current models cannot incorporate all the important components in the tropospheric ozone budget, but studies with simple models have shown that the response of ozone is complex and depends on the suite of changes in the global troposphere, including changes in methane and other VOCs, NO_x, water vapor, and carbon monoxide (Isaksen and Hov, 1987; Liu et al., 1987; Isaksen et al., 1988; Liu and Trainer, 1988; Prather, 1989; Thompson et al., 1989, 1990). Moreover, there is observational evidence that biomass burning (Delany et al., 1985; Logan and Kirchhoff, 1986; Kirchhoff et al., 1989) and regional pollution (Cox et al., 1975; Fishman et al., 1985) create extended layers of air in the troposphere with increased concentrations of ozone (see Figure 13-2) and could be an important part of the current budget for tropospheric ozone. Available predictions for future tropospheric ozone are varied, but they show limited increases over the next 50-100 years under the most extreme scenarios (not more than 50%, increases of 60-120 ppb).

Generally, increases in global tropospheric ozone will be expected to lead to a proportional rise in the ozone abundance of air entering metropolitan regions. However, the expected rise in ozone at a specific metropolitan region is dependent upon the local photochemistry and the initial ozone concentrations. Based on currently observed trends we might expect a systematic increase in local ozone of 10 ppb over the next 20 years. The feedback relationship between global air quality and local air quality could become more important in the future.

The abundance of global OH determines the global oxidative capacity of the lower atmosphere. OH, hydrogen peroxide (H_2O_2) and other oxidants are local quantities that respond to daily variations in ultraviolet sunlight, CO, O_3, CH_4 and NO_x (Levy, 1971; Sze, 1977; Thompson and Cicerone, 1986; Liu and Trainer, 1988; Thompson et al., 1990). The global average of OH is not directly influenced by the OH concentrations in the urban atmosphere because the air over cities constitutes such a small fraction of the global atmosphere. However, the abundance of OH in the nonurban atmosphere could be strongly affected by urban emissions of NO_x, because tropospheric chemistry over much of the globe is in the low-NO_x limit. When NO_x concentrations are low, e.g., less than 0.1 ppb, the abundance of OH responds almost linearly to NO_x, because HO_2 is effectively recycled to OH by reaction with NO. When NO_x concentrations are high, on the other hand, e.g., greater than 1 ppb, NO_x reactions such as formation of nitric acid reduce OH. A countering influence is the expected increase in CH_4, CO, and long-lived VOCs (such as C_2H_6) because of the increase in the emissions of these gases in the future. These species are the sink for OH in the remote atmosphere, which has low concen-

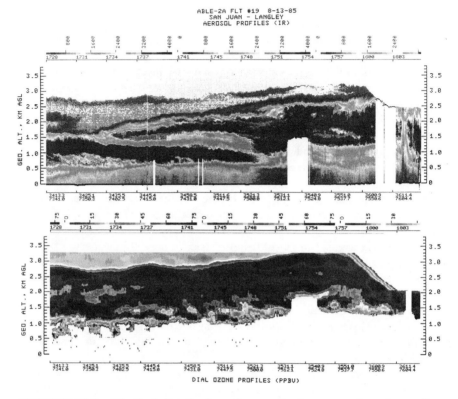

FIGURE 13-2   Vertical distribution of ozone in the troposphere immediately downwind of the east coast of the United States. The data were
obtained with an airborn lidar. The high concentration of ozone at
altitudes of 1-3 km are typical of polluted air masses moving off the
continental United States toward the North Atlantic Ocean.

trations of NO_x. Overall, the simple model studies mentioned above for
tropospheric ozone tend to predict decreases in global OH over the next 50
years of order tens of percent. Regional air quality will not respond directly
to changes in global OH, but the background abundances of biogenic VOCs
may be altered and regional ozone thus affected (Chameides et al., 1988;
Cardelino and Chameides, 1990).

Stratospheric NO_y will probably increase as aircraft emissions and N_2O (stratospheric source of NO_y) grow. Is this source of NO_x important anywhere other than the upper troposphere? Does it affect global tropospheric ozone production? These questions need to be addressed with global three-dimensional models, as for tropospheric ozone.

Stratospheric exchange, the rate of turnover of stratospheric air, is likely to change in a world where the concentration of CO_2 is doubled. If the rate of circulation in from the tropical tropopause, through the stratosphere, and out into the midlatitude troposphere were to speed up, then the stratospheric source of tropospheric ozone would increase. The only modeling study of this effect produced a measurable but modest increase of at most 15% in the residual circulation of the lower stratosphere (Rind et al., 1990). However, the implications for ozone and the lifetimes of other gases have not been examined.

Changes in atmospheric properties might lead to regional climate changes that could alter tropospheric concentrations (Hansen et al., 1989). Boundary layer exchange, in particular the rate of venting of the lowest atmospheric layers over the continents, has been predicted to change as temperatures rise (Rind, 1989). Storms, particularly the frequency and intensity of hurricanes, are hypothesized to increase as the sea surface temperature rises (Emanuel, 1987) and might lead to enhanced tropospheric mixing.

PREDICTING CHANGES IN TROPOSPHERIC OZONE

The global modeling of tropospheric ozone requires spatial and temporal resolution that can at best be achieved only with three-dimensional chemical transport models (CTMs). The scales of chemical heterogeneity critical to the modeling of global net production of tropospheric ozone occur over continental-maritime distances and on the much smaller scales of regional air pollution. In particular, the distributions of NO_x and reactive VOCs are patchy and likely to be correlated; much of the net production of ozone will come from the highly perturbed regions and will not be related to the longitudinally averaged concentrations.

High-resolution regional three dimensional CTMs for tropospheric ozone have been developed in response to the need to study air pollution as discussed in Chapters 9 and 10 and elsewhere (McRae and Seinfeld, 1983; Liu et al., 1984; Carmichael et al., 1986; Chang et al., 1987; McKeen et al., 1990). These models have detailed photochemical mechanisms, but they are extremely limited in that they must be initialized, they depend on boundary conditions, and they are used only for brief simulations (generally, no more than three days). Their grids (5 km up to 80 km) cannot be readily expanded to global size (400 km), and they cannot be used to simulate the annual global climatology of ozone. (Climatology refers to the distributions of means, patterns, and variability.) Nevertheless, the problems addressed by these models—heterogeneous distribution of NO_x and VOCs, and net ozone production—are important components of the ozone budget on a global scale (Liu et al., 1987). Even at the highest resolution, these regional models fail to resolve the individual pollution plumes associated with concentrated industrial sources that may represent an important nonlinear chemical processing of NO_x emissions (Sillman et al., 1990a). At the other extreme, global models will not resolve some of the processes in the regional CTMs; these processes must be accounted for in terms of larger-scale calculated variables such as wind speed and temperature gradients.

Some global two-dimensional, zonally averaged models for tropospheric ozone use realistic photochemical schemes but, by their nature, fail to resolve the continent-ocean differences in surface emissions and zonal transport (Isaksen and Hov, 1987; Hough and Derwent, 1990). Consequently, these models fail to account for the nonlinear dependence of ozone production on NO_x concentrations. Regional box models for tropospheric chemistry are a subset of the two-dimensional transport models and also have been applied to ozone (Thompson et al., 1989).

Three-dimensional models are far more difficult to design, initialize, and evaluate. Global CTMs depend on working, general circulation models for a complete and consistent picture of the physical climate system. It is not surprising therefore that none of the three-dimensional CTMs has presented a global tropospheric ozone simulation with realistic photochemistry. Early work with CTMs focused on stratospheric ozone chemistry (Hunt, 1969; Cunnold et al., 1975; Mahlman et al., 1980) and did not include the complexities of tropospheric ozone chemistry. More recent CTM studies have studied the climatology of tropospheric ozone (Levy et al., 1985) but include only a stratospheric source with a surface sink and no in situ chemistry. These studies have contributed to the tropospheric CTMs by better defining the stratospheric source of tropospheric ozone. Nevertheless, none of these CTM studies has been able to include the photochemical sources of ozone from urban regions or from tropical biomass burning (Fishman et al., 1985; 1990).

What steps are needed to devise a global CTM for tropospheric ozone? For example, the research at NOAA's General Fluid Dynamics Laboratory (GFDL) began with a seminal paper defining its tracer model (Mahlman and Moxim, 1978) and continued with a sequence of numerical experiments applicable to trace species with more complex sources and sinks: ozonelike (Mahlman et al., 1980), N_2O (Levy et al., 1982), tropospheric ozone with no chemistry (Levy et al., 1985), and tropospheric NO_x (Levy and Moxim, 1989). Similarly, the Goddard Institute for Space Studies/Harvard CTM began with a paper defining the model (Russell and Lerner, 1981) and then proceeded with a series of detailed studies that were meant to calibrate the model: CO_2 (Fung et al., 1983), CFCs (Prather et al., 1987), krypton (Jacob et al., 1987), radon (Balkanski and Jacob, 1990; Jacob and Prather, 1990), and CH_3CCl_3 (Spivakovsky et al., 1990a). Research groups at the Max Planck Institute in Hamburg (Heimann et al., 1990; Brost and Heimann, 1991), the National Center for Atmospheric Research (Rasch and Williamson, 1991) and the Lawrence Livermore National Laboratory (Penner et al., 1991) are similarly pursuing the development of global CTMs in studies of the calibration of continental and global transport and mixing, as well as the chemistry of tropospheric NO_x and hydrocarbons. The development of global CTMs will soon be applied to tropospheric ozone, and such tested and verified models will probably appear over the next five years.

A prediction of changes in global tropospheric ozone is needed by the U.S. Environmental Protection Agency (EPA) to study the effects of various policies on global chemistry and the climate. EPA has used results directly from the two-dimensional chemical or multibox budget models (Isaksen et al., 1988; Thompson et al., 1990) or has incorporated them into simple parameterized models that include a wider range of feedback couplings (e.g., Prather, 1989). Current models and emissions scenarios have not been consistently intercompared; however, most studies agree qualitatively, if not quantitatively (Isaksen and Hov, 1987; Liu et al., 1987; Isaksen et al., 1988; Liu and Trainer, 1988; Prather, 1989; Thompson et al., 1989, 1990). The response of ozone to global change is complex and will depend greatly on future global emissions. Under most circumstances, increases in tropospheric ozone are predicted by the middle of the twenty-first century, but they range from small to as much as 50%. A major limitation of these models is that they do not properly account for the nonlinear dependence of tropospheric chemistry on NO_x and VOC concentrations.

SUMMARY

The effect of global changes in the climate and atmospheric chemistry on tropospheric ozone are currently unpredictable, but they could lead to substantial increases in the number and duration of pollution episodes and in the size of the regions affected by high oxidant production. A warmer climate with less stratospheric ozone will enhance local photochemistry and probably local oxidant formation. Therefore, research efforts must elucidate the relationships of such effects with attainment of the ozone National Ambient Air Quality Standard (NAAQS). For example, it will be important to continue to develop global chemical transport models (CTMs) to predict changes in tropospheric ozone concentrations. A major synergism is potentially available between the global and regional CTMs in the simulation of ozone.

14

A Research Program
on Tropospheric Ozone

As discussed in this report, the scientific information used to develop regulations to control ozone is inadequate in many cases. It is difficult to predict accurately how ozone concentrations will respond to reductions in precursor emissions. It is also difficult to predict the effect of changes in atmospheric trace gas concentrations or climate on tropospheric ozone. A coherent and focused national research program with a long-term perspective is needed to provide government agencies with a better understanding of tropospheric ozone formation, transport, and accumulation. Progress toward reducing ozone concentrations in the United States has been severely hampered by the lack of such a program. This chapter addresses the need for a coordinated national research program directed at elucidating the chemical, physical, and meteorological processes that control ozone formation and concentrations over North America. The establishment of the North American Consortium for Atmospheric Modeling of Regional Air Quality (through a memorandum of understanding among more than a dozen funding agencies) represents an important step in coordinating research, but it does not answer the need for a focused research program directed from one program office.

A good analogy for the research program needed is the U.S. effort to address depletion of the stratospheric ozone layer by chlorofluorocarbons.[1] For this program, EPA is the relevant regulatory agency, but NASA's Upper

---

[1]This program is discussed as an example of one with many features that would be desirable in a tropospheric ozone research program. The committee does not recommend which agency should direct such a program.

Atmosphere Research Program is directed to "continue programs of research, technology, and monitoring of the phenomena of the stratosphere for the purpose of understanding the physics and chemistry of the stratosphere and for the early detection of potentially harmful changes in the ozone of the stratosphere" (Public Law 95-95, Clean Air Act Amendments, 1977). The partnership has worked well, and the basic research program has prepared the scientific foundation for international assessment (WMO, 1986, 1988, 1990) and for the 1987 Montreal Protocol on Substances that Deplete the Ozone Layer (1987). NASA has developed a basic research program of laboratory and field measurements, satellite data analysis, and theoretical modeling. The particular strengths of the program have been its broad participation base, which includes academic, government, industrial, and contract research groups, and its careful coordination with other federal and industrial programs and non-U.S. research efforts. The results of this comprehensive and coordinated research effort have been reported to Congress and to EPA. Its scientific assessments often include specific modeling studies that meet the regulatory and policy needs of EPA. A similar partnership that meets the needs of the research community and those of the regulatory agency will be necessary to establish a reliable scientific basis for the improvement of this nation's air quality.

The committee therefore recommends that a coordinated national program be established for the study of tropospheric ozone and for other related aspects of air quality in North America. This program should include coordinated field measurements, laboratory studies, and numerical modeling that will lead to a better predictive capability. In particular, it should elucidate the response of ambient ozone concentrations to possible regulatory actions or to natural changes in atmospheric composition or climate. To avoid conflict between the long-term planning essential for scientific research and the immediacy of requirements imposed on regulatory agencies, the research program should be managed independently of the EPA office that develops regulations under the Clean Air Act or other government offices that develop regulations. The research program must have a long-term commitment to fund research on tropospheric ozone. The direction and goals of this fundamental research program should not be subjected to short-term perturbations or other influences arising from ongoing debates over policy strategies and regulatory issues. The program should also be broadly based to draw on the best atmospheric scientists available in the nation's academic, government, industrial, and contract research laboratories. Further, the national program should foster international exchange and scientific evaluations of global tropospheric ozone and its importance in atmospheric chemistry and climate change. The recommended tropospheric ozone research program should be carefully coordinated with the Global Tropospheric Chemistry Program (UCAR, 1986) currently funded and coordinated by the National Science Foundation (NSF), and corresponding global change programs in the National Aeronautics and Space Administration (NASA), the National Oceanic and Atmospheric Administration (NOAA), the Department of Energy (DoE), and others.

References

Ackman, R.J. 1968. The flame ionization detector: Further comments on molecular breakdown and fundamental group response. J. Gas Chromatogr. 6:497—501.

Adams, R.M., S.A. Hamilton, and B.A. McCarl. 1985. An assessment of the economic effects of ozone on U.S. agriculture. J. Air Pollut. Control Assoc. 35:938—943.

Adams, R.M., J.D. Glyer, S.L. Johnson, and B.A. McCarl. 1989. A reassessment of the economic effects of ozone on U.S. agriculture. J. Air Waste Manage. 39:960—968.

Adams, R.P. 1989. Identification of Essential Oils by Ion Trap Mass Spectroscopy. San Diego, Calif.: Academic Press, Inc. 302 pp.

AFGL (Air Force Geophysics Laboratory). 1985. Atmospheric composition. Pp. 21-1 in Handbook of Geophysics and the Space Environment, A.S. Jursa, ed. Air Force Geophysics Laboratory, Air Force Systems Command, United States Air Force.

Aimedieu, P. 1983. Ozone profile intercomparison based on simultaneous observations between 20 and 40 km. Planet. Space Sci. 31:801—807.

Akimoto, H., K. Takagi, and F. Sakamaki. 1987. Photoenhancement of the nitrous acid formation in the surface reaction of nitrogen dioxide and water vapor: Extra radical source in smog chamber experiments. Int. J. Chem. Kinet. 19:539—551.

Albritton, D.L., S.C. Liu, and D. Kley. 1984. Global Nitrate Deposition from Lightning. Proceedings of the Conference on the Environmental Impact of

Natural Emissions, Research Triangle Park. Air Pollution Control Association, Pittsburgh, Penn.

Alson, A. 1988. The Emission Characteristics of Methanol and Compressed Natural Gas in Light Vehicles. Paper 88-99.3. Presented at the 81st Annual Meeting of the American Pollution Control Association, Dallas, Texas, June 19-24.

Altshuller, A.P. 1986. The role of nitrogen oxides in nonurban ozone formation in the planetary boundary layer over N America, W Europe and adjacent areas of ocean. Atmos. Environ. 20:245—268.

Altshuller, A.P. 1988. Some characteristics of ozone formation in the urban plume of St. Louis, Missouri, USA. Atmos. Environ. 22:499—510.

Altshuller, A.P. 1989. Nonmethane organic compound to nitrogen oxide ratios and organic composition in cities and rural areas. J. Air Waste Manage. Assoc. 39:936—943.

Altshuller, A.P., and J.J. Bufalini. 1971. Photochemical aspects of air pollution: A review. Environ. Sci. Technol. 5:39—64.

Ames, J., T.C. Myers, L.E. Reid, D.C. Whitney, S.H. Golding, S.R. Hayes, and S.D. Reynolds. 1985. Airshed Model Operations Manuals, Vol. I, User's Manual; Vol. II, Systems Manual. EPA-600/8-85/007a,b. U.S. Environmental Protection Agency, Atmospheric Sciences. Research Laboratory, Research Triangle Park, N.C. April.

Amiro, B.D., T.J. Gillespie, and G.W. Thurtell. 1984. Injury response of *phaseolus vulgaris* to ozone flux density. Atmos. Environ. 18:1207—1215.

Anderson, I.C., and J.S. Levine. 1987. Simultaneous field measurements of biogenic emissions of nitric oxide and nitrous oxide. J. Geophys. Res. 92:965—976.

Anderson, I.C., J.S. Levine, M.A. Poth, and J.P. Riggan. 1988. Enhanced biogenic emissions of nitric oxide and nitrous oxide following surface biomass burning. J. Geophys. Res. 93:3893—3898.

Anderson, L., J. Lanning, J., C. Machevec, and R. Meglen, R. 1989. The Effects of Colorado's Mandated High Oxygen Fuels Program on Amospheric Pollution in Denver. Paper 89-124.4. Presented at the 82nd Meeting and Exhibition of the Air and Waste Management Association, Anaheim, Calif., June 25-30.

Andreae, M.O., E.V. Browell, M. Garstang, G.L. Gregory, R.C. Harriss, G.F. Hill, D.J. Jacob, M.C. Pereira, G.W. Sasche, A.W. Setzer, P.L. Silva Dias, R.W. Talbot, A.L. Torres, and S.C. Wofsy. 1988. Biomass burning emissions and associated haze layers over Amazonia. J. Geophys. Res. 93:1509-—1527.

Aneja, V.P., C.S. Claiborn, Z. Li, and A. Murthy. 1990. Exceedances of the national ambient air quality standard for ozone occurring at a "pristine" area site. J. Air Waste Manage. Assoc. 40:217—20.

Angell, J.K., and J. Korshover. 1983. Global variations in total ozone and layer-mean ozone: An update through 1981. J. Clim. Appl. Meteor. 22:16-11—1627.

Anthes, R.A., and T.T. Warner. 1978. Development of hydrodynamic models suitable for air pollution and other mesometeorological studies. Mon. Wea. Rev. 106:1045—1078.

API (American Petroleum Institute). 1989. Detailed Analysis of Ozone State Implementation Plans in Seven Areas Selected for Retrospective Evaluation of Reasons for State Implementation Plan Failure, Vol. I, Executive Summary. No. 3708/89. Report prepared for the American Petroleum Institute by Pacific Environmental Services, Inc., Reston, Va., and Systems Applications, Inc., San Rafael, Calif. December.

Appel, B.R., Y. Tokiwa, J. Hsu, E.L. Kothny, and E. Hahn. 1985. Visibility as related to atmospheric aerosol constituents. Atmos. Environ. 19:1525—1534.

Arey, J., R. Atkinson, B. Zielinska, and P.A. McElroy. 1989. Diurnal concentrations of volatile polycyclic aromatic hydrocarbons and nitroarenes during a photochemical air pollution episode in Glendora, California. Environ. Sci. Technol. 23:321—327.

Arey, J., R. Atkinson, and S.M. Aschmann. 1990. Product study of the gas-phase reactions of monoterpenes with the OH radical in the presence of NO_x. J. Geophys. Res. 15:18539—18546.

Armerding, W., A. Herbert, T. Schindler, M. Spiekermann, and F.J. Comes. 1990. In-situ measurements of tropospheric OH radicals: A challenge for the experimentalist. Ber. Bunsenges. Phys. Chem. 94:776—781.

Arnts, R.R., and B.W. Gay, Jr. 1979. Photochemistry of Some Naturally Emitted Hydrocarbons. EPA/600/3-79-081. U.S. Environmental Protection Agency, Environmental Sciences Research Laboratories, Research Triangle Park, N.C.

Arnts, R.R., and S.A. Meeks. 1981. Biogenic hydrocarbon contribution to the ambient air of selected areas. Atmos. Environ. 15:1643—1651.

Ashbaugh, L.L., B.E. Croes, E.M. Fujita, and D.R. Lawson. 1990. Emission Characteristics of California's 1989 Random Roadside Survey. Paper presented at the 13th North American Motor Vehicle Emission Control Conference, Tampa, Fla., December 11-14.

Ashmore, M.R. 1984. Effects of ozone on vegetation in the United Kingdom. Pp. 92—104 in Proceedings of an International Workshop on the Evaluation and Assessment of the Effects of Photochemical Oxidants on Human Health, Agricultural Crops, Forestry, Materials and Visibility, P. Grennfelt, ed. Swedish Environmental Research Institute (IVL), Goteborg, Sweden.

Atkinson, R. 1988. Atmospheric transformations of automotive emissions. Pp. 99—132 in Air Pollution, the Automobile, and Public Health, A.Y. Watson, R.R. Bates, and D. Kennedy, eds. Washington DC: National Academy Press.

Atkinson, R. 1989. Kinetics and mechanisms of the gas-phase reactions of the hydroxyl radical with organic compounds. J. Phys. Chem. Ref. Data, Monograph 1:1—246.

Atkinson, R. 1990a. Gas-phase tropospheric chemistry of organic compounds: A review. Atmos. Environ. 24A:1—41.

Atkinson, R. 1990b. Tropospheric reactions of the haloalkyl radicals formed from hydroxyl radical reaction with a series of alternative fluorocarbons. Pp. 163—205 in Scientific Assessment of Stratospheric Ozone: 1989, Vol. II. Report No. 20. World Meteorological Organization Global Ozone Research and Monitoring Project. Geneva: World Meteorological Organization. 469 pp.

Atkinson, R. 1991. Kinetics and mechanisms of the gas-phase reactions of the NO_3 radical with organic compounds. J. Phys. Chem. Ref. Data 20:-459—507.

Atkinson, R., and W.P.L. Carter. 1984. Kinetics and mechanisms of the gas-phase reactions of ozone with organic compounds under atmospheric conditions. Chem. Rev. 84:437—470.

Atkinson, R. and A.C. Lloyd. 1984. Evaluation of kinetic and mechanistic data for modeling of photochemical smog. J. Phys. Chem. Ref. Data. 13:315—444.

Atkinson, R., A.M. Winer, and J.N. Pitts, Jr. 1986. Estimation of night-time N_2O_5 c concentrations from ambient NO_2 and NO_3 radical concentrations and the role of N_2O_5 in night-time chemistry. Atmos. Environ. 20:331—339.

Atkinson, R., S.M. Aschmann, and J.N. Pitts, Jr. 1988. Rate constants for the gas-phase reactions of the NO_3 radical with a series of organic compounds at 296 $\pm$ 2 K. J. Phys. Chem. 92:3454—3457.

Atkinson, R., D.L. Baulch, R.A. Cox, R.F. Hampson, Jr., J.A. Kerr, and J. Troe. 1989a. Evaluated kinetic and photochemical data for atmospheric chemistry: Supplement III. IUPAC Subcommittee on Gas Kinetic Data Evaluation for Atmospheric Chemistry. J. Phys. Chem. Ref. Data. 18:881—1097.

Atkinson, R., S.M. Aschmann, J. Arey, and W.P.L. Carter. 1989b. Formation of ring-retaining products from the OH radical-initiated reactions of benzene and toluene. Int. J. Chem. Kinet. 21:801—827.

Atkinson, R., S.M. Aschmann, E.C. Tuazon, J. Arey, and B. Zielinska. 1989c. Formation of 3-methylfuran from the gas-phase reaction of OH radicals with isoprene and the rate constant for its reaction with the OH radical. Int. J. Chem. Kinet. 21:593–604.

Atkinson, R., S.M. Aschmann, and J. Arey. 1990a. Rate constants for the gas-phase reactions of OH and NO_3 radicals and O_3 with sabinene and camphene at 296 ± 2 K. Atmos. Environ. 24A:2647–2654.

Atkinson, R., D. Hasegawa, and S.M. Aschmann. 1990b. Rate constants for the gas-phase reactions of O_3 with a series of monoterpenes and related compounds at 296 ± 2 K. Int. J. Chem. Kinet. 22:871–887.

Atkinson, R., E.C. Tuazon, and J. Arey. 1990c. Reactions of napthalene in N_2O_5-NO_3-NO_2-air mixtures. Int. J. Chem. Kinet. In press.

Attmannspacher, W., and H.U. Dütsch. 1981. 2nd International ozone sonde intercomparison at the Observatory Hohenpeissenberg. Ber. Dtsch. Wetterdienstes, 157, 1. April 5-20, 1978.

Atwater, M.A. 1984. Influence of Meteorology on High Ozone Concentrations. Air Pollution Control Association International Specialty Conference on Evaluation of the Scientific Basis for Ozone/Oxidants Standard, Houston, Texas, November 27-30.

Austin, T., R.G. Dulla, G. S. Rubenstein, and C.S. Weaver. 1989. Potential Emissions and Air Quality Effects of Alternative Fuels, Final Report. Report No. SR89-03-04. Sierra Research Inc., Sacramento, Calif. March.

Auto/Oil Air Quality Improvement Research Program. 1991. Air Quality Modeling Results for Reformulated Gasoline in Year 2005/2010. Technical Bulletin No. 3. Coordinating Research Council, Inc., Atlanta, Ga. May.

Ayers, G.P., and R.W. Gillett. 1988. Isoprene emissions from vegetation and hydrocarbon emissions from bush fires in tropical Australia. J. Atmos. Chem. 7:177–190.

Ayers, G.P., and R.W. Gillett. 1990. Tropospheric chemical composition: Overview of experimental methods in measurement. Rev. Geophys. 28:-297–314.

Bachmann, J. 1988. Potential effects of climate change and ozone depletion on air quality. Pp. 348–349 in Preparing for Climate Change, the Proceedings of First North American Conference on Climate Change: A Cooperative Approach, J.C. Topping, Jr., ed. Rockville, Md.: Government Institutes, Inc.

Balentine, H., C. Beskid, L. Edwards, R. Klausmeier, and S. Langevin. 1985. An Analysis of Chemistry Mechanisms and Photochemical Dispersion Models for Use in Simulating Methanol Photochemistry. EPA/460/3-85-008. U.S. Environmental Protection Agency, Emission Control Technology Division, Ann Arbor, Mich.

Balkanski, Y.J., and D.J. Jacob. 1990. Transport of continental air to the subantarctic Indian Ocean. Tellus 42B:62—75.

Barnes, R.A., A.R. Bandy, and A.L. Torres. 1985. Electrochemical concentration cell ozonesonde accuracy and precision. J. Geophys. Res. 90:-7881—7887.

Barnes, I., V. Bastian, K.H. Becker, and Z. Tong. 1990. Kinetics and products of the reactions of NO_3 with monoalkenes, dialkenes, and monoterpenes. J. Phys. Chem. 94:2413—2419.

Bauer, E. 1982. Natural and Anthropogenic Sources of Oxides of Nitrogen (NO_x) for the Troposphere. Institute for Defense Analysis Paper P-1619, FAA Technical Report No. FAA-EE-82-7. Federal Aviation Administration. Washington, D.C.: U.S. Department of Transportation.

Baugues, K. 1986. A Review of NMOC, NO_x and NMOC/NO_x Ratios Measured in 1984 and 1985. EPA-450/4-86-015. U.S. Environmental Protection Agency, Office of Air Quality, Planning and Standards, Research Triangle Park, N.C. Septembe.

Bechtold, R. and J.B. Pullman. 1980. Driving Cycle Economy, Emissions, and Photochemical reactivity using Alcohol Fuels and Gasoline. SAE Technical Paper Series No. 800260. Warrendale, Penn.: Society of Automotive Engineers.

Beck, M.B. 1987. Water quality modeling: A review of the analysis of uncertainty. Water Resources Research (WRERAQ) 23(8):1393—1442.

Bennett, J.H. 1979. Foliar exchange of gases. Pp. 10-1—10-29 in Methodology for the Assessment of Air Pollution Effects on Vegetation: A Handbook, W.W. Heck, S.V. Krupa, and S.N. Linzon, eds. Air Pollution Control Association, Pittsburgh, Penn.

Berglund, R.L., A.C. Dittenhoefer, H.M. Ellis, B.J. Watts, and J.L. Hansen. 1988. Evaluation of the stringency of alternative forms of a national ambient air quality standard for ozone. Pp. 343—369 in The Scientific and Technical Issues Facing Post-1987 Ozone Control Strategies, G.T. Wolff, J.L. Hanisch, and K. Schere, eds. Pittsburgh: Air and Waste Management Association.

Bernabo, J.C. 1989. The relationship of global climate change to other air quality issues. Pp. 224—228 in Coping with Climate Change, the Proceedings of the Second North American Conference on Preparing for Climate Change: A Cooperative Approach, J.C. Topping, Jr., ed. Climate Institute, Washington, D.C.

Betterton, E.A., and M.R. Hoffmann. 1988. Henry's law constants of some environmentally important aldehydes. Environ. Sci. Technol. 22:1415—1418.

Bicak, C. 1978. Plant Response to Variable Ozone Regimes of Constant Dosage, M.Sc. Thesis. Department of Plant Science, University of British Columbia, Vancouver, B.C., Canada. 115pp.

Bidleman, T.F. 1988. Atmospheric processes: Wet and dry deposition of organic compounds are controlled by their vapor-particle partitioning. Env. Sci. Technol. 22:361—367.

Bishop, G.A., and D.H. Stedman. 1990. On-road carbon monoxide emission measurement comparisons for the 1988-1989 Colorado oxy-fuels program. Environ. Sci. Technol. 24:843—847.

Bishop, G.A., J.R. Starkey, A. Ihlenfeldt, W.J. Williams, and D.H. Stedman. 1989. IR long-path photometry: A remote sensing tool for automobile emissions. Anal. Chem. 61:671A—677A.

Black, F.M. 1989. Motor vehicles as sources of compounds important to tropospheric and stratospheric ozone. Pp. 85—109 in Atmospheric Ozone Research and its Policy Implications, Proceedings of the 3rd US-Dutch International Symposium, T. Schneider, S.D. Lee, G.J.R. Wolters, and L.D. Grant, eds. Studies in Environmental Science 35. Amsterdam: Elsevier Science Publishers B.V.

Black, F. 1991. An overview of the technical implications of methanol and ethanol as highway motor vehicle fuels. Topical Technical (TOPTEC) Symposium on Alternative Liquid Transportation Fuels. Society of Automotive Engineers, Dearborn, Mich., August 28-29.

Black, F.M., and L.E. High. 1980. Passenger Car Hydrocarbon Emissions Speciation. EPA/ 600/2-80-085. U.S. Environmental Protection Agency, Environmental Sciences Rsearch Laboratories, Research Triangle Park, N.C. May.

Blanchard, C.L., P.M. Roth, and G.Z. Whitten. 1991. The Influence of NO$_x$ and VOC Emissions on Ozone Concentrations in Rural Environments. Electric Power Research Institute, Palo Alto, Calif. May.

Boekhaus, K.L., J.M. DeJovine, K.J. McHugh, D.A. Paulsen, L.A. Rapp, J.S. Segal, B.K. Sullivan, and D.J. Townsend. 1991. Reformulated Gasoline for California: EC-Premium Emission Control Gasoline and Beyond. SAE Technical Paper Series No. 911628. Warrendale, Penn.: Society of Automotive Engineers. August.

Bojkov, R.D. 1986. Surface ozone during the second half of the nineteenth century. J. Clim. Appl. Meteor. 25:343—352.

Bojkov, R.D. 1988. Ozone changes at the surface and in the free troposphere. Pp. 83-96 in Tropospheric Ozone: Regional and Global Scale Interactions, I.S.A. Isaksen, ed. NATO ASI Series C, Vol. 227. Dordrecht, Holland: D. Reidel Publishing Co.

Bollinger, M.J., C.J. Hahn, D.D. Parrish, P.C. Murphy, D.L. Albritton, and F.C. Fehsenfeld. 1984. NO$_x$ measurements in clean continental air and analysis of the contributing meteorology. J. Geophys. Res. 89:9623—9631.

Bonamassa, F., and H. Wong-Woo. 1966. Exhaust Hydrocarbon Composition of 1966 California Cars. Paper presented before the Division of Water, Air and Waste Chemistry, 152nd Annual Meeting, American Chemical Society, New York. September 11-16.

Boruki, W.J., and W. Chameides. 1984. Lightning: Estimates of the energy dissipation and nitrogen fixation. Rev. Geophys. Space Phys. 22:363–372.

Bottenheim, J.W., A.G. Gallant, and K.A. Brice. 1986. Measurement of NO species and O_3 at 82°N latitude. Geophys. Res. Lett. 13:113–117.

Bowne, N.E., P.M. Roth, S.D. Reynolds, C.L. Blanchard, and D.L. Shearer. 1990. Lake Michigan Ozone Study: Conceptual Design Plan. ENSR Consulting Engineering, Glastonbury, Conn.

Brewer, D.A., T.R. Augustsson, and J.S. Levine. 1983. The photochemistry of anthropogenic nonmethane hydrocarbons in the troposphere. J. Geophys. Res. 88:6683–6695.

Brorström-Lundé, E., and G. Lövblad. 1991. Deposition of soot-related hydrocarbons during long-range transport of pollution to Sweden. Atmos. Environ. 25A:207–215.

Brost, R.A. 1988. The sensitivity to input parameters of atmospheric concentrations simulated by a regional chemical model. J. Geophys. Res. 93:2371–2387.

Brost, R.A., and M. Heimann. 1991. The effect of the global background on a synoptic-scale simulation of tracer concentration. J. Geophys. Res. 96D:-15415–15425.

Brühl, C. and P. J. Crutzen. 1989. On the disproportionate role of tropospheric ozone as a filter against solar UV-B radiation. Geophys. Res. Lett. 16:703–706.

Bufalini, J.J. and M.C. Dodge. 1983. Ozone forming potential of light saturated hydrocarbons. Environ. Sci. Technol. 17:308–311.

Bufalini, J.J., T.A. Walter, and M.M. Bufalini. 1977. Contamination effects on ozone formation in smog chambers. Environ. Sci. Technol. 11:1181–1185.

Buhr, M.P., D.D. Parrish, R.B. Norton, F.C. Fehsenfeld, R.E. Sievers, and J.M. Roberts. 1990. Contribution of organic nitrates to the total reactive nitrogen budget at a rural eastern U. S. Site. J. Geophys. Res. 95:9809–9816.

Burns, V.R., J.D. Benson, A.M. Hochhauser, W.J. Koehl, W.M. Kreucher, and R.M. Reuter. 1991. Description of Auto/Oil Air Quality Improvement Research Program. SAE Technical Paper Series No. 912320. Warrendale, Penn.: Society of Automotive Engineers. October.

Cadle, S.H., D.P. Chock, J.M. Heuss, and P.R. Monson. 1976. Results of the General Motors Sulfate Dispersion Experiment. General Motors Research Publication GMR-2107. General Motors Research Laboratories, Warren,

Mich. 140pp.

Campbell, M.J., J.C. Farmer, C.A. Fitzner, M.N. Henry, J.C. Shepard, R.J. Hardy, J.F. Hopper, and V. Muralidhar. 1986. Radiocarbon tracer measurements of atmospheric hydroxyl radical concentrations. J. Atmos. Chem. 4:413–427.

Cantrell, C.A., and D.H. Stedman. 1982. A possible technique for the measurement of atmospheric peroxy radicals. Geophys. Res. Lett. 9:846–849.

Cantrell, C.A., D.H. Stedman, and G.J. Wendel. 1984. Measurement of atmospheric peroxy radicals by chemical amplification. Anal. Chem. 56:-1496–1502.

Cantrell, C.A., R.E. Shetter, A.H. McDaniel, J.A. Davidson, J.G. Calvert, D.D. Parrish, M. Buhr, F.C. Fehsenfeld, M. Trainer. 1988. Chain oxidation of NO as a measure of ambient peroxy radical content at an eastern U.S. site. Session A21A-05, EOS 69:1056.

CARB (California Air Resources Board). 1986. Methodology to Calculate Emission Factors for On-Road Motor Vehicles. Technical Support Section, Emissions Inventory Branch, Motor Vehicle Emissions and Projections Section, California Air Resources Board, Sacramento, Calif., November, 1986.

CARB (California Air Resources Board). 1989a. Definition of a Low-emission Motor Vehicle in Compliance with the Mandates of Health and Safety Code Section 39037.5. (Assembly Bill 234, Leonard, 1987). State of California Air Resources Board, Mobile Source Division, 9529 Telstar Ave., El Monte, Calif., May.

CARB (California Air Resources Board). 1989b. Methodology to Calculate Emissions Factors for On-road Motor Vehicles. Technical Support Division, California Air Resources Board, Sacramento, Calif.

CARB (California Air Resources Board). 1990. Proposed Regulations for Low-Emission Vehicles and Clean Fuels. Staff Report and Technical Support Document. State of California Air Resources board, Sacramento, Calif., August 13.

Cardelino, C.A., and W.L. Chameides. 1990. Natural hydrocarbons, urbanization, and urban ozone. J. Geophys. Res. 95:13971–3979.

Carroll, M.A., B.A. Ridley, D.D. Montzka, G. Hübler, J.G. Walega, R.B. Norton, B.J. Huebert, C.J. Hahn, and J.T. Merrill. Measurements of nitric oxide and nitrogen dioxide during MLOPEX. J. Geophys. Res., in press.

Carmichael, G.R., L.K. Peters, and T. Kitada. 1986. A second generation model for regional-scale transport/chemistry/deposition. Atmos. Environ. 20:173–18.

Carter, W.P.L. 1989. Ozone Reactivity Analysis of Emissions from Motor Vehicles. Draft report prepared for the Western Liquid Gas Association.

Carter, W.P.L. 1990a. A detailed mechanism for the gas-phase atmospheric reactions of organic compounds. Atmos. Environ. 24A:481–518.

Carter, W.P.L. 1990b. Ozone Reactivity Analysis of Emissions from Motor Vehicles. Report, University of California, Riverside, Calif.

Carter, W.P.L. 1991. Development of Ozone Reactivity Scales for Volatile Organic Compounds. EPA 600/3-91-050. U.S. Environmental Protection Agency, Research Triangle Park, N.C. August.

Carter, W.P.L., and R. Atkinson. 1987. An experimental study of incremental hydrocarbon reactivity. Environ. Sci. Technol. 21:670–679.

Carter, W.P.L., and R. Atkinson. 1988. Development and implementation of an up to date photochemical mechanism for use on airshed modeling. Final report, California Air Resource Board Contract A5-122-52. Sacramento, Calif. October.

Carter, W.P.L. and R. Atkinson. 1989a. Alkyl nitrate formation from the atmospheric photooxidation of alkanes: A revised estimation method. J. Atmos. Chem. 8:165–173.

Carter, W.P.L. and R. Atkinson. 1989b. Computer modeling study of incremental hydrocarbon reactivity. Environ. Sci. Technol. 23:864–880.

Carter, W.P.L., and F.W. Lurmann. 1991. Evaluation of a detailed gas-phase atmospheric reaction mechanism using environmental chamber data. Atmos. Environ. 25:2771–2806.

Carter, W.P.L., R. Atkinson, A.M. Winer, and J.N. Pitts, Jr. 1982a. Experimental investigation of chamber-dependent radical sources. Int. J. Chem. Kinet. 14:1071–1103.

Carter, W.P.L., A.M. Winer, and J.N. Pitts, Jr. 1982b. Effects of kinetic mechanisms and hydrocarbon composition on oxidant-precursor relationships predicted by the EKMA Isopleth Technique. Atmos. Environ. 16:113–120.

Carter, W.P.L., F.W. Lurmann, R. Atkinson, and A.C. Lloyd. 1986a. Development and Testing of a Surrogate Species Chemical Reaction Mechanism. EPA/600/3-86-031. U.S. Environmental Protection Agency, Research Triangle Park, N.C. 690 pp.

Carter, W.P.L., R. Atkinson, W.D. Long, L.N. Parker, and M.C. Dodd. 1986b. Effects of Methanol Fuel Substitution on Multi-day Air Pollution Episodes. Final Report. Contract No. A3-125-32. California Air Resources Board, Sacramento, Calif.

Chameides, W.L. 1984. The photochemistry of a remote marine stratiform cloud. J. Geophys. Res. 89:4739–4755.

Chameides, W.L. 1986. Possible role of NO_3 in the nighttime chemistry of a cloud. J. Geophys. Res. 91:5331–5337.

Chameides, W.L, and J. Rogers, 1988. Atmospheric Photochemical Oxidants:

A Southern Perspective. Summary of a workshop held in Atlanta, Ga., June 27-29.

Chameides, W.L., R.W. Lindsay, J. Richardson, and C.S. Kiang. 1988. The role of biogenic hydrocarbons in urban photochemical smog: Atlanta as a case study. Science 241:1473—1475.

Chan, C.Y., T.M. Hard, A.A. Mehrabzadeh, L.A. George, and R.J. O'Brien. 1990. Third-generation FAGE Instrument for Tropospheric Hydroxyl radical measurement. J. Geophys. Res. 95:18569—18576.

Chang, T. and Rudy, S.J. 1989. Urban air quality impact of methanol-fueled vehicles compared to gasoline-fueled vehicles. Paper presented at the Johns Hopkins University Conference on Methanol as a Fuel Choice: An Assessment. Washington, D.C., December 4-5.

Chang, T.Y., and S.J. Rudy, S.J. 1990. Ozone-forming Potential of Organic Emissions from Alternative-fueled Vehicles. Paper 90-96.3. Presented at the 83rd Annual Meeting and Exhibition of the Air and Waste Management Association, Pittsburgh, Penn., June 24-29.

Chang, J.S., R.A. Brost, I.S.A. Isaksen, S. Madronich, P. Middleton, W.R. Stockwell, and C.J. Walcek. 1987. A three-dimensional Eulerian acid deposition model: Physical concepts and formulation. J. Geophys. Res. 92:14681—14700.

Chang, T.Y., S.J. Rudy, G. Kuntasal, and R.A. Gorse, Jr. 1989. Impact of methanol vehicles on ozone air quality. Atmos. Environ. 23:1629—1644.

Charlson, R.J., J. Langner, H. Rodhe, C.B. Leovy, S.G. Warren. 1991. Perturbation of the northern hemisphere radiative balance by backscattering from anthropogenic sulfate aerosols. Tellus 43AB:152—163.

Chatfield, R., and H. Harrison. 1977. Tropospheric ozone 2. Variations along a meridianal band. J. Geophys. Res. 82:5969—5976.

Chatfield, R., and P.J. Crutzen. 1984. Sulfur dioxide in remote oceanic air: Cloud transport of reactive precursors. J. Geophys. Res. 89(D5):7111—7132.

Chen, W.Y. 1989. Estimate of dynamical predictability from NMC DERF Experiments. Mon. Wea. Rev. 117:1227—1236.

Ching, J.K.S., and A.J. Alkezweeny. 1986. Tracer study of vertical exchange by cumulus clouds. J. Clim. Appl. Meteor. 25:1702—1711.

Ching, J.K.S., S.T. Shipley, and E.V. Browell. 1988. Evidence for cloud venting of mixed layer ozone and aerosols. Atmos. Environ. 22:225—242.

Chock, D.P. 1988. A need for a more robust ozone air quality standard. Paper 88-121.5 in Proceedings of the 81st Annual Meeting and Exhibition of the Air Pollution Control Association, Dallas, Texas, June 19-24.

Chock, D.P. 1991. Issues regarding the ozone air quality standards. J. Air Waste Manage. Assoc. 41:148—152.

Chock, D.P., and J.M. Heuss. 1987. Urban ozone and its precursors. Environ. Sci. Technol. 21:1146—1153.

Chock, D.P., S. Kumar, and R.W. Herrmann. 1982. An analysis of trends in oxidant air quality in the south coast air basin of California. Atmos. Environ. 16:2615—2624.

Christian, G.D., and J.E. Reilly, eds. 1986. Instrumental Analysis, 2nd ed. Boston: Allyn and Bacon, Inc. 933 pp.

Clarke, J.F., and J.K.S. Ching. 1983. Aircraft observations of regional transport of ozone in the northeastern United States. Atmos. Environ. 17:1703—1712.

Clark, T.L., and J.F. Clarke. 1984. A Lagrangian study of the boundary layer transport of pollutants in the northeastern United States. Atmos. Environ. 18:287—297

Cleveland, W.S., B. Kleiner, J.E. McRae, J.L. Warner, and R.E. Pasceri. 1977. Geographical properties of ozone concentrations in the northeastern United States. J. Air Pollut. Control Assoc. 27:325—328.

Cofer, W.R., V.G. Collins, and R.W. Talbot. 1985. Improved aqueous scrubber for collection of soluble atmospheric trace gases. Environ. Sci. Technol. 19:557—560.

Corchnoy, S.B., and R. Atkinson. 1990. Kinetics of the gas-phase reactions of OH and NO$_3$ radicals with 2-carene, 1,8-cineole, p-cymene, and terpinolene. Environ. Sci. Technol. 24:1497—1502.

Cotton, W.R., and R.A. Anthes. 1989. Storm and Cloud Dynamics. International Geophysical Series, Vol. 44. San Diego: Academic Press, Inc. 883 pp.

Cox, R.A., A.E.J. Eggleton, R.G. Derwent, J.E.Lovelock, and D.H. Pack. 1975. Long-range transport of photochemical ozone in north-western Europe. Nature 225:118—121.

Crosley, D., and J. Hoell. 1985. Future directions for H$_x$O$_y$ detection. NASA Conference Publication CP-2448. U.S. National Aeronautics and Space Administration Workshop held at Menlo Park, Calif., August 12-15.

Crutzen, P.J. 1973. Gas-phase nitrogen and methane chemistry in the atmosphere. Pp. 110—124 in Physics and Chemistry of the Upper Atmosphere, Proceedings of a Symposium Organized by the Summer Advanced Study Institute, B.M. McCormac, ed. Dordrecht, Holland: D. Reidel Publishing Co.

Crutzen, P.J. 1979. The role of NO and NO$_2$ in the chemistry of the troposphere and stratosphere. Ann. Rev. Earth Planet Sci. 7:443—472.

Crutzen, P.J. 1988. Tropospheric ozone: An overview. Pp. 3-32 in Tropospheric Ozone: Regional and Global Scale Interactions, I.S.A. Isaksen, ed. NATO ASI Series C, Vol. 227. Dordrecht, Holland: D. Reidel Publishing

Co.

Crutzen, P.J., L.E. Heidt, J.P. Krasnec. W.H. Pollack, and W. Seiler. 1979. Biomass burning as a source of atmospheric gases CO, H_2, N_2Om NO, CH_3Cl and COS. Nature 282:253–256.

Crutzen, P.J., A.C. Delany, J. Greenberg, P. Haagenson, L. Heidt, R. Lueb, W. Pollack, W. Seiler, A. Wartburg, and P. Zimmerman. 1985. Tropospheric chemical composition measurements in Brazil during the dry season. J. Atmos. Chem. 2:233–256.

Cunnold, D.M., F. Alyea, N. Phillips, and R. Prinn. 1975. A three-dimensional dynamical-chemical model of atmospheric ozone. J. Atmos. Sci. 32:170–194.

Curran, T.C., and N.H. Frank. 1990. Ambient Ozone Trends Using Alternative Indicators. Presented at the Air and Waste Management Association Specialty Conference, Tropospheric Ozone and the Environment, Los Angeles, Calif. Air and Waste Management Transaction Series, Pittsburgh, Penn.

Darnall, K.R., A.C. Lloyd, A.M. Winer, and J.N. Pitts, Jr. 1976. Reactivity scale for atmospheric hydrocarbons based on reaction with hydroxyl radical. Environ. Sci. Technol. 10:692–696.

Davis, D.D., M.O. Rodgers, S.D. Fischer, and K. Asai. 1981. An experimental assessment of the O_2/H_2O interference problem in the detection of natural levels of OH via laser induced fluorescence. Geophys. Res. Lett. 8:69–72.

Decker, C.E., L.A. Ripperton, J.J.B. Worth, F.M. Vukovich, W.D. Bach, J.B.-Tommerdahl, F. Smith, and D.E. Wagoner. 1976. Formation and Transport of Oxidants Along Gulf Coast and in Northern U.S. EPA-450/-3-76-003. U.S. Environmental Protection Agency, Research Triangle Park, N.C.

DeJovine, J.M., K. J. McHugh, D.A. Paulsen, L.A. Rapp, J.S. Segal, B.K. Sullivan, and D.J. Townsend. Clean Fuels Report 91-06. ARCO Products Company, Anaheim, Calif.

Delany, A.C., P. Haagensen, S. Walters, A.F. Wartburg, and P.J. Crutzen. 1985. Photochemically produced ozone in the emission from large-scale tropical vegetation fires. J. Geophys. Res. 90: 2425–2429.

DeLuchi, M.A. 1989. Hydrogen vehicles: An evaluation of fuel storage, performance, safety, environmental impacts, and cost. Int. J. Hydrogen Energy 14:81–130.

DeLuchi, M.A., R.A. Johnston, and D. Sperling. 1988. Methanol vs. Natural Gas Vehicles: A Comparison of Resource Supply, Performance Emissions, Fuel Storage, Safety, Costs, and Transitions. SAE Technical Paper Series No. 881656. Warrendale, Penn.: Society of Automotive Engineers.

Demerjian , K.L., K.L. Schere, and J.T. Peterson. 1980. Theoretical estimates of actinic (spherically integrated) flux and photolytic rate constants of atmospheric species in the lower troposphere. Pp. 369—459 in Advances in Environmental Science and Technology Vol. 10., J.N. Pitts, Jr. and R.L. Metcalf, eds. New York: Wiley-Interscience Publication.

Demmy, J.L., W.M. Tax, and T.E. Warn. 1988. Area Source Documentation for the 1985 National Acid Precipitation Assessment Program Inventory. EPA/600/8-88-106. U.S. Environmental Protection Agency, Air and Energy Engineering Research Laboratory, Washington, D.C. 264 pp.

DeMore, W.B., M.J. Molina, S.P. Sander, D.M. Golden, R.F. Hampson, M.J. Kurylo, C.J. Howard, and A.R. Ravishankara. 1990. Chemical Kinetics and Photochemical Data for Use in Stratospheric Modeling. JPL Publ. 90-1, Evaluation No. 9. NASA Panel for Data Evaluation. Jet Propulsion Laboratory, Pasadena, Calif.

Dennis, R.L. 1986. Issues, design and interpretation of performance evaluations: Ensuring the emperor has clothes. Pp. 411—424 in Air Pollution Modeling and its Application V, C. De Wispelaere, F.A. Schiermeier, and N.V. Gillani, eds. New York: Plenum Press.

Dennis, R.L., and M.W. Downton. 1984. Evaluation of urban photochemical models for regulatory use. Atmos. Environ. 18:2055—2069.

Dennis, R.L., M.W. Downton, and R.S. Keil. 1983. Evaluation of Performance Measures for an Urban Photochemical Model. EPA 450/4-83-021. U.S. Environmental Protection Agency, Office of Air Quality Planning and Standards, Research Triangle Park, N.C.

Dennis, R.L., W.R. Barchet, T.L. Clark. 1990. Evaluation of Regional Acidic Deposition Models. NAPAP State of Science/Technology Report 5. National Acid Precipitation Assessment Program, 722 Jackson Place, NW, Washington, D.C.

Derwent, R.G. 1989. A comparison of model photochemical ozone formation potential with observed regional scale ozone formation during a photochemical episode over the United Kingdom in April 1987. Atmos. Environ. 23:1361—1371.

Derwent, R., and Ö. Hov. 1988. Application of sensitivity and uncertainty analysis techniques to a photochemical ozone model. J. Geophys. Res. 93:5185—5199.

Dickerson, R.R., G.J. Huffman, W.T. Luke, L.J. Nunnermacker, K.E. Pickering, A.C.D. Leslie, C.G. Lindsey, W.G.N. Slinn, T.J. Kelly, P.H. Daum, A.C. Delany, J.P. Greenberg, P.R. Zimmerman, J.F. Boatman, J.D. Ray, and D.H. Stedman. 1987. Thunderstorms: An important mechanism in the transport of air pollutants. Science 235:460—465.

Diederen, H., R. Guicherit, and J. Hollander. 1985. Visibility reduction by air pollution in the Netherlands. Atmos. Environ. 19:377–384.

Dimitriades, B. 1974. Proceedings of the Solvent Reactivity Conference. EPA-650/3-74-010. U.S. Environmental Protection Agency, National Environmental Research Center, Research Triangle Park, N.C.

Dimitriades, B., and S.B. Joshi. 1977. International Conference on Photochemical Oxidant Pollution and its Control: Proceedings, Vol. II, B. Dimitriades, ed. EPA/600/3-77-001b. U.S. Environmental Protection Agency, Environmental Sciences Research Laboratory, Research Triangle Park, N.C.

Dodge, M.C. 1977. Combined use of modeling techniques and smog chamber data to derive ozone--precursor relationships. Pp. 881–889 in International Conference on Photochemical Oxidant Pollution and its Control: Proceedings, Vol. II., B. Dimitriades, ed. EPA/600/3-77-001b. U.S. Environmental Protection Agency, Environmental Sciences Research Laboratory, Research Triangle Park, N.C.

Dodge, M.C. 1984. Combined effects of organic reactivity and nonmethane hydrocarbon to nitrogen oxides ratio on photochemical oxidant formation: A modeling study. Atmos. Environ. 18:1657–1666.

Dodge, M.C. 1989. A comparison of three photochemical oxidant mechanisms. J. Geophys. Res. 94:5121–5136.

Dodge, M.C. 1990. Formaldehyde production in photochemical smog as predicted by three state-of-the-science chemical oxidant mechanisms. J. Geophys. Res. 95:3635–3648.

Dorn, H.-P., J. Callies, U. Platt, and D.H. Ehhalt. 1988. Measurement of tropospheric OH concentrations by laser long-path absorption spectroscopy. Tellus 40B:437–445.

Downton, M.W., and R.L. Dennis. 1985. Evaluation of urban air quality models for regulatory use: Refinement of an approach. J. Clim. Appl. Meteor. 24:161–173.

Drummond, J.W., D.H. Ehhalt, and A. Volz. 1988. Measurements of nitric oxide between 0-12 km altitude and 67°N to 60°S latitude obtained during STRATOZ III. J. Geophys. Res. 93: 15831–15849.

Duce, R.A., V.A. Mohnen, P.R. Zimmerman, D. Grosjean, W. Cautreels, R. Chatfield, R. Jaenicke, J.A. Ogren, E.D. Pellizzari, and G.T. Wallace. 1983. Organic material in the global troposhere. Rev. Geophys. Space Phys. 21:921–952.

Duffy, L., I.E. Galbally, and C.M. Elsworth. 1988. Biogenic NO, emissions in the Latrobe Valley. Clean Air (Australia and New Zealand) 22:196–198.

Dunker, A.M. 1980. The response of an atmospheric reaction-transport model to changes in input functions. Atmos. Environ. 14:671–679.

Dunker, A.M. 1984. The decoupled direct method for calculating sensitivity coefficients in chemical kinetics. J. Chem. Phys. 81:2385—93.

Dunker, A. M. 1990. Relative reactivity of emissions from methanol-fueled and gasoline-fueled vehicles in forming ozone. Environ. Sci. Technol. 24:853—862.

Dunker, A.M., S. Kumar, and P.H. Berzins. 1984. A comparison of chemical mechanisms used in atmospheric models. Atmos. Environ. 18:311—321.

Eisele, F.L., and D.J. Tanner. 1990. Rapid ion-assisted tropospheric OH measurements. Session AllC-9, EOS Transactions 71:1229.

EPA (Environmental Protection Agency). 1981. Procedures for Emission Inventory Preparation, Vols I-V. EPA 450/4-81-026 (a-e). United States Environmental Protection Agency, Research Triangle Park, N.C. September.

EPA (Environmental Protection Agency). 1983. 1982 Ozone SIP Data Base and Summary Report. U.S. Environmental Protection Agency, Office of Air Quality Planning and Standards, Research Triangle Park, N.C.

EPA (Environmental Protection Agency). 1985a. Compilation of Air Pollutant Emission Factors, Vol. I. Stationary Point and Area Sources. No. AP-42. U.S. Environmental Protection Agency, Research Triangle Park, N.C.

EPA (Environmental Protection Agency). 1985b. Compilation of Air Pollutant Emission Factors, Vol 2. Mobile Sources. No. AP-42. U.S. Environmental Protection Agency, Research Triangle Park, N.C.

EPA (Environmental Protection Agency). 1986a. Air Quality Criteria for Ozone and Other Photochemical Oxidants, Vol. II. EPA-600/8-84/020bF. U.S. Environmental Protection Agency, Environmental Criteria and Assessment Office, Research Triangle Park, N.C. August.

EPA (Environmental Protection Agency). 1986b. Guideline on Air Quality Models. EPA/450/2-78-027R. U.S. Environmental Protection Agency, Office of Air Quality Planning and Standards, Research Triangle Park, N.C.

EPA (Environmental Protection Agency). 1986c. Development of the 1980 NAPAP Emissions Inventory. EPA/600/4-85-038. U.S. Environmental Protection Agency, Research Triangle Park, N.C.

EPA (Environmental Protection Agency). 1987. AIRS (Aeronometric Information Retrieval System. Users Guide. Volume 1 (of 7), internal report. U.S. Environmental Protection Agency, National Air Data Branch, Research Triangle Park, N.C. December. Update, 1991.

EPA (Environmental Protection Agency). 1988. Regional Oxidant Model for Northeast Transport (ROMNET) Project, Alliance Technologies Corporation correspondence to Emissions Committee. Subject: Responses to the ROMNET Existing Control Questionnaires, October 14, 1988.

EPA (Environmental Protection Agency). 1989a. The 1985 NAPAP Emissions Inventory. (Version 2): Development of the Annual Data and Modeler's Tapes. EPA-600/7-89-012a. U.S. Environmental Protection Agency, Office of Research and Development, Research Triangle Park, N.C. November. 196 pp.

EPA (Environmental Protection Agency). 1989b. User's Guide for MOBILE4 (Mobile Source Emissions Factor Model). EPA-AA-TEB-89-01. U.S. Environmental Protection Agency, Office of Mobile Sources, Ann Arbor, Mich. February.

EPA (Environmental Protection Agency). 1989c. NEDS Source Classification Codes and Emission Factor Listing. U.S. Environmental Protection Agency.

EPA (Environmental Protection Agency). 1989d. National Air Pollutant Emission Estimates 1940-1987. EPA-450/4-88-022. U.S. Environmental Protection Agency, Research Triangle Park, N.C.

EPA (Environmental Protection Agency). 1990a. National Air Quality and Emission Trends Report, 1988. EPA-450/4-90-002. U.S. Environmental Protection Agency, Office of Air Quality Planning and Standards, Research Triangle Park, N.C. March.

EPA (Environmental Protection Agency). 1990b. EPA Announces 96 Areas Failing to Meet Smog Standards, U.S. Environmental Protection Agency, Office of Public Affairs, Washington, D.C. August.

EPA (Environmental Protection Agency). 1990c. State and Local Air Monitoring Stations (SLAMS) Network. 1989 Status Report. U.S. Environmental Protection Agency, Office of Air Quality Planning and Standards, Research Triangle Park, N.C. March.

EPA (Environmental Protection Agency). 1990d. National Air Pollutant Emission Estimates-1940—1988. EPA/450/4-90-001. U.S. Environmental Protection Agency, Office of Air Quality Planning and Standards, Research Triangle Park, N.C. March.

EPA (Environmental Protection Agency). 1990e Quality Review Guidelines for Post-1987 State Implementation Plan (SIP) Base Year Emission Inventories. 450/Draft Report. U.S. Environmental Protection Agency, Office of Air Quality Planning and Standards, Research Triangle Park, N.C. February.

EPA (Environmental Protection Agency). 1990f. Volatile Organic Compounds from On-Road Vehicles: Sources and Control Options. Draft, May, 1990. Prepared for: Working Group on Volatile Organic Compounds, Long-range Transboundary Air Pollution Convention, United Nations-- Economic Commission for Europe.

EPA (Environmental Protection Agency). 1991a. National Air Quality and Emissions Trends Report, 1989. EPA/450/4-91-003. Environmental Protection Agency, Office of Air Quality Planning and Standards, Research Triangle Park, N.C. February.

EPA (Environmental Protection Agency). 1991b. Guideline for Regulatory Application of the Urban Airshed Model. EPA450/4-91-013. U.S. Environmental Protection Agency, Research Triangle Park, N.C. July.

EPA/OAQPS (Environmental Protection Agency). 1991. Ozone and Carbon Monoxide Areas Designated Nonattainment. Environmental Protection Agency, Office of Air Quality Planning and Standards, Research Triangle Park, N.C.

Evans, G.F. 1985. National Air Pollution Background Network, 1976-1984: Final Project Report. EPA/600/4-85-038. U.S. Environmental Protection Agency, Environmental Monitoring Systems Laboratories, Cincinnati, Ohio.

Evans, G.F., P. Finkelstein, B. Martin, N. Possiel, and M. Graves. 1983. Ozone measurements from a network of remote sites. J. Air Pollut. Control Assoc. 33:291-296.

Fahey, D.W., G. Hübler, D.D. Parrish, E.J. Williams, R.B. Norton, B.A. Ridley, H.G. Singh, S.C. Liu, and F.C. Fehsenfeld. 1986. Reactive nitrogen species in the troposphere: Measurements of NO, NO_2, HNO_3, particulate nitrate, peroxyacetyl nitrate (PAN), O_3, and total reactive odd nitrogen (NO_y) at Niwot Ridge, Colorado. J. Geophys. Res. 91:9781–9793.

Farmer, J.C., D.R. Cronn, D.W. Arlander, and F.A. Menzia. 1987. Remote marine tropospheric measurements of organic acids and aldehydes. Session A12-07, EOS 68:1212.

Federal Register. 1991. Air Quality Designations and Classifications; Final Rule. 56(215):56694– 40 CFR Part 81. November 6.

Feister, U., and W. Warmbt. 1987. Long-term measurements of surface ozone in the German Democratic Republic. J. Atmos. Chem. 5:1–21.

Fehsenfeld, F.C. 1988. Chemistry of the rural troposphere. Session A21A, EOS:69:1055–1057.

Fehsenfeld, F.C., M.J. Bollinger, S.C. Liu, D.D. Parrish, M. McFarland, M. Trainer, D. Kley, P.C. Murphy, and D.L. Albritton, and D.H. Lenschow. 1983. A study of ozone in the Colorado mountains. J. Atmos. Chem. 1:87–105.

Fehsenfeld, F.C., R.R. Dickerson, G. Hübler, W.T. Luke, L.J. Nunnermacker, E.J. Williams, J.M. Roberts, J.G. Calvert, C.M. Curran, A.C. Delany, C.S. Eubank, D.W. Fahey, A. Fried, B.W. Gandrud, A.O. Langford, P.C. Murphy, R.B. Norton, K.E. Pickering, and B.A. Ridley. 1987. A ground-based intercomparison of NO, NO_x, and NO_y measurement techniques. J. Geophys. Res. 92:14710–14722.

Fehsenfeld, F.C., J.W. Drummond, U.K. Roychowdhury, P.J. Galvin, E.J. Williams, M.P. Buhr, D.D. Parrish, G. Hübler, A.O. Langford, J.G. Calvert, B.A. Ridley, F. Grahek, B. Heikes, G. Kok, J. Shetter, J. Walega, C.M. Elsworth, R.B. Norton, D.W. Fahey, P.C. Murphy, C. Hovermale, V.A. Mohnen, K.L. Demerjian, G.I. Mackay, and H.I. Schiff. 1990. Intercomparison of NO_2 measurement techniques. J. Geophys. Res. 95:3579—3597.

Finlayson-Pitts, B.J. 1983. Reaction of NO_2 with NaCl and atmospheric implications of NOCl formation. Nature 306:676—677.

Finlayson-Pitts, B.J., and J.N. Pitts, Jr. 1977. The chemical basis of air quality: Kinetics and mechanism of photochemical air pollution and application to control strategies. Pp. 75-162 in Advances in Environmental Science and Technology, Vol. 7., J.N. Pitts, Jr. and R.L. Metcalf, eds. New York: Wiley-Interscience Publication.

Finlayson-Pitts, B.J. and J.N. Pitts, Jr. 1986. Atmospheric Chemistry: Fundamentals and Experimental Techniques. New York: Wiley-Interscience Publication. 1098pp.

Finlayson-Pitts, B.J., M.J. Ezell, and J.N. Pitts, Jr. 1989a. Formation of chemically active chlorine compounds by reactions of atmospheric NaCl particles with gaseous N_2O_5 and $ClONO_2$. Nature 337:241—244.

Finlayson-Pitts, B.J., F.E. Livingston, and H.N. Berko. 1989b. Synthesis and identification by infrared spectroscopy of gaseous nitryl bromide, $BrNO_2$. J. Phys. Chem. 93:4397—4400.

Finlayson-Pitts, B.J., F.E. Livingston, and H.N. Berko. 1990. Ozone destruction and bromine photochemistry at ground level in the arctic spring. Nature 343:622—625.

Fishman, J., and P.J. Crutzen. 1977. A numerical study of tropospheric photochemistry using a one-dimensional model. J. Geophys. Res. 82:5897—5906.

Fishman, J., F.M. Vukovich, and E.V. Browell 1985. The photochemistry of synoptic scale ozone synthesis: implications for the global tropospheric ozone budget. J. Atmos. Chem. 3:299—320.

Fishman, J., C.E. Watson, J.C. Larsen, and J.A. Logan. 1990. Distribution of tropospheric ozone determined from satellite data. J. Geophys. Res. 95:3599—3617.

Foreman, W.T., and T.F. Bidleman. 1990. Semivolatile organic compounds in the ambient air or Denver, Colorado. Atmos. Environ. 24A:2405—2416.

Fox, D.G. 1981. Judging air quality model performance. Bull. Am. Meteorol. Soc. 62:599—609.

Fujita, E.M., B.E. Croes, and F.W. Lurmann. 1990. Comparisons of Ambient and Emission Inventory Data in the South Coast Air Basin. Pp. 4/105—4/-

135 in Proceedings of the CRC-APRAC Vehicle Emissions Modeling Workshop, S. Cadle, ed. Newport Beach, Calif., October 30-31. Coordinating Research Council, Inc., Atlanta, Ga.

Fung, I., K. Prentice, E. Matthews, J. Lerner, and G. Russell. 1983. Three-dimensional tracer model study of atmospheric CO_2: Response to seasonal exchanges with the terrestrial biosphere. J. Geophys. Res. 88:1281—1294.

Gabele, P.A. 1990. Characterization of emissions from a variable gasoline/-methanol fueled car. J. Air Waste Manag. Assoc. 40:296—304.

Galbally, I.E. 1989. Factors controlling NO, emissions from soils. Pp. 23—37 in Exchange of Trace Gases Between Terrestrial Ecosystems and the Atmosphere, M.O. Andreae and D.S. Schimel, eds. New York: Wiley-Interscience Publication.

Galbally, I.E., and C.R. Roy. 1978. Loss of fixed nitrogen from soils by nitric oxide exhalation. Nature 275:734—735.

Galbally, I.E., C.R. Roy, C.M. Elsworth, and H.A.H. Rabich. 1985. The Measurement of Nitrogen Oxide (NO, NO_2) Exchange Over Plant/Soil Surfaces. Technical Paper No 8. Division of Atmospheric Research, Commonwealth Scientific and Industrial Research Organization (CSIRO), Aspendale, Australia. 23 pp.

Galbally, I.E., J.R. Freney, W.A. Muirhead, J.R. Simpson, A.C.F. Trevitt, and P.M. Chalk. 1987. Emission of nitrogen oxides (NO,) from a flooded soil fertilized with urea: Relation to other nitrogen loss process. J. Atmos. Chem. 5:343—365.

Geophysical Research Letters. 1990. 17:4.

Gertler, W.W., W.R. Pierson, and J.G. Watson. 1990. Review of on-road emission factors from the SCAQS tunnel experiment. Pp. 6-37—6-41 in Proceedings of the CRC-APRAC Vehicle Emissions Modeling Workshop, S. Cadle, ed., Newport Beach, Calif., October 30-31. Coordinating Research Council, Inc, Atlanta, Ga.

Gery, M.W., G.Z. Whitten, and J.P. Killus. 1988a. Development and Testing of the CBM-IV for Urban and Regional Modeling. EPA-600/3-88-012. U.S. Environmental Protection Agency, Research Triangle Park, N.C.

Gery, M. W., R. D. Edmund, and G.Z. Whitten. 1988b. Stratospheric ozone modification and ground level ozone. Pp. 334-347 in Preparing for Climate Change, Proceedings of the First North American Conference on Climate Change: A Cooperative Approach, J.C. Topping, Jr., ed. Rockville, Md.: Government Institutes, Inc.

Gery, M.W., G.Z. Whitten, J.P. Killus, and M.C. Dodge. 1989. A photochemical kinetics mechanism for urban and regional scale computer modeling. J. Geophys. Res. 94:12925—12956.

Gidel, L.T. 1983. Cumulus cloud transport of transient tracers. J. Geophys.

Res. 88C:6587—6599.

Glasson, W.A., and C.S. Tuesday. 1970a. Hydrocarbon reactivity and the kinetics of the atmospheric photooxidation of nitric oxide. J. Air Pollut. Control Assoc. 20:239—243.

Glasson, W.A., and C.S. Tuesday. 1970b. Hydrocarbon reactivities in the atmospheric photooxidation of nitric oxide. Environ. Sci. Technol. 4:916—924.

Glasson, W.A., and C.S. Tuesday. 1971. Reactivity relationships of hydrocarbon mixtures in atmospheric photooxidation. Environ. Sci. Technol. 5:151-—154.

Glendening, J.W., B.L. Ulrickson, and J.A. Businger. 1986. Mesoscale variability of boundary layer properties in the Los Angeles Basin. Mon. Wea. Reṽ. 114:2537—2549.

Gold, M.D., and C.E. Moulis. 1988. Effects of Emission Standards on Methanol Vehicle-related Ovone, Frmaldehyde, and Mthanol Exposure. Paper 88-41.4. Presented at the 81st Annual Meeting of American Pollution Control Association, Dallas, Texas, June 19-24.

Gorse, R.A., Jr. 1984. On-road emission rates of carbon monoxide, nitrogen oxides, and gaseous hydrocarbons. Environ. Sci. Technol. 18:500—507.

Gorse, R.A., Jr., and J.M. Norbeck. 1981. CO emission rates for in-use gasoline and diesel vehicles. J. Air Pollut. Control Assoc. 31:1094—1096.

Greenberg, J.P., and P.R. Zimmerman. 1984. Nonmethane hydrocarbons in remote tropical continental, and marine atmospheres. J. Geophys. Res. 89:4764—4778.

Greenberg, J.P., P.R. Zimmerman, L. Heidt, and W. Pollock. 1984. Hydrocarbon and carbon monoxide emissions from biomass burning in Brazil. J. Geophys. Res. 89:1350—1354.

Greenhut, G.K. 1986. Transport of ozone between boundary layer and cloud layer by cumulus clouds. J. Geophys. Res. 91D:8613—8622.

Greenhut, G.K., J.K.S. Ching, R. Pierson, Jr., and T.P. Repoff. 1984. Transport of ozone by turbulence and clouds in an urban boundary layer. J. Geophys. Res. 89D:4757—4766.

Gregory, G.L., E.V. Browell, and L.S. Warren. 1988. Boundary layer ozone: An airborne survey above the Amazon Basin. J. Geophys. Res. 93:1452—1468.

Gregory, G.L., J.M. Hoell, Jr., M.A. Carroll, B.A. Ridley, D.D. Davis, J. Bradshaw, M.O. Rodgers, S.T. Sandholm, H.I. Schiff, D.R. Hastie, D.R. Karecki, G.I. MacKay, G.W. Harris, A.L. Torres, and A. Fried. 1990. An intercomparison of airborne nitrogen dioxide instruments. J. Geophys. Res. 95:10103—10127.

Groblicki, P.J., G.T. Wolff, and R.J. Countess. 1981. Visibility reducing species in the Denver brown cloud. I. Relationships between extinction and chemical composition. Atmos. Environ. 15:2473–2484.

Grosjean, D. 1982. Formaldehyde and other carbonyls in Los Angeles ambient air. Envir. Sci. Technol. 16:254–262.

Grosjean, D. 1989. Organic acids in southern California air: Ambient concentrations, mobile source emissions, in situ formation and removal processes. Environ. Sci. Technol. 23:1506–1514.

Grosjean, D. 1990. Liquid chromatography analysis of chloride and nitrate with "negative" ultraviolet detection: Ambient levels and relative abundance of gas-phase inorganic and organic acids in southern California. Environ. Sci. Technol. 24:77–81.

Grosjean, D., and J.H. Seinfeld. 1989. Parameterization of the formation potential of secondary organic aerosols. Atmos. Environ. 23:1733–1747.

Grosjean, D., and S.S. Parmar. 1990. Interferences from aldehydes and peroxyacetyl nitrate when sampling urban air organic acids on alkaline traps. Environ. Sci. Technol. 24:1021-1026.

Grosjean, D., E.C. Tuazon, and E. Fujita. 1990a. Ambient formic acid in southern California air: A comparison of two methods, Fourier transform infrared spectroscopy and alkaline trap-liquid chromatography with UV detection. Environ. Sci. Technol. 24:144–146.

Grosjean, D., A.H. Miguel, and T.M. Tavares. 1990b. Urban air pollution in Brazil: Acetaldehyde and other carbonyls. Atmos. Environ. 24B:01–06.

Grotch, S.L. 1988. Regional intercomparisons of general circulation model predictions and historical climate data. Contract No. DOE/NBB-0084. 291 pp.

Gu, C.-L., C.M. Rynard, D.G. Hendry, and T. Mill. 1985. Hydroxyl radical oxidation of isoprene. Environ. Sci. Technol. 19: 151–155.

Guensler, R. 1990. Evaluating the Effectiveness of State Implementation Plan Strategies. Paper 90-93.12. Presented at the 83rd Annual Meeting and Exhibition of the Air and Waste Management Association, Pittsburgh, Penn., June 24-29.

Haagen-Smit, A.J. 1952. Chemistry and physiology of Los Angeles smog. Indust. Eng. Chem. 44:1342–1346.

Haagen-Smit, A.J., and M.M. Fox. 1954. Photochemical ozone formation with hydrocarbons and automobile exhaust. J. Air Pollut. Control Assoc. 4:105–109.

Haagen-Smit, A.J., and M.M. Fox. 1955. Automobile exhaust and ozone formation. SAE Technical Series No. 550277. Warrendale, Penn.: Society of Automotive Engineers,

Haagen-Smit, A.J., and M.M. Fox. 1956. Ozone formation in photochemical oxidation of organic substances. Indust. Eng. Chem. 48:1484–.

Haagen-Smit, A.J., E.F. Darley, M. Zaitlin, H. Hull, and W. Noble. 1951. Investigation on injury to plants from air pollution in the Los Angeles area. Plant Physiol. 27:18—.

Haagen-Smit, A.J., C.E. Bradley, and M.M. Fox. 1953. Ozone formation in photochemical oxidation of organic substances. Indust. Eng. Chem. 45:- 2086—.

Hameed, S., O.G. Paidoussis, and R.W. Stewart. 1981. Implications of natural sources for the latitudinal gradients of NO, in the unpolluted troposphere. Geophys. Res. Lett. 8:591—594.

Hansen, J., and S. Lebedeff. 1988. Global surface air temperatures: Update through 1987. Geophys Res. Lett. 15:323—326.

Hansen, J., A. Fung, A. Lacis, D. Rind, S. Lebedeff, R. Reudy, G. Russell, and P. Stone. 1988. Global climate changes as forecast by Goddard Institute for Space Studies (GISS) 3-dimensional model. J. Geophys Res. 93:9341—9364.

Hansen, J., D. Rind, A. DelGenio, A. Lacis, S. Lebedeff, M. Prather, R. Ruedy, and T. Karl. 1989. Regional greenhouse climate effects. Pp. 68-81 in Coping with Climate Change, Proceedings of the Second North American Conference on Preparing for Climate Change: A Cooperative Approach, J.C. Topping, Jr., ed. Climate Institute, Washington, D.C.

Hard, T.M., C.Y. Chan, A.A. Mehrabzadeh, W.H. Pan, and R.J. O'Brien. 1986. Diurnal cycle of tropospheric OH. Nature 322:617—620.

Harris, G.W., W.P.L. Carter, A.M. Winer, J.N. Pitts, Jr., U. Platt, and D. Perner. 1982. Observations of nitrous acid in the Los Angeles atmosphere and implications for predictions of ozone-precursor relationships. Environ. Sci. Technol. 16:414—419.

Harris, S.J., and J.A. Kerr. 1989. A kinetic and mechanistic study of the formation of alkyl nitrates in the photo-oxidation of n-heptane studied under atmospheric conditions. Int. J. Chem. Kinet. 21:207—218.

Hatakeyama, S. and M.-T. Leu. 1989. Rate constants for reactions between atmospheric reservoir species. 2. H_2O. J. Phys. Chem. 93:5784—5789.

Hatakeyama, S., K. Izumi, T. Fukuyama, and H. Akimoto. 1989. Reactions of ozone with α-pinene and β-pinene in air: Yields of gaseous and particulate products. J. Geophys. Res. 94:13013—13024.

Hatekeyama, S., K. Izumi, T. Fukuyama, H. Akimoto, and N. Washida. 1991. Reactions of OH with α-pinene and β-pinene in air: Estimate of global CO production from the atmospheric oxidation of terpenes. J. Geophys. Res. 96:947—958.

Hayes, E.M., and J.M. Skelly. 1977. Transport of ozone from the Northeast U.S. into Virginia and its effect on Eastern white pines. Plant Dispos. Rep. 61:778—782.

Heck, W.W., and D.T. Tingey. 1971. Ozone. Time-concentration model to predict acute foliar injury. Pp. 249–255 in Proceedings of the 2nd International Clear Air Congress, H.M. Englund and W.T. Beery, eds. New York: Academic Press.

Heck, W.W., J.A. Dunning, and I.J. Hindawi. 1966. Ozone: Nonlinear relation of dose and injury in plants. Science 151:577–578.

Heck, W.W., O.C. Taylor, R. Adams, G. Bingham, J. Miller, E. Preston, and L. Weinstein. 1982. Assessment of crop loss from ozone. J. Air Pollut. Control Assoc. 32:353–361.

Heicklen, J., K. Westberg, and N. Cohen. 1969. Conversion of Nitrogen Oxide to Nitrogen Dioxide in Polluted Atmospheres. Report No. 115-69. Center for Air Environment Studies, Pennsylvania State University, University Park, Penn.

Heikes, B.G., and A.M. Thompson. 1983. Effects of heterogeneous processes on NO_3, HONO, and HNO_3 chemistry in the troposphere. J. Geophys. Res. 88:10883–10895.

Heimann, M., P. Monfray, and G. Polian. 1990. Modeling the long-range transport of ^{222}Rn to subantarctic and antarctic areas. Tellus 42B:83–99.

Hellpointner, E., and S. Gäb. 1989. Detection of methyl, hydroxymethyl and hydroxyethyl hydroperoxides in air and precipitation. Nature 337:631–634.

Hellman, K.H., R.I. Bruetsch, G.K. Piotrowski, and W.D. Tallent. 1989. Resistive Materials Applied to Quick Light-off Catalysts. SAE Technical Paper Series 890799. Warrendale, Penn: Society of Automotive Engineers.

Hempel, L.C., D. Press, D. Gregory, J.M. Hough, and M.E. Moore, M.E. 1989. Curbing Air Pollution in Southern California: The Role of Electric Vehicles. Policy Clinic Report, Claremont Graduate School, Claremont, Calif., April.

Henderson, W.R., and R.A. Reinert. 1979. Yield response of four fresh market tomato cultivars after acute ozone exposure in the seedling stage. J. Am. Soc. Hort. Sci. 104:754–759.

Hering, S.V., D.R. Lawson, I. Allegrini, J.E. Sickles II, K.G. Anlauf, A. Wiebe, B.R. Appel, W. John, S. Wall, R.S. Braman, R. Sutton, G.R. Cass, P.A. Solomon, D.J. Eatough, N.L. Eatough, E.C. Ellis, D. Grosjean, B.B. Hicks, J.D. Womack, J. Horrocks, K.T. Knapp, T.B. Ellestad, R.J. Paur, W.J. Mitchell, M. Pleasant, E. Peake, A. MacLean, W.R. Pierson, W. Brachaczek, H.I. Schiff, G.I. Mackay, C.W. Spicer, D.H. Stedman, H.W. Biermann, A.M. Winer, and E.C. Tuazon. 1988. The nitric acid shootout: Field comparison of measurement methods. Atmos. Environ. 22:1519–1539.

Herron, J.T., and R.E. Huie. 1978. Stopped-flow studies of the mechanisms of ozone-alkene reactions in the gas phase. Propene and isobutene. Int. J. Chem. Kinet. 10:1019–1041.

Heuss, J.M. and W.A. Glasson. 1968. Hydrocarbon Reactivity and Eye Irritation. Environ. Sci. Technol. 2:1109—1116.

Hilsenrath, E., W. Attmannspacher, A. Bass, W. Evans, R. Hagemeyer, R.A. Barnes, W. Komhyr, K. Mauersberger, J. Mentall, M. Proffitt, D. Robbins, S. Taylor, A. Torres, and E. Weinstock. 1986. Results from the balloon ozone intercomparison campaign (BOIC). J. Geophys. Res. 90:13137—13152.

Hlavlinka, M.W., and J.A. Bullin. 1988. Validation of mobile source emission estimates using mass balance techniques. J. Air Pollut. Cont. Assoc. 38:-1035—1039.

Hochhauser, A.M., J.D. Benson, V. Burns, R.A. Gorse, W.J. Koehl, L.J. Painter, B.H. Rippon, R.M. Reuter, and J.A. Rutherford. 1991. The Effect of Aromatics, MTBE, Olefins, and T_{90} on Mass Exhaust Emissions from Current and Older Vehicles--The Auto/Oil Air Quality Improvement Research Program. SAE Technical Paper Series No. 912322. Warrendale, Penn: Society of Automotive Engineers. August.

Hoell, J.M., Jr., G.L. Gregory, D.S. McDougal, M.A. Carroll, M. McFarland, B.A. Ridley, D.D. Davis, J. Bradshaw, M.O. Rodgers, and A.L. Torres. 1985. An intercomparison of nitric oxide measurement techniques. J. Geophys. Res. 90:12843—12851.

Hoell, J.M., Jr., G.L. Gregory, D.S. McDougal, G.W. Sachse, G.F. Hill, E.P. Condon, and R.A. Rasmussen. 1987a. Airborne intercomparison of carbon monoxide measurement techniques. J. Geophys. Res. 92:2009—2019.

Hoell, J.M., G.L. Gregory, D.S. McDougal, A.L. Torres, D.D. Davis, J. Bradshaw, M.O. Rodgers, B.A. Ridley, and M.A. Carroll. 1987b. Airborne intercomparison of nitric oxide measurement techniques. J. Geophys. Res. 92:1995—2008.

Hofzumahaus, A., H.-P. Dorn, and U. Platt. 1990a. Tropospheric OH radical measurement techniques: Recent developments. Pp. 103—108 in Physico-Chemical Behavior of Atmospheric Pollutants, G. Restelli and G. Angeletti, eds. Dordrecht: Kluwer Academic Publishers.

Hofzumahaus, A., H.-P. Dorn, J. Callies, U. Platt, and D.H. Ehhalt. 1990b. Tropospheric OH concentrations measurements by laser long-path absorption spectroscopy. Atmos. Environ. 25:2017—2022.

Hogsett, W.E., D.T. Tingey, and S.R. Holman. 1985a. A programmable exposure control system for determination of the effects of pollutant exposure regimes on plant growth. Atmos. Environ. 19:1135—1145.

Hogsett, W.E., M. Plocher, V. Wildman, D.T. Tingey, and J.P. Bennett. 1985b. Growth response of two varieties of slash pine seedlings to chronic ozone exposures. Can. J. Bot. 63:2369—2376.

Hogsett, W.E., D.T. Tingey, and E.H. Lee. 1988. Exposure indices: Concepts for development and evaluation of their use. Pp. 107—138 in Assessment of Crop Loss from Air Pollutants, W.W. Heck, O.C. Taylor, and D.T. Tingey, eds. London, U.K.: Elsevier Applied Science.

Holdren, M.W., H.H. Westberg, and H.H. Hill, Jr. 1979. Analytic Methodology for the Identification and Quantification of Vapor Phase Organic Pollutants. Interim Report, CRC-APRAC-CAPA-11-71. Coordinating Research Council, Air Pollution Research Advisory Committee, New York.

Horn J.C., and S.K. Hoekman. 1989. Methanol-fueled Light Duty Vehicle Exhaust Emissions. Paper 89-9.3. Presented at the 82nd Annual Meeting and Exhibition of the American Waste Management Association, Anaheim, Calif., June 25-30.

Hough, A.M. 1988. An intercomparison of mechanisms for the production of photochemical oxidants. J. Geophys. Res. 93:3789—3812.

Hough, A.M., and C. Reeves. 1988. Photochemical oxidant formation and the effects of vehicle exhaust emission controls in the U. K.: The results from 20 different chemical mechanisms. Atmos. Environ. 22:1121—1136.

Hough, A.M., and R.G. Derwent. 1990. Changes in the global concentration of tropospheric ozone due to human activities. Nature 344:645—648.

Hov, Ö. 1984a. Modelling of the long-range transport of peroxyacetyl nitrate to Scandinavia. J. Atmos. Chem. 1:187—202.

Hov, Ö. 1984b. Ozone in the troposphere: High level pollution. Ambio 13:73—79.

Hov, Ö., E. Hesstvedt, and I.S.A. Isaksen. 1978. Long-range transport of tropospheric ozone. Nature 273:341—344.

Hov, Ö., J. Schjoldager, and B.M. Wathne. 1983. Measurement and modeling of the concentrations of terpenes in coniferous forest air. J. Geophys. Res. 88:10679—10688.

Hübler, G., D. Perner, U. Platt, A. Tönnissen, and D.H. Ehhalt. 1984. Groundlevel OH radical concentration: new measurements by optical absorption. J. Geophys. Res. 89:1309—1319.

Hull, L.A. 1981. Terpene ozonolysis products. Pp. 161—186 in Atmospheric Biogenic Hydrocarbons, Vol. 2, J. J. Bufalini and R. R. Arnts, eds. Ann Arbor, Mich.: Ann Arbor Science Publishers, Inc.

Hunt, B.G. 1969. Experiments with a stratospheric general circulation model, III. Large-scale diffusion of ozone including photochemistry. Mon. Wea. Rev. 97:287—306.

Hwang, H., and P.K. Dasgupta. 1986. Fluorometric flow injection determination of aqueous peroxides at nanomolar level using membrane reactors. Anal. Chem. 58:1521—1524.

Ingalls, M.N. 1989. On-road Vehicle Emission Factors from Measurements in a Los Angeles Area Tunnel. Paper 89-137.3. Presented at the 82nd

Annual Meeting and Exhibition of the Air and Waste Management Association Association, Anaheim, Calif., June 25-30.

Ingalls, M.N., L.R. Smith, and R.E. Kirksey. 1989. Measurement of On-road Vehicle Emission Factors in the California South Coast Air Basin, Vol. 1. Regulated Emissions. SCAQS-1, Final Report. No. SwRI-1604. Southwest Research Institute, San Antonio, Texas. June.

Ioffe, B.V., V.A. Isidorov, and I.G. Zenkevich. 1979. Certain regularities in the composition of volatile organic pollutants in the urban atmosphere. Environ. Sci. Technol. 13:864—868.

IPCC (Intergovernmental Panel on Climate Change). 1990. Climate Change: The IPCC Scientific Assessment. Intergovernmental Panel on Climate Change, World Meteorological Organization/United Nations Environment Programme, J.T. Houghton, G.J. Jenkins, and J.J. Ephraums, eds. Cambridge, U.K.: Cambridge University Press.

Isaksen, I.S.A., and O. Hov. 1987. Calculation of trends in the tropospheric concentration of ozone, hydroxyl, carbon monoxide, methane, and nitrogen oxides. Tellus 39B:271—285.

Isaksen, I.S.A., O.Hov, and E. Hesstvedt. 1978. Ozone generation over rural areas. Environ. Sci. Technol. 12:1279—1284.

Isaksen, I.S.A., T. Berntsen, and S. Solberg. 1988. Calculated changes in the tropospheric distribution of long-lived primary trace species and in secondary species resulting from releases of pollutants, report to the EPA, 1988.

Izumi, K., K. Murano, M. Mizuochi, and T. Fukuyama. 1988. Aerosol formation by the photooxidation of cyclohexene in the presence of nitrogen oxides. Environ. Sci. Technol. 22:1207—1215.

Jacob, D.J. 1986. Chemistry of OH in remote clouds and its role in the production of formic acid and peroxymonosulfate. J. Geophys. Res. 91:-9807—9826.

Jacob, D.J., and M.J. Prather. 1990. Radon-222 as a test of the convective transport in a general circulation model. Tellus 42B:118—134.

Jacob, D.J., M.J. Prather, S.C. Wofsy, and M.B. McElroy. 1987. Atmospheric distribution of ^{85}Kr simulated with a general circulation model. J. Geophys. Res. 92:6614—6626.

Jacob, D.J., E.W. Gottlieb, and M.J. Prather. 1989. Chemistry of a polluted cloudy boundary layer. J. Geophys. Res. 94:12975—13002.

Janach, W.E. 1989. Surface ozone: Trend details, seasonal variations, and interpretation. J. Geophys. Res. 94:18289—18295.

Japar, S.M., C.H. Wu, and H. Niki. 1976. Effect of molecular oxygen on the gasphase kinetics of the ozonolysis of olefins. J. Phys. Chem. 80:2057—2062.

Japar, S.M., T.J. Wallington, J.F.O. Richert, and J.C. Ball. 1990. The atmospheric chemistry of oxygenated fuel additives: t-butyl alcohol, dimethyl
ether and methyl t-butyl ether. Int. J. Chem. Kinet. 22:1257—1269.

Jeffries, H.E., K.G. Sexton, and M.S. Holleman. 1985. Outdoor Smog Cham-
ber Experiments: Reactivity of Methanol Exhaust. EPA 460/3-85-009a.
U.S. Environmental Protection Agency, Emission Control Technology
Division, Ann Arbor, Mich.

Jeffries, H.E., K.G. Sexton, and J.R. Arnold. 1989. Validation Testing of
New Mechanisms with Outdoor Chamber Data, Vol. 2: Analysis of VOC
Data for the CB4 and CAL Photochemical Mechanisms. EPA/600/3-89/-
010b. U.S. Environmental Protection Agency, Research Triangle Park,
N.C.

Jenkin, M.E., R.A. Cox, and D.J. Williams. 1988. Laboratory studies of the
kinetics of formation of nitrous acid from the thermal reaction of nitrogen
dioxide and water vapor. Atmos. Environ. 22:487—498.

Johansson, C. 1984. Field measurements of nitric oxide emissions from
fertilized and unfertilized forest soils in Sweden. J. Atmos. Chem. 1:429—
442.

Johansson, C., and L. Granat. 1984. Emission of nitric oxide from arable
land. Tellus 36B:25—37.

Johansson, C., and E. Sanhueza. 1988. Emission of NO from savanna soils
during the rainy season. J. Geophys. Res. 93:14193—14198.

Johansson, C., H. Rodhe, and E. Sanhueza. 1988. Emission of NO in a
tropical savanna and a cloud forest during the dry season. J. Geophys. Res.
93:7180—7192.

Jones, P.D. 1988. Hemispheric surface air temperature variations: Recent
trends and an update to 1987. J. Clim. 1:654—660.

Jones, K., L. Militana, and J. Martini. 1989. Ozone Trend Analysis for Selected Urban Areas in the Contintental U.S. Paper 89-3.6. Presented at the
82nd Annual Meeting and Exhibition of the Air and Waste Management
Association, Anaheim, Calif., June 25-30.

Joshi, S.B, M.C. Dodge, and J.J. Bufalini. 1982. Reactivities of selected
organic compounds and contamination effects. Atmos. Environ. 16:1301—
1310.

Kamens, R.M., M.W. Gery, H.E. Jeffries, M. Jackson, and E.I. Cole. 1982.
Ozone-isoprene reactions: Product formation and aerosol potential. Int.
J. Chem. Kinet. 14:955—975.

Kaplan, W.A., S.C. Wofsy, M. Keller, and J.M. Da Costa. 1988. Emission of
NO and deposition of O_3 in a tropical forest system. J. Geophys. Res.
93:1389—1395.

Karamchandani, P.K., G. Kuntasal, and A. Venkatram. 1988. Results from a comprehensive acid deposition model: Predictions of regional ozone concentrations. Presented at the 17th International Technical Meeting of NATO-CCMS on Air Pollution Modeling and its Application. September 19-22, Downing College, Cambridge, England.

Keene, W.C., J. N. Galloway, and J.D. Holdren, Jr. 1983. Measurement of weak organic acidity in precipitation from remote areas of the world. J. Geophys. Res. 88:5122–5130.

Keene, W.C., A.A.P. Pszenny, J.R. Maben, J.N. Galloway, J.C. Farmer, H. Westberg, S.M. Li, M.O. Andreae, J.W. Munger, M.R. Hoffman, R.B. Norton, R.W. Talbot, W. Winiwarter, and H. Puxbaum. 1986. An intercomparison of measurement systems for vapor and particulate phase concentrations of formic and acetic acids. EOS 67:884.

Kelly, T.J., D.H. Stedman, J.A. Ritter, and R.B. Harvey. 1980. Measurements of oxides of nitrogen and nitric acid in clean air. J. Geophys. Res. 85:7417–7425.

Kelly, N.A., G.T. Wolff, and M.A. Ferman. 1982. Background pollution measurements in air masses affecting the eastern half of the United States. Atmos. Environ. 16:1077–1088.

Kelly, N.A., R.L. Tanner, L. Newman, P.J. Galvin, and J.A. Kadlecek. 1984. Trace gas and aerosol measurements at a remote site in the northeast United States. Atmos. Environ. 18:2565–2576.

Kelly, N.A., M.A. Ferman, and G.T. Wolff. 1986. The chemical and meteorological conditions associated with high and low ozone concentrations in Southeastern Michigan and nearby areas of Ontario. J. Air Pollut. Control Assoc. 36:150–158.

Kessler, R.C. 1988. What Techniques are Available for Generating Windfields? Paper presented at Conference on Photochemical Modeling as a Tool for Decision Makers, State of California Air Resources Board, Pasadena, Calif.

Khalil, M.A.K., and R.A. Rasmussen. 1988. Carbon monoxide in the earth's atmosphere: Indications of a global increase. Nature 332:242–245.

Killus, J.P., and G.Z. Whitten. 1990. Background reactivity in smog chambers. Int. J. Chem. Kinet. 22:547–575.

Kinney, P.L., J.H. Ware, and J.D. Spengler. 1988. A critical evaluation of acute ozone epidemiology results. Arch. Environ. Health 43:168–173.

Kirchoff, V.W.J.H. 1988. Surface ozone measurements in Amazonia. J. Geophys. Res. 93:1469–1476.

Kirchhoff, V.W.J.H., A.W. Setzer, and M.C. Pereira. 1989. Biomass burning in Amazonia: seasonal effects on atmospheric O_3 and CO. Geophys. Res. Lett. 16:469–472.

Kirchner, F., F. Zabel, and K.H. Becker. 1990. Determination of the rate constant ratio for the reaction of the acetylperoxy radical with nitric oxide and nitrogen dioxide. Ber. Bunsen-Ges Phys. Chem. 94:1379—1382.

Kleindienst, T.E., P.B. Shepson, C.M. Nero, R.R. Arnts, S.B. Tejada, G.I. Mackay, L.K. Mayne, H.I. Schiff, J.A. Lind, G.L. Kok, A.L. Lazrus, P.K. Dasgupta, and S. Dong. 1988a. An intercomparison of formaldehyde measurement techniques at ambient concentration. Atmos. Environ. 22:-1931—1940.

Kley, D., and M. McFarland. 1980. Chemiluminescence detector for NO and NO_2. Atmos. Technol. 12:63—69.

Kley, D., J.W. Drummond, M. McFarland, and S.C. Liu. 1981. Tropospheric profiles of NO_x. J. Geophys. Res. 86:3153—3161.

Knispel, R., R. Roch, M. Siese, and C. Zetzsch. 1990. Adduct formation of OH radicals with benzene, toluene, and phenol and consecutive reactions of the adducts with nitrogen oxide and oxygen. Ber. Bunsen-Ges Phys. Chem. 94:1375—1379.

Knoerr, K.R., and F.L. Mowry. 1981. Energy balance/Bowen ratio technique for estimating hydrocarbon fluxes. Pp. 35-52 in Atmospheric Biogenic Hydrocarbons, Vol. 1, J. Bufalini and R. Arnts, eds. Ann Arbor, Mich.: Ann Arbor Science Publishers, Inc.

Kohout, E.H., D.J. Miller, L. Nieves, D.S. Rothman, C.L. Saricks, F. Stodolsky, and D.A. Hanson. 1990. Current Emission Trends for Nitrogen Oxides, Sulfur Dioxide, and Volatile Organic Compounds by Month and State: Methodology and Results. ANL/EAIS/TM-25. Environmental Assessment and Information Sciences Division, Argonne National Laboratory, Argonne, Ill. August.

Kolaz, D.J., and R.L. Swinford. 1990. How to Remove the Influence of Meteorology from the Chicago Area Ozone Trend. Paper 90-97.5. Presented at the 83rd Annual Meeting and Exhibition of the Air and Waste Management Association, Pittsburgh, Penn., June 24-29.

Komhyr, W.D. 1969. Electrochemical concentration cells for gas analysis. Ann. Geophys. 25:203—210.

Korsog, P.E. and G.T. Wolff. 1991. An examination of tropospheric ozone trends in the northeastern U.S. (1973-1983) using a robust statistical method. Atmos. Environ. 25B:47—57.

Kotzias, D., J.L. Hjorth, and H. Skov. 1989. A chemical mechanism for dry deposition - the role of biogenic hydrocarbon (terpene) emissions in the dry deposition of ozone, sulfur dioxide, and nitrogen oxides in forest areas. Toxicol. Environ. Chem. 20/21:95—99.

Kowalczyk, M.L., and E. Bauer. 1982. Lightning as a Source of Nitrogen Oxides in the Troposphere. FAA-EE-82-4. Federal Aviation Administration, U.S. Department of Transportation. Washington, D.C.: U.S. Government Printing Office.

Krupnick, A., H. Dolawtabadi, and A. Russell. 1990. Electric Vehicles and the Environment: Consequences for Emissions and Air Quality in Los Angeles and U.S. Regions. Washington, D.C.: Resources for the Future.

Kumar, S., and D.P. Chock. 1984. An update on oxidant trends in the south coast air basin of California. Atmos. Environ. 18:2131–2134.

Kuntasal, G., and T.Y. Chang. 1987. Trends and relationships of O3, NOx and HC in the south coast air basin of California. J. Air Pollut. Control Assoc. 37:1158-1163.

Laity, J.L., I.G. Burstain, and B.R. Appel. 1973. Photochemical smog and the atmospheric reactions of solvents. Advan. Chem. Ser. 124:95–112.

Lamb, R.G. 1983. Regional Scale (1000km) Model of Photochemical Air Pollution, Part 1. Theoretical Formulation. EPA/600/3-83-035. U.S. Environmental Protection Agency, Environmental Sciences Research Laboratories, Research Triangle Park, N.C.

Lamb, R.G. 1984. Regional Scale (1000km) Model of Photochemical Air Pollution, Part 2. Input Processor Network Design. EPA 600/3-84-085. U.S. Environmental Protection Agency, Environmental Sciences Research Laboratories, Research Triangle Park, N.C.

Lamb, B., H. Westberg, G. Allwine, and T. Quarles. 1985. Biogenic hydrocarbon emissions from deciduous and coniferous trees in the United States. J. Geophys. Res. 90:2380–2390.

Lamb, B., H. Westberg, and G. Allwine. 1986. Isoprene emission fluxes determined by an atmospheric tracer technique. Atmos. Environ. 20:1–8.

Lamb, B., A. Guenther, D. Gay, and H. Westberg. 1987. A national inventory of biogenic hydrocarbon emissions. Atmos. Environ. 21:1695–1705.

Lammel, G., and D. Perner. 1988. The atmospheric aerosol as a source of nitrous-acid in the polluted atmosphere. J. Aeros. Sci. 19:1199–1202.

Larsen, L.C., R.A. Bradley, and G. L. Honcoop. 1990. A New Method of Characterizing the Variability of Ozone Air Quality-Related Indicators. Presented at the Air and Waste Management Association Specialty Conference, Tropospheric Ozone and the Environment, Los Angeles, Calif. Air and Waste Management Transactions Series, Pittsburgh, Penn.

Lavoie, R.L. 1972. Mesoscale numerical model of lake-effect storms. J. Atmos. Sci. 29:1025–1049.

Lawson, D.R. 1990. The southern California air quality study. J. Air Waste Manage. Assoc. 40:156–165.

Lawson, D.R., P.J. Groblicki, D.H. Stedman, G.A. Bishop, and P.L. Guenther. 1990a. Emissions from in-use motor vehicles in Los Angeles: A pilot study of remote sensing and the inspection and maintenance program. J.

Air Waste Manage. Assoc. 40:1096—1105.

Lawson, D.R., H.W. Biermann, E.C. Tuazon, A.M. Winer, G.I. Mackay, H.I. Schiff, G.L. Kok, P.K. Dasgupta, and K. Fung. 1990b. Formaldehyde measurement methods evaluation and ambient concentrations during the carbonaceous species methods comparison study. Aerosol Sci. Tech. 12:-64—76.

Lazrus, A.L., G.L. Kok, J.A. Lind, S.N. Gitlin, B.G. Heikes, and R.E. Shetter. 1986. Automated fluorometric method for hydrogen peroxide in air. Anal. Chem. 58:594—597.

Lee, E.H., D.T. Tingey, and W.E. Hogsett. 1988. Evaluation of ozone exposure indices in exposure-response modeling. Environ. Pollut. 53:43—62.

Lee, E.H., D.T. Tingey, and W.E. Hogsett. 1989. Interrelation of Experimental Exposure and Ambient Air Quality Data for Comparison of Ozone Exposure Indices and Estimating Agricultural Losses. EPA/600/3-89-047. U.S. Environmental Protection Agency, NSI Technology Services Corporation, Corvallis Environmental Research Laboratory, Corvallis, Oregon. June.

Lefohn, A.S., and H.M. Benedict. 1982. Development of a mathematical index that describes ozone concentration, frequency and duration. Atmos. Environ. 16:2529—2532.

Lefohn, A.S., and V.A. Mohnen. 1986. The characterization of ozone, sulfur dioxide, and nitrogen dioxide for selected monitoring sites in the Federal Republic of Germany. J. Air Pollut. Control Assoc. 36:1329—1337.

Lefohn, A.S., and V.C. Runeckles. 1987. Establishing standards to protect vegetation--ozone exposure/dose considerations. Atmos. Environ. 21:561—568.

Lefohn, A.S. and J.E. Pinkerton. 1988. High resolution characterization of ozone data for sites located in forested areas of the United States. J. Air Pollut. Control Assoc. 38:1504—1511.

Lefohn, A.S., J.A. Lawrence, and R.J. Kohout. 1988. A comparison of indices that describe the relationship between exposure to ozone and reduction in the yield of agricultural crops. Atmos. Environ. 22:1229—1240.

Leighton, P.A. 1961. Photochemistry of Air Pollution. New York: Academic Press.

Lelieveld, J., and P.J. Crutzen. 1990. Influences of cloud photochemical processes on tropospheric ozone. Nature 343:227—233.

Leuenberger, C. M.P. Ligocki, and J.F. Pankow. 1985. Trace organic compounds in rain. 4. Identities, concentrations and scavenging mechanisms for phenosl in urban air and rain. Environ. Sci. Technol. 19:1053—1058.

Levy, H., II. 1971. Normal atmosphere: Large radical and formaldehyde concentrations predicted. Science 173:141—143.

Levy, H., II. 1972. Photochemistry of the troposphere. Planet. Space Sci. 20:919—935.

Levy, H., II, and W.J. Moxim. 1989. Simulated global distribution and deposition of reactive nitrogen emitted by fossil fuel combustion. Tellus 41B:-256—271.

Levy, H., II, J.D. Mahlman, and W.J. Moxim. 1982. Tropospheric N_2O variability. J. Geophys. Res. 87:3061—3080.

Levy, H., II, J.D. Mahlman, W.J. Moxim, and S.C. Liu. 1985. Tropospheric ozone: The role of transport. J. Geophys. Res. 90:3753—3772.

Lindsay, R.W., and W.L. Chameides. 1988. High-ozone events in Atlanta, Georgia, in 1983 and 1984. Environ. Sci. Technol. 22:426—431.

Lindsay, R.W., J.L. Richardson, and W.L. Chameides. 1989. Ozone trends in Atlanta, Georgia: Have emission controls been effective? J. Air Waste Manage. Assoc. 39:40—43.

Linn, W.S., E.L. Avol, D.A. Shamoo, R.C. Peng, L.M. Valencia, E.E. Little, and J.D. Hackney. 1988. Repeated laboratory ozone exposures of volunteer Los Angeles residents: An apparent seasonal variation in response. Toxicol. Indust. Health 4:505—520.

Lioy, P.J., J. Vollmuth, J., and M. Lippmann. 1985. Persistence of peak flow decrements in children following ozone exposures exceeding the NAAQS. J. Air Pollut. Control Assoc. 35:1068—1071.

Lioy, P.J., and R.V. Dyba. 1989. Tropospheric ozone: The dynamics of human exposure. Tox. Ind. Health 5:493—504.

Lipfert, F.W., S.C. Morris, and R.E. Wyzga. 1989. Acid aerosols: The next criteria air pollutant? Environ. Sci. Technol. 23:1316—1322.

Lippmann, M. 1989. Health effects of ozone: A critical review. J. Air Waste Manage. Assoc. 39:672—695.

Liu, S. C., and M. Trainer. 1988. Responses of the tropospheric ozone and odd hydrogen radicals to column ozone change. J. Atmos. Chem. 6:221—233.

Liu, M.-K., R.E. Morris, and J.P. Killus. 1984. Development of a regional oxidant model and application to the northeastern United States. Atmos. Environ. 18:1145—1161.

Liu, S.C., M. Trainer, F.C. Fehsenfeld, D.D. Parrish, E. J. Williams, D.W. Fahey, G. Hübler, and P.C. Murphy. 1987. Ozone production in the rural troposphere and the implications for regional and global ozone distributions. J. Geophys. Res. 92:4191—4207.

Livingston, F.E., and B.J. Finlayson-Pitts. 1991. The reaction of gaseous N_2O_5 with solid NaCl at 298-K - Estimated lower limit to the reaction probability and its potential role in tropospheric and stratospheric chemistry. Geophys. Res. Lett. 18:17—20.

Lloyd, A.C., R. Atkinson, F.W. Lurmann, and B. Nitta. 1983. Modeling potential ozone impacts from natural hydrocarbons. 1. Development and testing of a chemical mechanism for the nitrogen oxides-air photooxidations of isoprene and α-pinene under ambient conditions. Atmos. Environ. 17:1931—1950.

Logan, J.A. 1983. Nitrogen oxides in the troposphere: Global and regional budgets. J. Geophys. Res. 88:10785—10807.

Logan, J.A. 1985. Tropospheric ozone: Seasonal behavior, trends, and anthropogenic influence. J. Geophys. Res. 90:10463—10482.

Logan, J.A. 1988. The ozone problem in the rural areas of the United States. Pp. 327-344 in Tropospheric Ozone: Regional and Global Scale Interactions, I.S.A. Isaksen, ed. NATO ASI Series C, Vol. 227. Dordrecht, Holland: D. Reidel Publishing Co.

Logan, J.A. 1989. Ozone in rural areas of the United States. J. Geophys. Res. 94:8511—8532.

Logan, J.A., and V.W.J.H. Kirchhoff. 1986. Seasonal variations of tropospheric ozone at Natal, Brazil. J. Geophys. Res. 91:7875—7881.

Logan, J.A., M.J. Prather, S.C. Wofsy and M.B. McElroy. 1981. Tropospheric chemistry: A global perspective. J. Geophys. Res. 86:7210—7254.

Lonneman, W.A., T.A. Bellar, and A.P. Altshuller. 1968. Aromatic hydrocarbons in the atmosphere of the Los Angeles basin. Environ. Sci. Technol. 2:1017—1020.

Lonneman, W.A., S.L. Kopczynski, P.E. Darley, and F.D. Sutterfield. 1974. Hydrocarbon composition of urban air pollution. Environ. Sci. Technol. 8:229—236.

Lonneman, W.A., R.L. Seila, and S.A. Meeks. 1986. Non-methane organic composition in the Lincoln Tunnel. Environ. Sci. Technol. 20:790—796.

Lovelace, B. 1990. EMFAC - past, present, and future. Pp. 3-73—3-79 in Proceedings of the CRC-APRAC Vehicle Emissions Modeling Workshop, S. Cadle, ed., Newport Beach, Calif., October 30-31. Coordinating Research Council, Inc., Atlanta, Ga.

Lübkert, B., and S. De Tilly. 1989. The OECD-Map emission inventory for sulphur dioxide, nitrogen oxide, and volatile organic compounds in Western Europe. Atmos. Environ. 23:3—15.

Lübkert, B., and K.-H. Zierock. 1989. European emission inventories--A proposal of international worksharing. Atmos. Environ. 23:37—48.

Ludwig, F., E. Reiter, E. Shelar, and W.B. Johnson. 1977. The Relation of

Oxidant Levels to Precursor Emissions and Meteorological Features., Vol. 1. Analysis and Findings. EPA-50/3-77-022a. Environmental Protection Agency, Office of Air Quality Standards and Planning, Research Triangle Park, N.C. October.

Lurmann, F.W., A.C. Lloyd, and R. Atkinson. 1986a. Suggested Gas-phase Species Measurements for the Southern California Air Quality Study. Environmental Research and Technology Document No. P-E124-001. Report to Coordinating Research Council, Atlanta, Ga. Environmental Research & Technology, Inc., Westlake Village, Calif. September.

Lurmann, F.W., A.C. Lloyd, and R. Atkinson. 1986b. A chemical mechanism for use in long-range transport/acid deposition computer modeling. J. Geophys. Res. 91:10905—10936.

Lurmann, F.W., W.P.L. Carter, and L.A. Coyner. 1987. A Surrogate Species Chemical Reaction Mechanism for Urban-scale Air Quality Simulation Models, Vol. I, Adaptation of the Mechanism; Vol. II, Guidelines for Using the Mechanism. EPA-600/3-87-014a,b. Environmental Protection Agency, Atmospheric Sciences Lab, Research Triangle Park, N.C.

Lyons, W.A., R.H. Calby, and C.S. Keen. 1986. The impact of mesoscale convective systems on regional visibility and oxidant distributions during persistent elevated pollution episodes. J. Clim. Appl. Meteor. 25:1518—1531.

Machiele, P.A. 1987. Flammability and Toxicity Trade-offs with Methanol Fuels. SAE Technical Paper Series 872064. Warrendale, Penn.: Society of Automotive Engineers.

Mahlman, J.D., and W.J. Moxim. 1978. Tracer simulation using a global general circulation model: Results from a mid-latitude instantaneous source experiment. J. Atmos. Sci. 35:1340—1374.

Mahlman, J.D., H. Levy, II, and W.J. Moxim. 1980. Three-dimensional tracer structure and behavior as simulated in two ozone precursor experiments. J. Atmos. Sci. 37:655—685.

Mann, L.K., S.B. McLaughlin, and D.S. Shriner. 1980. Seasonal physiological responses of white pine under chronic air pollution stress. Environ. Exp. Bot. 20:99—105.

Martin, R.S., H. Westberg, E. Allwine, L. Ashman, J.C. Farmer, and B. Lamb. 1991. Measurement of isoprene and its atmospheric oxidation-products in a central Pennsylvania deciduous forest. J. Atmos. Chem. 13:1—32.

Martinez, J.R., and H.B. Singh. 1979. Survey of the Role of NO, in Nonurban Ozone Formation. EPA/450/4-79-035. United States Environmental Protection Agency, Research Triangle Park, N. C.

Mayrsohn, H., and J.H. Crabtree. 1976. Source reconciliation of atmospheric hydrocarbons. Atmos. Environ. 10:137—143.

McCurdy, T. 1990. Relationships among alternative formulations of an ozone NAAQS. Paper 90-113.2 in Proceedings of the 83rd Annual Meeting and Exhibition of the Air and Waste Management Association, Pittsburgh, Penn., June 24-29.

McCurdy, T., and R. Atherton. 1990. Variability of ozone air quality indicators in selected metropolitan statistical areas. J. Air Waste Manage. Assoc. 40:477–486.

McKeen, S.A., M. Trainer, E.-Y. Hsie, R. Tallamraju, and S.C. Liu. 1990. On the indirect determination of atmospheric OH radical concentrations from reactive hydrocarbon measurements. J. Geophys. Res. 95:7493–7500.

McKeen, S.A., E.-Y. Hsie, M. Trainer, R. Tallamraju, and S.C. Liu. 1991a. A regional model study of the ozone budget in the eastern United States. J. Geophys. Res. 96:10809–10846.

McKeen, S.A., E.-Y. Hsie, and S.C. Liu. 1991b. A study of the dependence of rural ozone on ozone precursors in the eastern United States. J. Geophys. Res. 96:15377–15394.

McRae, G.J., and J.H. Seinfeld. 1983. Development of a second-generation mathematical model for urban air pollution--II: Evaluation of model performance. Atmos. Environ. 17:501–522.

McRae, G.J., W.R. Goodin, and J.H. Seinfeld. 1982. Development of a second-generation mathematical model for urban air pollution--I. Model formulation. Atmos. Environ. 16:679–696.

Meyer, E.L. 1987. A Review of 1986 Philadelphia NMOC, NO_x and NMOC/NO_x Ratio Data, Interim Report. U.S. Environmental Protection Agency, Research Triangle Park, N.C.

Middleton, P., and J.S. Chang. 1990. Analysis of RADM gas concentration prediction using OSCAR and NEROS monitoring data. Atmos. Environ. 24A:2113–2125.

Middleton, P., J.S. Chang, J.C. Del Corral, H. Geiss, and J.M. Rosinski. 1988. Comparison of RADM and OSCAR precipitation chemistry data. Atmos. Environ. 22:1195–1208.

Middleton, P., W.R. Stockwell, and W.P.L. Carter. 1990. Aggregation and analysis of volatile organic compound emissions for regional modeling. Atmos. Environ. 24A:1107–1134.

Milford, J.B., A.G. Russell, and G.J. McRae. 1989. A new approach to photochemical pollution control: Implications of spatial patterns in pollutant responses to reductions in nitrogen oxides and reactive organic gas emissions. Environ. Sci. Technol. 23:1290–1301.

Miller, P.R., J.R. McBride, S.L. Schilling, and A.P. Gomez. 1989. Trend of Ozone Damage to Conifer Forests 1984-1988 in the San Bernadino Mountains of Southern California. Paper 89-129.6 in Proceedings of the 82nd

Annual Meeting and Exhibition of the American Waste Management Association Meeting, Anaheim, Calif., June 25-30.

Mohnen, V.A., A. Hogan, and P. Coffey. 1977. Ozone measurements in rural areas. J. Geophys. Res. 82:5889—5895.

Monson, R.K., and R. Fall. 1989. Isoprene emission from aspen leaves: Influence of environment and relation to photosynthesis and photorespiration. Plant Physiol. 90:267—274.

Montreal Protocol on Substances that Deplete the Ozone Layer. 1987. Final Act. UNEP NA-6106. Conference on the Protocol on Chlorofluorocarbons to the Vienna Convention for the Protection of the Ozone Layer, United Nations Environment Programme. Montreal, Canada. September 14-16.

Montzka, D.D., M.A. Carroll, G. Hübler, C. Hahn, B.A. Ridley, J.G. Walega, P. Zimmerman, L. Heidt, E. Atlas, S. Schauffler, B. Heikes. 1989. In-situ measurements of NO_x During the MLOPEX Program. Session A12B-9, EOS 70:1014.

Moortgat, G.K., B. Veyret, and R. Lesclaux. 1989. Kinetics of the reaction of HO_2 with $CH_3C(O)O_2$ in the temperature range 253-368 K. Chem. Phys. Lett. 160:443—447.

Morris, R. 1990. Selection of Cities for Assessing the Effects of Test Fuels on Air Quality. Systems Applications International, Inc., San Rafael, Calif. January 17.

Morris, R., S. Reynolds, M. Yocke, and M.-K. Liu. 1988. The Systems Applications Inc. regional transport model: Current status and future needs. Pp. 257—287 in The Scientific and Technical Issues Facing Post-1987 Ozone Control Strategies , G.T. Wolff, J.L. Hanisch, and K. Schere, eds. Pittsburgh, Penn.: Air and Waste Management Association.

Mozurkewich, M. and J.G. Calvert. 1988. Reaction probability of N_2O_5 on aqueous aerosols. J. Geophys. Res. 93:15889—15896.

Mueller, P.K. 1982. Eastern Regional Air Quality Measurements. EPRI Report EA-1914, Vol. 1. Electric Power Research Institute, Palo Alto, Calif.

Mueller, P.K. and G.M. Hidy. 1983. The Sulfate Regional Experiment: Report of Findings. EPRI Report EA-1901 (3 vols.). Electric Power Research Institute, Palo Alto, Calif.

Musselman, R.C., R.J. Oshima, and R.E. Gallavan. 1983. Significance of pollutant concentration distribution in the response of red kidney beans (Phaseolus vulgaris) to ozone. J. Am. Soc. Hortic. Sci. 108:347—351.

Musselman, R.C., A.J. Huerta, P.M. McCool, and R.J. Oshima. 1986. Response of beans to simulated ambient and uniform ozone distribution with equal peak concentrations. J. Am. Soc. Hortic. Sci. 111:470—473.

Mylonas, D.T., D.T. Allen, S.H. Ehrman, and S.E. Pratsinis. 1991. The sources and size distributions of organnonitrates in Los Angeles aerosol. Atmos. Environ. 25A:2855–2861.

NAS (National Academy of Sciences). 1991. Policy Implications of Greenhouse Warming-- Synthesis Panel. Washington, D.C.: National Academy Press. 127 pp.

Nichols, R.J., and J.M. Norbeck. 1985. Assessment of Emissions from Methanol-fueled Vehicles: Implications for Ozone Air Quality. Paper 85-38.3. Presented at the 78th Annual Meeting of the Air Pollution Control Association, Detroit, Mich., June 16-21.

Niemann, B.L. 1988. Trends in Ozone Concentrations in the Major Ozone Transport Areas: The Influence of Precursor Emissions and Abnormal Meteorology. Paper 88-50.7. Presented at the 81st Annual Meeting and Exhibition of the Air Pollution Control Association, Dallas, Texas, June 19-24.

Niki, H., P.D. Maker, C.M. Savage, and L.P. Breitenbach. 1983. Atmospheric ozone-olefin reactions. Environ. Sci. Technol. 17:312A–322A.

Norton, R.B. 1987. The measurement of tropospheric organic gas phase acids with NA_2CO_3 coated denuder tubes. EOS 68:1216.

Nouchi, I., and K. Aoki. 1979. Morning glory as a photochemical oxidant indicator. Environ. Pollut. 18:289–303.

Novak, J.H., and J.A. Reagan. 1986. A comparison of natural and man-made hydrocarbon emission inventories necessary for regional acid deposition and oxidant modeling. Paper 86-30. Proceedings of the 79th Annual Air Pollution Control Association Meeting, Air Pollution Control Association, Pittsburgh, Penn, June 23-27.

NRC (National Research Council). 1981. Formaldehyde and Other Aldehydes. Washington, D.C.: National Academy Press. 340pp.

NRC (National Research Council). 1982. Causes and Effects of Stratospheric Ozone Reduction: An Update. Washington, D.C.: National Academy Press. 339 pp.

NRC (National Research Council). 1984. Causes and Effects of Changes in Stratospheric Ozone: Update 1983. Washington, D.C.: National Academy Press. 254 pp.

NRC (National Research Council), 1990. Advancing the Understanding and Forecasting of Mesoscale Weather in the United States. Washington, D.C.: National Academy Press. 56 pp.

Oke, T.R. 1973. City size and the urban heat island. Atmos. Environ. 7:769-–779.

Olson, R. 1980. Geoecology: A County-level Environmental Data Base for the Conterminous United States. Oak Ridge National Laboratory. Publication No. 1537. Oak Ridge, Tenn.

Oltmans, S.J., and W.D. Komhyr. 1986. Surface ozone distributions and variations from 1973-1984 measurements at the NOAA Geophysical Monitoring for Climate Change Baseline Observatories. J. Geophys. Res. 91:-5229–5236.

Oltmans, S.J., W.D. Komhyr, P.R. Franchois, and W.A. Matthews. 1989. Tropospheric ozone: Variations from surface and ECC ozonesonde observations. Pp. 539–543 in Ozone in the Atmosphere, Proceedings of the Quadrennial Ozone Symposium, 4-13 August, 1988, Gottingen, Fed. Rep. Germany, R.D. Bojkov and P. Fabian, eds. Hampton, Virg.: A. Deepak Publishing.

Ouimette, J.R., and R.C. Flagan. 1982. The extinction coefficient of multicomponent aerosols. Atmos. Environ. 16:2405–2420.

Oshima, R.J. 1975. Final Report to the California Air Resources Board. Development of a System for Evaluating and Reporting Economic Crop Losses Caused by Air Pollution in California. III. Ozone Dosage--Crop Loss Conversion Function--Alfalfa, Sweet Corn. IIIA. Procedures for Production, Ozone Effects on Alfalfa, Sweet Corn and Evaluation of These Systems. State of California, Department of Food and Agriculture, Sacramento, California. 107 pp.

OTA (Office of Technology Assessment, U.S. Congress). 1989. Catching Our Breath--Next Steps for Reducing Urban Ozone. OTA-O-412. Congressional Board of the 101st Congress. Washington, DC: U.S. Government Printing Office. July.

O'Toole, R.P., E. Dutzi, R. Gershman, R. Heft, W. Kalema, and D. Maynard. 1983. California Methanol Assessment, Vol. II, Technical Report. Chapter 6 in Air Quality Impact of Methanol Use in Vehicles, JPL Publ. 83-18. Jet Propulsion Laboratory and Division of Chemistry and Chemical Engineering, California Institute of Technology, Pasadena, Calif.

Pandis, S.N., and J.H. Seinfeld. 1989. Mathematical modeling of acid deposition due to radiation fog. J. Geophys. Res. 94:12911–12923.

Pandis, S.N., S.E. Paulson, J.H. Seinfeld, and R.C. Flagan. 1991. Aerosol formation in the photooxidation of isoprene and β-pinene. Atmos. Environ. 25A:997–1008.

Parrish, D.D. 1990. Tropospheric Chemistry--Continental Measurements. Fall American Geophysical Union meeting in San Francisco, D.D. Parrish, presiding. Session A11C, EOS 71:1228.

Parrish, D.D., E. Williams, R.B. Norton, F.C. Fehsenfeld. 1985. Measurement of odd-nitrogen species and O_3 at Point Arena, California. Session A21A-09, EOS 66:820.

Parrish, D.D., M. Trainer, E.J. Williams, D.W. Fahey, G. Hübler, C.S. Eubank, S.C. Liu, P.C. Murphy, D.L. Albritton, and F.C. Fehsenfeld. 1986. Measurements of the NO_x-O_3 photostationary state at Niwot Ridge, Colorado. J. Geophys. Res. 91:5361—5370.

Parrish, D., M. Buhr, R. Norton, and F. Fehsenfeld. 1988. Study of atmospheric oxidation processes at a rural, eastern U.S. site. EOS 69:1056.

Parrish, D.D., C.H. Hahn, D.W. Fahey, E.J. Williams, M.J. Bollinger, G. Hubler, M.P. Buhr, P.C. Murphy, M. Trainer, E.Y. Hsie, S.C. Liu, and F.C. Fehsenfeld. 1990. Systematic variations in the concentration of NO_x (NO plus NO_2) at Niwot Ridge, Colorado. J. Geophys. Res. 95:1817—1836.

Parrish, D.D., M. Trainer, M.P. Buhr, B.A. Watkins, and F.C. Fehsenfeld. 1991. Carbon monoxide concentrations and their relation to concentrations of total reactive oxidized nitrogen at two rural U.S. sites. J. Geophys. Res. 96:9309—9320.

Paul, R.A., W.F. Biller, T. McCurdy. 1987. National estimates of population exposure to ozone. Paper 87-42.7 in Proceedings of the 80th Annual Meeting and Exhibition of the Air Pollution Control Association, American Pollution Control Association, Pittsburgh, Penn.

Paulson, S.E., R.C. Flagan, and J.H. Seinfeld. 1992a. Atmospheric photooxidation of isoprene. Part 1: The hydroxyl radical and ground state atomic oxygen reactions. Int. J. Chem. Kinet. 24:79—101.

Paulson, S.E., R.C. Flagan, and J.H. Seinfeld. 1992b. Atmospheric photo-oxidation of isoprene. Part 2: The ozone-isoprene reaction. Int. J. Chem. Kinet., in press.

Pefley, R.K., J.B. Pullman, and G.Z. Whitten. 1984. The Impact of Alcohol Fuels on Urban Air Pollution: Methanol Photochemistry Study. Final Report. DOE/84/CE/50036-1. U.S. Department of Energy, Office of Vehicles and Engine Research and Development, Washington, D.C. Washington, D.C.: U.S. Government Printing Office.

Pechan, E.H. & Associates, Inc. 1990. National Assessment of VOC, CO, and NO_x Controls, Emissions, and Costs. EPA Contract 68-W8-0038, U.S. Environmental Protection Agency, Office of Policy Planning and Evaluation, Washington, D.C.

Penner, J.E., C.S. Atherton, J. Dignon, S.J. Ghan, and J.J. Walton, and S. Hameed. 1991. Tropospheric nitrogen: A three-dimensional study of sources, distributions, and deposition. J. Geophys. Res. 96D:959—990.

Perner, D., U. Platt, M. Trainer, G. Hübler, J. Drummond, W. Junkermann, J. Rudolph, B. Schubert, A. Volz, D.H. Ehhalt, K.J. Rumpel, and G. Helas. 1987. Measurements of tropospheric OH concentrations: A comparison of field data with model prediction. J. Atmos. Chem. 5:185—216.

Peterson, J.T. 1976. Calculated actinic fluxes (290-700 nm) for air pollution photochemistry application. EPA 600/4-76-025. U.S. Environmental Protection Agency. 55 pp.

Petersson, G. 1988. High ambient concentrations of monoterpenes in a Scandinavian pine forest. Atmos. Environ. 22:2617—2620.

Pickering, K.E., R.R. Dickerson, G.J. Huffman, J.F. Boatman, and A. Schanot. 1988. Trace gas transport in the vicinity of frontal convective clouds. J. Geophys. Res. 93D:759—773.

Pickering, K.E., R.R. Dickerson, W.T. Luke, and L.J. Nunnermacker. 1989. Clear-sky vertical profiles of trace gases as influenced by upstream convective activity. J. Geophys. Res. 94D:14879—14892.

Pickering, K.E., A.M. Thompson, R.R. Dickerson, W.T. Luke, D.P. McNamara, J.P. Greenberg, and P.R. Zimmerman. 1990. Model calculations of tropospheric ozone production potential following observed convective events. J. Geophys. Res. 95D:14049—14062.

Pickle, T., D.T. Allen, and S.E. Pratsinis. 1990. The sources and size disbributions of aliphatic and carbonyl carbon in Los Angeles aerosol. Atmos. Environ. 24A:2221—2228.

Pielke, R.A. 1984. Mesoscale Meteorological Modeling. Orlando, Fla.: Academic Press. 612 pp.

Pierce, T.E., and P.S. Waldruff. 1991. PC-BEIS: A personal computer version of the Biogenic Emissions Inventory System. J. Air Waste Manage. Assoc. 41:937—941.

Pierce, T.E., B.K. Lamb, and A.R. Van Meter. 1990. Development of a Biogenic Emissions Inventory System for Regional Air Pollution Models. Paper 90-94.3. Presented at the 83rd Annual Meeting and Exhibition of the Air and Waste Management Association, Pittsburgh, Penn., June 24-29.

Pierotti, D., S.C. Wofsy, D. Jacob, and R.A. Rasmussen. 1990. Isoprene and its oxidation products: Methacrolein and methyl vinyl ketone. J. Geophys. Res. 95:1871—1881.

Pierson, W.R., W.W. Brachaczek, R.H. Hammerle, D.E. McKee, and J.W. Butler. 1978. Sulfate emissions from vehicles on the road. J. Air Pollut. Control Assoc. 28:123—132.

Pierson, W.R. 1990. Memorandum to CRC-APRAC Vehicles Emissions Modeling Workshop attendees, Newport Beach, Calif., October 30-31.

Pilar, S., and W.F. Graydon. 1973. Benzene and toluene distribution in Toronto atmosphere. Environ. Sci. Technol. 7:628—631.

Pitts, J.N. Jr. 1987. Nitration of gaseous polycyclic aromatic hydrocarbons in simulated and ambient urban atmospheres: A source of mutagenic nitroarenes. Atmos. Environ. 21:2531—2547.

Pitts, J.N., Jr., A.M. Winer, K.R. Darnall, A.C. Lloyd, and G.J. Doyle. 1977. Pp. 687—704 in International Conference on Photochemical Oxidant Pollution and Its Control: Proceedings, Vol. II, B. Dimitriades, ed. EPA/-600/3-77-001b. U.S. Environmental Protection Agency, Environmental Sciences Research Laboratory, Research Triangle Park, N.C.

Pitts, J.N., Jr., H.W. Biermann, R. Atkinson, and A.M. Winer. 1984a. Atmospheric implications of simultaneous nighttime measurements of NO_3 radicals and HONO. Geophys. Res. Lett. 11:557—560.

Pitts, J.N., Jr., E. Sanhueza, R. Atkinson, W.P.L. Carter, A.M. Winer, G.W. Harris, and C.N. Plum. 1984b. An investigation of the dark formation of nitrous acid in environmental chambers. Int. J. Chem. Kinet. 16:919—939.

Pitts, J.N., Jr., H.W. Biermann, A.M. Winer, and E.C. Tuazon. 1984c. Spectroscopic identification and measurement of gaseous nitrous acid in dilute auto exhaust. Atmos. Environ. 18:847—854.

Pitts, J.N., Jr., H.W. Biermann, E.C. Tuazon, M. Green, W.D. Long, and A.M. Winer. 1989. Time-resolved identification and measurement of indoor air pollutants by spectroscopic techniques: Gaseous nitrous acid, methanol, formaldehyde and formic acid. J. Air Waste Manage. 39:1344—1347.

Placet, M., R.E. Battye, F.C. Fehsenfeld, and G.W. Bassett. 1990. Emissions Involved in Acidic Deposition Processes. State-of-Science/Technology Report 1. National Acid Precipation Assessment Program. Washington, D.C.: U.S. Government Printing Office. December.

Platt, U., D. Perner, J. Schröeder, C. Kessler, and A. Toennissen. 1981. The diurnal variation of NO_3. J. Geophys. Res. 86:11965—11970.

Platt, U.F., A.M. Winer, H.W. Biermann, R. Atkinson, and J.N. Pitts, Jr. 1984. Measurement of nitrate radical concentrations in continental air. Environ. Sci. Technol. 18:365—369.

Platt, U., M. Rateike, W. Junkermann, J. Rudolph, and D.H. Ehhalt. 1988. New tropospheric OH measurements. J. Geophys. Res. 93:5159—5166.

Plum, C.N., E. Sanhueza, R. Atkinson, W.P.L. Carter, and J.N. Pitts, Jr. 1983. OH radical rate constants and photolysis rates of α-dicarbonyls. Environ. Sci. Technol. 17:479—484.

Pollack, A.K. 1986. Application of a simple meteorological index of ambient ozone potential to ten cities. Paper 86-19.5 in Proceedings of the 79th Annual Meeting and Exhibition of the Air Pollution Control Association, Minneapolis, Minn., June 23-27.

Pollack, A.K., T.E. Stockenius, J.L. Haney, T.S. Stocking, J.L. Fieber, and M. Moezzi. 1988. Analysis of Historical Ozone Concentrations in the Northeast, Volume I. Main Report. SYSAPP-88-/192a. Systems Applications, Inc., San Rafael, Calif. November.

Possiel, N.C., and W.M. Cox. 1990. The Relative Effectiveness of NO_x and

VOC Strategies in Reducing Northeast Ozone Concentrations. Paper presented at the symposium on The Role and Importance of Nitrogen Oxides Emissions Controls in Ozone Attainment Strategies for Eastern North America, sponsored by NESCAUM (Northeast States for Coordinated Air Use Management) meeting. Cambridge, Mass., May 30-June 1.

Possiel, N.C., D.C. Doll, K.A. Bauges, E.W. Baldridge, and R.A. Wayland. 1990. Impacts of Regional Control Strategies on Ozone in the Northeastern United States. Paper 90-93.3. Presented at the 83rd Annual Meeting and Exhibition of the Air and Waste Management Association, Pittsburgh, Penn., June 24-29.

Prather, M.J., ed. 1988. An Assessment Model for Atmospheric Composition. NASA Conference Publication, CP-3203. Proceedings of a workshop held at NASA-Goddard Institute for Space Studies, New York, N.Y., January 10-13. U.S. National Aeronautics and Space Administration. 64 pp.

Prather, M.J., and R.T. Watson. 1990. Stratospheric ozone depletion and future levels of atmospheric chlorine and bromine. Nature 344:729–734.

Prather, M., M. McElroy, S. Wofsy, G. Russell, and D. Rind. 1987. Chemistry of the global troposphere: Fluorocarbons as tracers of air motion. J. Geophys. Res. 92:6579–6613.

Pratt, G.C., R. C. Hendrickson, B.I. Chevone, D.A. Christopherson, M.V. O'Brien, and S.V. Krupa. 1983. Ozone and oxides of nitrogen in the rural upper-midwestern USA. Atmos. Environ. 17:2013–2023.

Prinn, R., D. Cunnold, R. Rasmussen, P. Simmonds, F. Alyea, A. Crawford, P. Fraser, and R. Rosen. 1987. Atmospheric trends in methylcholoroform and the global average for the hydroxyl radical. Science 238:945–950.

Ramanathan, V., R.D. Cess, E.F. Harrison, P. Minnis, B.R. Barkstrom, E. Ahmad, and D. Hartmann. 1989. Cloud-radiative forcing and climate: Results from the Earth Radiation Budget Experiment. Science 243:57–63.

Ranzieri, A.J., and R. Thullier. 1991. San Joaquin Valley Air Quality Study (SJVAQS) and Atmospheric Utility Signatures, Predictions, and Experiments (AUSPEX): A Collaborative Modeling Program. Paper 91-70.5. Presented at the 84th Annual Meeting and Exhibition of the Air and Waste Management Association, Vancouver, BC, June 16-21.

Rao, S.T. 1987. Application of the Urban Airshed Model to the New York Metropolitan Area. EPA-450/4-87-011. Environmental Protection Agency, Office of Air Quality Planning and Standards, Research Triangle Park, N.C. May.

Rao, S.T., and G. Sistla, G. 1990. The efficacy of nitrogen oxides emissions control in ozone attainment strategies as predicted by the urban airshed model. Intl. J. Water, Air, and Soil Poll.

Rao, S.T., G. Sistla, J.Y. Ku, K. Schere, R. Scheffe, and J. Godowitch. 1989. Nested Grid Modeling Approach for Assessing Urban Ozone Air Quality. Paper 89-42A.2. Presented at the 82nd Annual Meeting and Exhibition of the Air and Waste Management Association, Anaheim, Calif., June 25-30.

Rapoport, R.D. 1990. Analysis of Volatile Organic Compound Emission Control Methods to Reduce Urban Ozone. Paper 90-93.11. Presented at the 83rd Annual Meeting and Exhibition of the Air and Waste Management Management Association, Pittsburgh, Penn., June 24-29.

Rasch, P.J., and D.L. Williamson. 1991. The sensitivity of a general circulation model climate to the moisture transport function. J. Geophys. Res. 96D:13123–13137.

Rasmussen, R.A. 1970. Isoprene: Identified as a forest-type emission to the atmosphere. Environ. Sci. Technol. 4:667–671.

Rasmussen, R.A. 1972. What do the hydrocarbons from trees contribute to air pollution? J. Air Pollut. Cont. 22:537–543.

Rasmussen, R.A., and M.A.K. Khalil. 1988. Isoprene over the Amazon Basin. J. Geophys. Res. 93:1417–1421.

Raval, A., and V. Ramanathan. 1989. Observational determination of the greenhouse effect. Nature 342:758–761.

Ravishankara, A.R. 1988. Kinetics of radical reactions in the atmospheric oxidation of CH_4. Ann. Rev. Phys. Chem. 39:367–394.

Reinert, R.A., and P.V. Nelson. 1979. Sensitivity and growth of twelve elatior begonia-hiemalis cultivars to ozone. Hort. Sci. 14:747–748.

Reynolds, S.D. 1977. The Systems Applications, Inc. Urban Airshed Model: An Overview of Recent Developmental Work. International Conference on Photochemical Oxidant Pollution and Its Control. EPA-600/3-77-001b. U.S. Environmental Protection Agency, Research Triangle Park, N.C.

Reynolds, S.D., P.M. Roth, and J.H. Seinfeld. 1973. Mathematical modeling of photochemical air pollution-I. Formulation of the model. Atmos. Environ. 7:1033–1061.

Reynolds, S.D., T.W. Tesche, and L.E. Reid. 1979. An Introduction to the SAI Airshed Model and Its Usage. SAI-EF79-31. Systems Applications, Inc., San Rafael, Calif.

Ridley, B.A. 1989. Recent Measurement of Oxidized Nitrogen Compounds in the Troposphere. Paper presented at the International Conference on the Generation of Oxidants on Regional and Global Scales, University of East Anglia, Norwich, England. July 3-7.

Ridley, B.A., M.A. Carroll, and G.L. Gregory. 1987. Measurements of nitric oxide in the boundary layer and free troposphere over the Pacific Ocean. J. Geophys. Res. 92:2025–2047.

Ridley, B.A., M.A. Carroll, A.L. Torres, E.P. Condon, G.W. Sachse, G.F. Hill, and G.L. Gregory. 1988. An intercomparison of results from ferrous sulphate and photolytic converter techniques for measurements of NO_x made during the NASA GTE/CITE 1 aircraft program. J. Geophys. Res. 93:15803—15811.

Ridley, B.A., M.A. Carroll, D.D. Dunlap, M. Trainer, G.W. Sachse, G.L. Gregory, and E.P. Condon. 1989. Measurements of NO_x over the eastern Pacificozone in Ocean and southwestern United States during the spring 1984 NASA GTE aircraft program. J. Geophys. Res. 94:5043—5067.

Rind, D., R. Suozzo, N.K. Balachandran, and M.J. Prather. 1990. Climate change and the middle atmosphere, Part 1: The doubled CO_2 climate. J. Atmos. Sci. 47:475—494.

Robbins, D.E. 1983. NASA-Johnson Space Center (JSC) measurements during "La campagne d'intercomparison d'ozonemetres, Gap France, 1981. Planet. Space Sci. 31:761—765.

Roberts, J.M. 1990. The atmospheric chemistry of organic nitrates. Atmos. Environ. 24A:243—287.

Roberts, J.M., F.C. Fehsenfeld, S.C. Liu, M.J. Bollinger, C. Hahn, D.L. Albritton, and R.E. Sievers. 1984. Measurement of aromatic hydrocarbon ratios and NO_x concentrations in the rural troposphere: Observation of air mass photochemical aging and NO_x removal. Atmos. Environ. 18:2421—2432.

Rogers, J.D. 1990. Ultraviolet absorption cross sections and atmospheric photodissociation rate constants of formaldehyde. J. Phys. Chem. 94:4011—4015.

Rodgers, M.O. 1986. Development and Application of a Photofragmentation/Laser-Induced Fluorescence Detection System for Atmospheric Nitrous Acid. Ph.D. Thesis. School of Geophysical Sciences, Georgia Institute of Technology.

Rodgers, M.O., and D.D. Davis. 1989. A UV-photofragmentation/laser-induced fluorescence sensor for the atmospheric detection of HONO. Environ. Sci. Technol. 23:1106—1112.

Rodgers, M.O., J.D. Bradshaw, S.T. Sandholm, S. KeSheng, and D.D. Davis. 1985. A 2-λ laser- induced fluorescence field instrument for ground based and airborne measurements of atmospheric OH. J. Geophys. Res. 90:12819—12834.

Rogozen, M.B., and R.A. Ziskind. 1984. Formaldehyde: A Survey of Airborne Concentrations and Sources. SAI Publication No. 84/1642. Systems Applications, Inc., San Rafael, Calif.

Roselle, S.J., and K.L. Schere. 1990. Sensitivity of the EPA Regional Oxidant

Model to Biogenic Hydrocarbon Emissions. Paper 90-94.4. Presented at the 83rd Annual Meeting and Exhibition of the Air and Waste Management Association, Pittsburgh, Penn., June 24-29.

Roselle, J.S., T.E. Pierce, and K.L. Schere. 1991. The sensitivity of regional ozone modeling to biogenic hydrocarbons. J. Geophys. Res. 96(D4):7371–7394.

Roth, P.M. 1988. A Proposed Concept and Scope for the San Joaquin Valley Air Quality Study, Vol. 1-Scoping Report Study. Sonoma Technology, Inc.

Roth, P.M., S.D. Reynolds, T.W. Tesche, P.D. Gutfreund, and C. Seigneur. 1983. An appraisal of emissions control requirements in the California south coast air basin. Environ. Int. 9:549–571.

Roth, P.M., C.E. Blanchard, S.D. Reynolds. 1989. The Role of Grid-Based Reactive Air Quality Modeling in Policy Analysis: Perspectives and Implications as Drawn from a Case Study. EPA/600/3-89-082. U.S. Environmental Protection Agency, Atmospheric Research and Exposure Assessment Laboratory, Research Triangle Park, N.C.

RTI (Research Triangle Institute). 1975. Investigation of Rural Oxidant Levels as Related to Urban Hydrocarbon Control Strategies. EPA-450/3-75-036. U.S. Environmental Protection Agency, Research Triangle Park, N.C. 359 pp.

Rubino, R.A., L. Bruckman, and J. Magyar. 1976. Ozone transport. J. Air Pollut. Control Assoc. 26:972–975.

Russell, A.G. 1989. Methanol Fuel Use for Photochemical Smog Control. Paper presented at Methanol as an Alternative Fuel Choice: An Assessment. Johns Hopkins University, Washington, D.C., December 4-5.

Russell, A.G. 1990. Air Quality Modeling of Alternative Fuel Use in Los Angeles, CA: Sensitivity of Pollutant Formation to Individual Pollutant Compounds. Paper 90-96.9. Presented at the 83rd Annual Meeting and Exhibition of the Air and Waste Management Association., Pittsburgh, Penn., June 24-29.

Russell, G.L., and J.A. Lerner. 1981. A new finite-differencing scheme for the tracer transport equation. J. Appl. Meteorol. 20:1483—1498.

Russell, A.G., and G.R. Cass. 1986. Verification of a mathematical model for aerosol nitrate and nitric acid formation and its use for control measure evaluation. Atmos. Environ. 20:2011—2025.

Russell, A.G., G.J. McRae, and G.R. Cass. 1985. The dynamics of nitric acid production and the fate of nitrogen oxides. Atmos. Environ. 19:893—903.

Russell, A.G., K.F. McCue, and G.R. Cass. 1988a. Mathematical modeling of the formation of nitrogen-containing air pollutants. 1. Evaluation of an Eulerian photochemical model. Environ. Sci. Technol. 22:263—271.

Russell, A.G., K.F. McCue, and G.R. Cass. 1988b. Mathematical modeling of the formation of nitrogen-containing air pollutants. 2. Evaluation of the effect of emission controls. Environ. Sci. Technol. 22:1336—1347.

Russell, A.G., J.N. Harris, D. St. Pierre, and J.B. Milford. 1989. Quantitative Estimate of the Air Quality Impacts of Methanol Fuel Use. Prepared for the California Air Resources Board and South Coast Air Quality Management District under ARB Agreement No. A6-048-32, Carnegie Mellon University, Pittsburgh, Pa. April.

Russell, A.G., D. St. Pierre, and J.B. Milford. 1990. Ozone control and methanol fuel use. Science 247:201—205.

Russell, A.G., L.A. McNair, M.T. Odman, and N. Kumar. 1991. Organic Compound Reactivities and the Use of Alternative Fuels. Report to the California Air Resources Board, Sacramento, Calif.

Russell, A.G., L.A. McNair, and M.T. Odman. 1992. Airshed calculation of the sensitivity of pollutant formation to organic compound classes and oxygenates associated with alternative fuels. J. Air Waste Manage., in press.

SAI (Systems Applications, Inc.). 1990. Protocol for Modeling the Air Quality Impact of Fuel Composition Changes in Light Duty Vehicles. SYSAPP-90/056. Systems Applications, Inc., San Rafael, Calif.

SAE (Society of Automotive Engineers). 1991. Topical Technical (TOPTEC) Symposium on Alternative Liquid Transportation Fuels, Alternative Fuel Series, TOPTEC, Dearborn, Mich., August 28-29.

Sakamaki, F., S. Hatakeyama, and H. Akimoto. 1983. Formation of nitrous acid and nitric oxide in the heterogeneous dark reaction of nitrogen dioxide and water vapour in a smog chamber. Int. J. Chem. Kinet. 15:1013—1029.

Sakugawa, H., and I.R. Kaplan. 1990. Observation of the diurnal variation of gaseous H_2O in Los Angeles using a cryogenic collection method. Aerosol. Sci. Technol 12:77—85.

Samson, P.J., and B. Shi. 1988. A Meteorological Investigation of High Ozone Values in American Cities. Report prepared for the United States

Congress, Office of Technology Assessment. Washington, D.C.: U.S. Government Printing Office.

SCAQMD (South Coast Air Quality Management District). 1989. Air Quality Management Plan. South Coast Air Quality Management District and Southern California Association of Governments, El Monte, Calif., March.

Scheffe, R.D. 1990. Urban Airshed Model Study of Five Cities. EPA-450/4-90-006a. U.S. Environmental Protection Agency, Office of Air Quality Planning and Standards, Research Triangle Park, N.C.

Schere, K., and E. Wayland, E. 1989. EPA Regional Oxidant Model (ROM 2.0). Evaluation on 1980 NEROS Data Bases. EPA-600/S3-89/057. U.S. Environmental Protection Agency, Research Triangle Park, N.C. 351 pp.

Schnell, R.C., S.C. Liu, S.J. Oltmans, R.S. Stone, D.J. Hofmann, E.G. Dutton, T. Deshler, W.T. Sturges, J.W. Harder, S.D. Sewell, M. Trainer, and J.M. Harris. 1991. Decrease of summer tropospheric ozone concentrations in Antarctica. Nature 351:726–729.

Schönbein, C.F. 1840. Boebachtungen u;ber den bei der elektrolysation des wassers and dom Ausstro;men der gewo;hnlichen electrizitat aus spitzen eich entwichelnden geruch. Ann. Phys. Chem. 50:616.

Schönbein, C.F. 1854. Ube;r verschiedene zusta;nde des sauerstoffs, liebigs. Ann. Chem. 89:257–300.

Schuetzle, D., and R.A. Rasmussen. 1978. The molecular composition of secondary aerosol particles formed from terpenes. J. Air Pollut. Control Assoc. 28:236–240.

Schwartz, W.E. 1974. Chemical Characterization of Model Aerosols. EPA/650/3-74-001. U.S. Environmental Protection Agency, National Environmental Research Center, Research Triangle Park, N.C. 80 pp.

Seigneur, C., T.W. Tesche, P.M. Roth, and L.E. Reid. 1981. Sensitivity of a complex urban air quality model to input data. J. Appl. Meteorol. 20:1020–1040.

Seiler, W., and P.J. Crutzen. 1980. Estimates of gross and net fluxes of carbon between the biosphere and the atmosphere from biomass burning. Clim. Change 2:207–247.

Seiler, W., H. Geihl, and H. Ellis. 1976. A Method for Monitoring of Background CO and first Results of Continuous Monitoring of CO Resisterations on Mauna Loa Observatory. Pp. 31-39 in Special Environmental Report No. 10. Geneva: World Meteorological Organization.

Seinfeld, J.H. 1986. Atmospheric Chemistry and Physics of Air Pollution. New York: Wiley-Interscience Publication.

Seinfeld, J.H. 1988. Ozone air quality models: A critical review. J. Air

Pollut. Control Assoc. 38:616—645.

Seinfeld, J.H. 1989. Urban air pollution: State of science. Science 243:745—753.

Sexton, K. 1983. Evidence of an additive effect for ozone plumes from small cities. Environ. Sci. Technol. 17:402—407.

Sexton, K. and H. Westberg. 1980. Elevated ozone concentrations measured downwind of the Chicago-Gary urban complex. J. Air Pollut. Control Assoc. 30:911—914.

Shafer, T.B., and J.H. Seinfeld. 1986. Comparative analysis of chemical reaction mechanisms for photochemical smog--II. Sensitivity of EKMA to chemical mechanism and input parameters. Atmos. Environ. 20:487—499.

Shah, J.J., J.G. Watson, Jr., J.A. Cooper, and J.J. Huntzicker. 1984. Aerosol chemical composition and light scattering in Portland, Oregon: The role of carbon. Atmos. Environ. 18:235—240.

Shareef, G.S., W.A. Butler, L.A. Bravo, and M.B. Stockton. 1988. Air Emissions Species Manual, Vol. I. Volatile Organic Compound Species Profiles. EPA 450/2-88-003a. U.S. Environmental Protection Agency, Research Triangle Park, N.C. 641 pp.

Shreffler, J.H., and R. B. Evans. 1982. The surface ozone record from the Regional Air Pollution Study, 1975-1976. Atmos. Environ. 16:1311—1322.

Sicre, M.A., J.C. Marty, A. Saliot, X. Aparicio, J. Grimalt, and J. Albaiges. 1987. Aliphatic and aromatic hydrocarbons in different sized aerosols over the Mediterranean Sea: Occurrence and origin. Atmos. Environ. 21:2247—2259.

Sigsby, J.E., Jr., S. Tejada, W. Ray, J.M. Lang, and J.W. Duncan. 1987. Volatile organic compound emissions from 46 in-use passenger cars. Environ. Sci. Technol. 21:466—475.

Sillman, S., and P.J. Samson. 1989. Impact of Methanol-fueled Vehicles on Rural and Urban Ozone Concentrations During a Region-wide Ozone Episode in the Midwest. Paper presented at Methanol as an Alternative Fuel Choice: An Assessment. Johns Hopkins University, Washington, D.C., December 4-5.

Sillman, S., and P.J. Samson. 1990. A Regional-Scale Analysis of the Sources of Elevated Ozone Concentrations in Southeastern Michigan. Paper 90-95.1. Presented at the 83rd Annual Meeting and Exhibition of the Air and Waste Management Association, Pittsburgh, Penn., June 24-29.

Sillman, S., J.A. Logan, and S.C. Wofsy. 1990a. A regional scale model for ozone in the United States with subgrid representation of urban and power plant plumes. J. Geophys. Res. 95:5731—5748.

Sillman, S., J.A. Logan, and S.C. Wofsy. 1990b. The sensitivity of ozone to nitrogen oxides and hydrocarbons in regional ozone episodes. J. Geophys.

Res. 95:1837—1851.

Simoneit, BR.T.   1986.   Characterization of organic constituents in aerosols in relation to their origin and transport: A review.   Int. J. Environ. Anal. Chem. 23:207—237.

Simoneit, BR.T., G. Sheng, X. Chen, J. Fu, J. Zhang, and Y. Xu.   1991. Molecular marker study of extractable organic matter in aerosols from urban areas of China.   Atmos. Environ. 25A:2111—12129.

Singh, H.B.   1980.   Guidance for the Collection and Use of Ambient Hydrocarbon Species Data in Development of Ozone Control Strategies. EPA 450/4-80-008.   U.S. Environmental Protection Agency.   64 pp.

Singh, H.B., and P.L. Hanst.   1981.   Peroxyacetyl nitrate (PAN) in the unpolluted atmosphere: An important reservoir for nitrogen oxides. Geophys. Res. Lett. 8:941—944.

Singh, H.B., F.L. Ludwig, and W.B. Johnson.   1978.   Tropospheric ozone: Concentrations and variabilities in clean remote atmospheres.   Atmos. Environ. 12:2185—2196.

Singh, H.B., L.J. Salas, B.A. Ridley, J.D. Shetter, N.M. Donahue, F.C. Fehsenfeld, D.W. Fahey, D.D. Parrish, E.J. Williams, S.C. Liu, G. Hübler, and P.C. Murphy.   1985.   Relationship between peroxyacetyl nitrate (PAN) and nitrogen oxides in the clean troposphere.   Nature 318:347—349.

Slemr, R., and W. Seiler.   1984.   Field measurements of NO and NO$_2$ emissions from fertilized and unfertilized soils.   J. Atmos. Chem. 2:1—24.

Slemr, F., G.W. Harris, D.R. Hastie, G.I. Mackay, and H.I. Schiff.   1986. Measurement of gas phase hydrogen peroxide in air by tunable diode laser absorption spectroscopy.   J. Geophys. Res. 91:5371—5378.

Smith, B.   1982.   Aldehyde and Other Emissions From the Use of High Alcohol Fuels.   Liquid Fuels Trust Board, Project 610/01/4, Report No. LF 2018.   Wellington, New Zealand.   June.

Smith, G.P., and D.R. Crosley.   1990.   A photochemical model of ozone interference effects in laser detection of tropospheric OH.   J. Geophys. Res. 95:16427—16442.

Snow, R., L. Baker, W. Crews, C.O. Davis, J. Duncan, N. Perry, P. Siudak, F. Stump, W. Ray, and J. Braddock.   1989.   Characterization of emissions from a methanol fueled motor vehicle.   J. Air Waste Manage. Assoc. 39:48—54.

Sperling, D.   1988.   New Transportation Fuels: A Strategic Approach to Technological Change.   Berkeley, Calif: University of California Press.

Sperling, D.   1991.   An incentive-based transition to alternative transportation fuels.   Pp. 251—264 in Energy and the Environment in the 21st Century, J.W. Tester, D.O. Wood, and N.A. Ferrari, eds.   Cambridge, Mass.: The

MIT Press.

Sperling, D., and M.A. DeLuchi. 1989. Transportation energy futures. Ann. Rev. Energy 14:375–424.

Spicer, C.W. 1977. The fate of nitrogen oxides in the atmosphere. Pp. 163-261 in Advances in Environmental Science and Technology, Vol. 7., J.N. Pitts, Jr. and R.L. Metcalf, eds. New York: Wiley-Interscience Publication.

Spicer, C.W. 1982. Nitrogen oxide reactions in the urban plume of Boston. Science 215:1095–1097.

Spicer, C.W., and G.M. Sverdrup. 1981. Tracer Nitrogen Chemistry During the Philadelphia Oxidant Enhancement Study, 1979. A Battelle-Columbus report (Contract No. 68-02-2957) to the Office of Air Quality Planning and Standards, U.S. Environmental Protection Agency, Research Triangle Park, N.C.

Spicer, C.W., D.W. Joseph, and P.R. Sticksel. 1982. An investigation of the ozone plume from a small city. J. Air Pollut. Cont. Assoc. 32:278–281.

Spicer, C.W., D.W. Joseph, and G.F. Ward. 1978. Investigations of Nitrogen Oxides Within the Plume of an Isolated City. CAPA-9-77, Battelle Columbus report to Coordinating Research Council.

Spicer, C.W., D.W. Joseph, P.R. Sticksel, and G.F. Ward. 1979. Ozone sources and transport in the northeastern United States. Environ. Sci. Technol. 13:975–985.

Spicer, C.W., J.R. Koetz, G.W. Keigley, G.M. Sverdrup, and G.F. Ward. 1983. Nitrogen oxides reactions within urban plumes transported over the ocean. EPA 600/3-83-028. U.S. Environmental Protection Agency, Environmental Sciences Research Laboratory, Research Triangle Park, N.C.

Spivakovsky, C.M., R. Yevich, J.A. Logan, S.C. Wofsy, M.B. McElroy, and M.J. Prather. 1990a. Tropospheric OH in a three-dimensional chemical tracer model: An assessment based on observations of CH_3CCl_3. J. Geophys. Res. 95:18441–18471.

Spivakovsky, C.M., S.C. Wofsy, and M.J. Prather. 1990b. A numerical method for parameterization of atmospheric chemistry: Computation of tropospheric OH. J. Geophys. Res. 95:18433–18439.

Stan, H.-J., S. Schicker, and H. Kassner. 1981. Stress ethylene evolution of bean plants--a parameter indicating ozone pollution. Atmos. Environ. 15:391–395.

Stasiuk, W.N., Jr., and P.E. Coffey. 1974. Rural and urban ozone relationships in New York state. J. Air Pollut. Control Assoc. 24:564–568.

Stedman, D.H., E.D. Morris, Jr. E.E. Daby, H. Niki, and B. Weinstock. 1970. The role of OH radicals in photochemical smog reactions. Paper No.

WATR-26. 160th National Meeting of the American Chemical Society, Chicago, Illinois. September 14-18.

Stephens, E.R., and F.R. Burleson. 1967. Analysis of the atmosphere for light hydrocarbons. J. Air Pollut. Control Assoc. 17:147—153.

Stephens, E.R., and F.R. Burleson. 1969. Distribution of light hydrocarbons in ambient air. J. Air Pollut. Control Assoc. 19:929—936.

Stephens, R.D., and S.H. Cadle. 1991. Remote sensing measurements of carbon monoxide emissions from on-road vehicles. J. Air Waste Manage. Assoc. 41:39—46.

Stockwell, W.R. 1986. A homogeneous gas phase mechanism for use in a regional acid deposition model. Atmos. Environ. 20:1615—1632.

Stockwell, W.R. 1988. The RADM-II gas-phase chemical mechanism, interim report. Environmental Protection Agency, Research Triangle Park, N.C.

Stockwell, W.R., P. Middleton, and J.S. Chang. 1990. The RADM2 chemical mechanism for regional air quality modeling. J. Geophys. Res. 95D:16343—16367.

Stoeckenius, T. 1989. Comparison of Techniques for Adjusting Ozone Exceedances to Account for Variations in Meteorological Conditions. Paper 89-35.2. Presented at the 82nd Annual Meeting and Exhibition of the Air and Waste Management Association, Anaheim, Calif., June 25-30.

Stoeckenius, T. 1990. Adjustment of Ozone Trends for Meteorological Variation. Presented at the Air and Waste Management Specialty Conference, Tropospheric Ozone and the Environment, Los Angeles, Calif., March.

Stolarski, R.S., P. Bloomfield, R.D. McPeters, and J.R. Herman. 1991. Total ozone trends deduced from Nimbus 7 TOMS data. Geophys. Res. Lett. 18:1015—1018.

Strutt, R.J. 1918. Ultraviolet transparency of the lower atmosphere and its relative poverty in ozone. Proc. Roy. Soc. A94:260—268.

Svensson, R., E. Ljungstrom, and O. Lindqvist. 1987. Kinetics of the reaction between nitrogen dioxide and water vapour. Atmos. Environ. 21:1529—1539.

Sze, N.D. 1977. Anthropogenic CO emissions: Implications for the atmospheric CO-OH-CH$_4$ cycle. Science 195:673—675.

Talbot, R.W., K.M. Beecher, R.C. Harriss, and W.R. Cofer, II. 1988. Atmospheric geochemistry of formic and acetic acids at a midlatitude temperate site. J. Geophys. Res. 93:1638—1652.

Talbot, R.W., A.S. Vijgen, and R.C. Harriss. 1990. Measuring tropospheric nitric acid: Problems and prospects for nylon filter and mist chamber techniques. J. Geophys. Res. 95:7553—7561.

Tanner, R.L., A.H. Miguel, J.B. de Andrade, J.S. Gaffney, and G.E. Streit. 1988. Atmospheric chemistry of aldehydes: Enhanced peroxyacetyl nitrate formation from ethanol-fueled vehicular emissions. Environ. Sci. Technol. 22:1026–1034.

Tesche, T.W. 1983. Photochemical dispersion modeling: Review of model concepts and applications studies. Environ. Int. 9:465–490.

Tesche, T.W. 1987. Photochemical Modeling of 1984 SCCCAMP Oxidant Episodes: Protocol for Model Selection, Adaptation, and Performance Evaluation. Report prepared for the U.S. Environmental Protection Agency by Radian Corporation, Sacramento, Calif.

Tesche, T.W. 1988. Accuracy of ozone air quality models. J. Environ. Eng. 114:739–752.

Tesche, T.W., and D.E. McNally. 1989. A Three-Dimensional Photochemical Aerosol Model for Episodic and Long-Term Simulation: Formulation and Initial Application in the Los Angeles Basin. Proceedings of the American Chemical Society Symposium on Atmospheric Modeling, Miami Beach, Fla., September 10-15.

Tesche, T.W., and D. McNally. 1990. Photochemical modeling of two 1984 SCCCAMP ozone episodes. J. Appl. Meteor. 30:745–763.

Tesche, T.W., C. Seigneur, L.E. Reid, P.M. Roth, and W.R. Oliver. 1981. The Sensitivity of Complex Photochemical Model Estimates to Detail in Input Information. EPA 450/4-81-031A. U.S. Environmental Protection Agency, Office of Air Quality Planning and Standards, Research Triangle Park, N.C.

Tesche, T.W., D.E. McNally and J.G. Wilkinson. 1988a. Airshed Modeling of San Francisco Bay Region: Evaluation of Ozone Impact of Proposed Crockett Cogeneration Facility under Four Different Meteorological Scenarios. Radian Corporation Report to California Energy Commission, Sacramento, Calif.

Tesche, T.W., J.G. Wilkinson, D.E. McNally, R. Kapahi, and W.R. Oliver. 1988b. Photochemical Modeling of Two SCCCAMP 1984 Oxidant Episodes. Vol. II: Modeling Procedures and Evaluation Results. Report prepared by Radian Corporation for U.S. Environmental Protection Agency, Sacramento, Calif., January.

Tesche, T.W., D.E. McNally, and J.G. Wilkinson. 1988c. Importance of Boundary Layer Measurements for Urban Airshed Modeling. Paper 88-48.5. Paper presented at the 81st Annual Meeting of Air Pollution Control Association, Dallas, Texas, June 19-24.

Tesche, T.W., P. Georgopoulos, J.H. Seinfeld, G. Cass, F.L. Lurmann, and P.M. Roth. 1990. Improvement of Procedures for Evaluating

Photochemical Models. Report prepared by Radian Corporation for the State of California Air Resources Board, Sacramento, Calif.

Thompson, A.M., and R.J. Cicerone. 1986. Possible perturbations to atmospheric CO, CH_4, and OH. J. Geophys. Res. 91:10853–10864.

Thompson, A.M., R.W. Stewart, M.A. Owens, and J.A. Herwehe. 1989. Sensitivity of tropospheric oxidants to global chemical and climate change. Atmos. Environ. 23:519–532.

Thompson, A.M., M.A. Huntley, and R.W. Stewart. 1990. Perturbations to tropospheric oxidants, 1985-2035. 1. Calculations of ozone and OH in chemically coherent regions. J. Geophys. Res. 95:9829–9844.

Tiao, G.C., G.C. Reinsel, J.H. Pedrick, G.M. Allenby, C.L. Mateer, A.J. Miller, and J.J. DeLuisi. 1986. A statistical trend analysis of ozonesonde data. J. Geophys. Res. 91:13121–13136.

Tilden, J.W., V. Costanza, G.J. McRae, and J.H. Seinfeld. 1981. Sensitivity analysis of chemically reacting systems. Pp. 69-91 in Modelling of Chemical Reaction Systems, Proceedings of an International Workshop, K.H. Ebert, P. Deuflhard, and W. Jäger, eds. Springer-Verlag Series in Chemical Physics 18. Berlin: Springer-Verlag.

Tingey, D.T. 1981. The effect of environmental factors on the emission of biogenic hydrocarbons from live oak and slash pine. Pp. 53–79 in Atmospheric Biogenic Hydrocarbons, Vol. I, J.J. Bufalini and R.R. Arnts, eds. Ann Arbor, Mich.: Ann Arbor Science Publishers, Inc.

Tingey, D.T., M. Manning, L.C. Grothaus, and W.F. Burns. 1979. The influence of light and temperature on isoprene emission rates from live oak. Physiol. Plant. 47:112–118.

Tingey, D.T., M. Manning, L.C. Grothaus, and W.F. Burns. 1980. Influence of light and temperature on monoterpene emission rates from slash pine. Plant Physiol. 65:797–801.

Tingey, D.T., W.E. Hogsett, and E.H. Lee. 1989. Analysis of crop loss for alternative ozone exposure indices. Pp 219-227 in Atmospheric Ozone Research and Its Policy Implications, Proceedings of the 3rd US-Dutch International Symposium, T. Schneider, S.D. Lee, G.J.R. Wolters, and L.D. Grant, eds. Studies in Environmental Science 35. Amsterdam: Elsevier Science Publishers.

Tonneijck, A.E.G. 1984. Effects of peroxyacetyl nitrate (PAN) and ozone on some plant species. Pp. 118-127 in Proceedings of the International Workshop on an Evaluation and Assessment of the Effects of Photochemical Oxidants on Human Health, Agricultural Crops, Forestry, Materials and Visibility, P. Grennfelt, ed. Swedish Environmental Research Institute (IVL), Goteborg, Sweden.

Torres, A.L., and H. Buchan. 1988. Tropospheric nitric oxide measurements over the Amazon Basin. J. Geophys. Res. 93:1396—1406.

Townsend, D. 1990. Roads to Alternative Transportation Fuels. Presented at Asilomar, California, July 12.

Trainer, M., E.T. Williams, D.D. Parrish, M.P. Buhr, E.J. Allwine, H.H. Westberg, F.C. Fehsenfeld, and S.C. Liu. 1987. Models and observations of the impact of natural hydrocarbons on rural ozone. Nature 329:705—707.

Tuazon, E.C., and R. Atkinson. 1989. A product study of the gas-phase reaction of methyl vinyl ketone with the OH radical in the presence of NO_x. Int. J. Chem. Kinet. 21:1141—1152.

Tuazon, E.C., and R. Atkinson. 1990a. A product study of the gas-phase reaction of isoprene with the OH radical in the presence of NO_x. Int. J. Chem. Kinet. 22:1221—1236.

Tuazon, E.C., and R. Atkinson. 1990b. A product study of the gas-phase reaction of methacrolein with the OH radical in the presence of NO_x. Int. J. Chem. Kinet. 22:591—602.

Tuazon, E.C., W.P.L. Carter, and R. Atkinson. 1991. Thermal decomposition of peroxyacetyl nitrate and reactions of acetyl peroxy radicals with NO and NO_2 over the temperature range 283-313 K. J. Phys. Chem. 95:2434—2437.

Turman, B.N., and B.C. Edgar. 1982. Global lightning distributions at dawn and dusk. J. Geophys. Res. 87:1191—1206.

Turpin, B.J., and J.J. Huntzicker. 1991. Secondary formation of organic aerosols in the Los Angeles basin: A descriptive analysis of organic and elemental carbon concentrations. Atmos. Environ. 25A:207—216.

UCAR (University Cooperation for Atmospheric Research). 1986. Global Tropospheric Chemistry - Plans for the U.S. Research Effort. OIES Report 3. 110 pp.

U.S. Senate Committee on Environment and Public Works. 1989. Clean Air Act Amendments of 1989. U.S. Senate Committee on Environment and Public Works. S-1630, Report 101-228, Calendar No. 427. Washington, DC. December.

Vaghjiani, G.L., and A.R. Ravishankara. 1991. New measurement of the rate coefficient for the reaction of OH with methane. Nature 350:406—409.

van den Dool, H.M., and S. Saha. 1990. Frequency dependence in forecast skill. Mon. Wea. Rev. 118:128—137.

Vaughan, W.M. 1985. Transport of Pollutants in Plumes and PEPEs. EPA-600/3-85/033. U.S. Environmental Protection Agency, Research Triangle Park, N.C.

Venkatram, A., P.K. Karamchandani, and P.K. Misra. 1988. Testing a comprehensive acid deposition model. Atmos. Environ. 22:737—747.

Viterito, A. 1989. Implications of urbanization for local and regional temperatures in the United States. Pp. 115-119 in Coping with Climate Change, Proceedings of the Second North American Conference on Preparing for Climate Change: A Cooperative Approach, J.C. Topping, Jr., ed. Climate Institute, Washington, D.C.

Volz, A., and D. Kley. 1988. Evaluation of the Montsouris series of ozone measurements in the nineteenth century. Nature 332:240—242.

Volz, A., D. Mihelcic, P. Müsgen, H.W. Pätz, G. Pilwat, H. Geiss, and D. Kley. 1988. Ozone production in the Black Forest: Direct measurements of RO_2, NO_x and other relevant parameters. Pp. 293—302 in Tropospheric Ozone: Regional and Global Scale Interactions, I.S.A. Isaksen, ed. NATO ASI Series C, Vol. 227. Dordrecht, Holland: D. Reidel Publishing Co.

Vukovich, F.M., and J. Fishman. 1986. The climatology of summertime O_3 and SO_2 (1977-1981). Atmos. Environ. 20:2423—2433.

Vukovich, F.M., W.D. Bach, Jr., B.W. Crissman, and W.J. King. 1977. On the relationship between high ozone in the rural surface layer and high pressure systems. Atmos. Environ. 11:967—983.

Vukovich, F.M., J. Fishman, and E.V. Browell. 1985. The reservoir of ozone in the boundary layer of the eastern United States and its potential impact on the global tropospheric ozone budget. J. Geophy. Res. 90:5687—5698.

Wackter, D.J., and P.V. Bayly. 1988. The effectiveness of emission controls on reducing ozone levels in Connecticut from 1976 through 1987. Pp. 398-415 in The Scientific and Technical Issues Facing Post-1987 Ozone Control Strategies, G.T. Wolff, J.L. Hanisch, and K. Schere, eds. Pittsburgh, Penn.: Air and Waste Management Association.

Wagner, K.K. 1988. Photochemical Modeling for the South Coast Air Basin. Technical Support Division, California Air Resources Board, Sacramento, Calif.

Wagner, K.K., and A. Ranzieri. 1984. Model Performance Evaluations for Regional Photochemical Models in California. Paper 84-47.6. Presented at the 77th Annual Meeting of the Air Pollution Control Association, San Francisco, Calif., June 24-27.

Wagner, K.K., and B.E. Croes. 1986. Evaluation of Predicted Hydrocarbons to Oxides of Nitrogen Ratios in Ozone Modeling for Emission Control Strategy Development. Proceedings of the 5th Joint Conference on Applications of Air Pollution Meteorology. Boston: American Meteorological Society.

Wagner, K.K., and N.J. Wheeler. 1989. Variability of Photochemical Modeling Emission Sensitivity with Respect to Meteorology. Proceedings of the 6th Joint Conference on Applications of Air Pollution Meteorology. Boston: American Meteorological Society.

Wakim, P.G. 1989. Temperature-adjusted Ozone Trends for Houston, New

York, and Washington, 1981-1987. Paper 89-35.1. Presented at the 82nd Annual Meeting and Exhibition of the Air and Waste Management Association, Anaheim, Calif., June 25-30.

Wakim, P.G. 1990. 1981 to 1988 Ozone Trends Adjusted to Meteorological Conditions for 13 Metropolitan Areas. Paper 90-97.9. Presented at the 83rd Annual Meeting and Exhibition of the Air and Waste Management Association, Pittsburgh, Penn., June 24-29.

Walker, H.M. 1985. Ten-year ozone trends in California and Texas. J. Air Pollut. Control Assoc. 35:903—912.

Wang, C.C., L.I. Davis, Jr., P.M. Selzer, and R. Munoz. 1981. Improved airborne measurements of OH in the atmosphere using the technique of laser-induced fluorescence. J. Geophys. Res. 86:1181—1186.

Warneck, P. 1988. Chemistry of the Natural Atmosphere. New York: Academic Press. 757 pp.

Watson, R.T., M.J. Prather, and M.J. Kurylo. 1988. Present State of Knowledge of the Upper Atmosphere 1988: An Assessment Report. NASA Reference Publication 1208. U.S. National Aeronautics and Space Administration, Office of Space Science and Applications, Washington, D.C. August. 208pp.

Went, F.W. 1960. Organic matter in the atmosphere and its possible relation to petroleum formation. Proc. Natl. Acad. Sci. 46:212—221.

Westberg, H. and B. Lamb. 1985. Ozone Production and Transport in the Atlanta, Georgia Region. EPA/600/3-85/013. U.S. Environmental Protection Agency, Research Triangle Park, N.C. 244 pp.

Whitby, R.A., and E.R. Altwicker. 1978. Acetylene in the atmosphere: Sources, representative ambient concentrations and ratios to other hydrocarbons. Atmos. Environ. 12:1289—1296.

White, W.H., and P.T. Roberts. 1977. On the nature and origins of visibility-reducing aerosols in the Los Angeles air basin. Atmos. Environ. 11:803—812.

White, W.H., J.A. Anderson, D.L. Blumenthal, R.B. Husar, N.V. Gillani, J.D. Husar, and W.E. Wilson, Jr. 1976. Formation and transport of secondary air pollutants: Ozone and aerosols in the St. Louis urban plume. Science 194:187-189.

Whitten, G.Z., and H. Hogo. 1983. Impact of Methanol on Smog: A Preliminary Estimate. SAI Publication No. 83044, Systems Applications, Inc., San Rafael, Calif.

Whitten, G.Z., and M.W. Gery. 1986. Development of the CBM-X Mechanisms for Urban and Regional AQSMs. EPA-600/3-86-012. U.S. Environmental Protection Agency, Office of Mobile Sources, Ann Arbor, Mich.

Whitten, G.Z., N. Yonkow, and T.C. Myers. 1986. Photochemical Modeling of Methanol-use Scenarios in Philadelphia. EPA/460/3-86-001. U.S. Environmental Protection Agency, Emission Control Technology Division, Ann Arbor, Mich.

Williams, E.J., and F.C. Fehsenfeld. 1991. Measurement of soil nitrogen oxide emissions at three North American ecosystems. J. Geophys. Res. 96:1033—1042.

Williams, E.J., D.D. Parrish, and F.C. Fehsenfeld. 1987. Determination of nitrogen oxide emissions from soils: Results from a grassland site in Colorado, United States. J. Geophy. Res. 92:2173—2179.

Williams, E.J., D.D. Parrish, M.P. Buhr, F.C. Fehsenfeld, and R. Fall. 1988. Measurement of soil NO. emissions in Central Pennsylvania. J. Geophys. Res. 93:9539—9546.

Williams, R.L., F. Lipari, and R.A. Potter. 1989. Formaldehyde, Methanol, and Hydrocarbon Emissions from Methanol-Fueled Cars. Paper No. 89-124.3. Prepared for presentation at the 82nd Annual Meeting of the Air and Waste Management Association, Anaheim, Calif., June 25-30.

Williams, E.J., A. Guenther, and F.C. Fehsenfeld. 1992a. An inventory of nitric oxide emissions from soils in the United States. J. Geophys. Res., submitted.

Williams, E.J., D.D. Parrish, and F.C. Fehsenfeld. 1992b. An empirical representation of nitric oxide emissions from soil. J. Geophys. Res., submitted.

Wilson, K.W., and G.J. Doyle. 1970. Investigation of photochemical reactivities of organic solvents. Final Report, SRI Project PSU-8029. SRI International, Menlo Park, Calif.

Wilson, W.E., Jr., W.E. Schwartz, and G.W. Kinzer. 1972. Haze Formation: Its Nature and Origin. CRC-APRAC-CAPA-6-68-3. Coordinating Research Council, Air Pollution Advisory Committee, New York.

Wilson, J.H. Jr., E.J. Laich, and F.L. Bunyard. 1990. Ozone Nonattainment-An Analysis of Proposed Legislation. Paper 90—93.5. Presented at the 83rd Annual Meeting and Exhibition of the Air and Waste Management Association, Pittsburgh, Penn., June 24-29.

Winer, A.M. 1986. Air pollution chemistry. Pp. 133—154 in Handbook of Air Pollution Analysis, 2nd ed., R.M. Harrison and R. Perry, eds. London: Chapman and Hall.

Winer, A.M., K.R. Darnall, R. Atkinson, and J.N. Pitts, Jr. 1979. Smog chamber study of the correlation of hydroxyl radical rate constants with ozone formation. Environ. Sci. Technol. 13:822—826.

Winer, A.M., R. Atkinson, and J.N. Pitts, Jr. 1984. Gaseous nitrate radical:

Possible nighttime atmospheric sink for biogenic organic compounds. Science 224:156—159.

Winer, A.M., R. Atkinson, J. Arey, H.W. Biermann, W.P. Harger, E.C. Tuazon, and B. Zielinska. 1987. The Role of Nitrogenous Pollutants in the Formation of Atmospheric Mutagens and Acid Deposition. Final Report to California Air Resources Board, Contract No. A4-081-32. Statewide Air Pollution Research Center, University of California, Riverside, Calif. March. 348 pp.

Winer, A.M., J. Arey, S.M. Aschmann, R. Atkinson, W.D. Long, L.C. Morrison, and D.M. Olszyk. 1989. Hydrocarbon Emissions from Vegetation Found in California's Central Valley. Final Report. Contract No. A732-155. California Air Resources Board, Sacramento, Calif. November. 456 pp.

WMO (World Meteorological Organization). 1986. Atmospheric Ozone 1985. Atmospheric Ozone Assessment of Our Understanding of the Processes Controlling its Present Distribution and Change. Report No.16, Vol. 1. Global Ozone Research and Monitoring Project. Geneva: World Meteorological Organization.

WMO (World Meteorological Organization). 1988. Ozone Trends Panel Report. Global Ozone Research and Monitoring Project, Report No. 18, R.T. Watson, ed. Geneva: World Meteorological Organization.

WMO (United Nations Environment Program and World Meteorological Organization). 1990. Scientific Assessment of Stratospheric Ozone: 1989. Vol. I. World Meteorological Organization Global Ozone Research and Monitoring Project, Report No. 20. Geneva: Geneva: World Meteorological Organization. 486 pp.

Wolff, G.T., and P.J. Lioy. 1978. An empirical model for forecasting maximum daily ozone levels in the northeastern U.S. J. Air Pollut. Control Assoc. 28:1034—1038.

Wolff, G.T., and P.J. Lioy. 1980. Development of an ozone river associated with synoptic scale episodes in the eastern United States. Environ. Sci. Technol. 14:1257—1260.

Wolff, G.T., P.J. Lioy, G.D. Wight, R.E. Meyers, and R.T. Cederwall. 1977. An investigation of long-range transport of ozone across the midwestern and eastern United States. Atmos. Environ. 11:797—802.

Wolff, G.T., M.A. Ferman, N.A. Kelley, D.P. Stroup, and M.S. Ruthkosky. 1982. The relationships between the chemical composition of fine particles and visibility in the Detroit metropolitan area. J. Air. Pollut. Control Assoc. 32:1216—1220.

Wolff, G.T., N.A. Kelly, and M.A. Ferman. 1982. Source regions for summertime ozone and haze episodes in the eastern United States. Water

Air Soil Pollut. 18:65—81.

Yamartino, R.J., J.S. Scire, S.R. Hanna, G.R. Carmichael, and Y.S. Chang. 1989. CALGRID: A Mesoscale Photochemical Grid Model: Model Formulation Document. Report A049-1, Sigma Research Corporation. Prepared for the State of California Air Resources Board, Sacramento, Calif. June.

Yokouchi, Y., and Y. Ambe. 1985. Aerosols formed from the chemical reaction of monoterpenes and ozone. Atmos. Environ. 19:1271—1276.

Zeldin, M.D., and W. Meisel. 1978. Use of Meteorological Data in Air Quality Trend Analysis. EPA-450/3-78-024. Technology Service Corporation, Santa Monica, Calif., March. 101 pp.

Zeldin, M.D., J.C. Cassmassi, and M. Hoggan. 1990. Ozone Trends in the South Coast Air Basin: An Update. Paper presented at the Air and Waste Management Association Specialty Conference, Tropospheric Ozone and the Environment., Los Angeles, Calif., March. Air and Waste Management Transaction Series, Pittsburgh, Penn.

Zellner, R., B. Fritz, and M. Preidel. 1985. A cw UV laser absorption study of the reactions of the hydroxy-cyclohexadienyl radical with NO_2 and NO. Chem. Phys. Lett. 121:412—416.

Zetzsch, C., R. Koch, M. Siese, F. Witte, and P. Devolder. 1990. Adduct formation of OH with benzene and toluene and reaction of the adducts with NO and NO_2. Pp. 320—327 in the Proceedings of the 5th European Symposium on the Physio-Chemical Behavior of Atmospheric Pollutants. Dordrecht: Reidel Publishing Co.

Zimmerman, P.R. 1979. Determination of Emission Rates of Hydrocarbons from Indigenous Species of Vegetation in the Tampa/St. Petersburg, Florida Area. Tampa Bay Area Photochemical Oxidant Study. EPA 904/9-77-028. U.S. Environmental Protection Agency, Atlanta, Ga.

Zika, R.G., and E.S. Saltzman. 1982. Interaction of ozone and hydrogen peroxide in water: Implications for analysis of H_2O_2 in air. Geophys. Res. Lett. 9:231—234.

Zimmerman, P.R. 1981. Testing of hydrocarbon emissions from vegetation and methodology for compiling biogenic emission inventories. Pp. 15—33 in Atmospheric Biogenic Hydrocarbons, Vol. I, J. Bufalini and R. Arnts, eds. Ann Arbor, Mich.: Ann Arbor Science Publishers, Inc.

Zimmerman, P.R., W. Tax, M. Smith, J. Demmy, and R. Battye. 1988a. Anthropogenic Emissions Data for the 1985 NAPAP Inventory. EPA 600/7-88-022. U.S. Environmental Protection Agency, Washington, D.C. 295 pp.

Zimmerman, P.R., J.P. Greenberg, and C.E. Westberg. 1988b. Measurements of atmospheric hydrocarbons and biogenic emission fluxes in the Amazon boundary layer. J. Geophys. Res. 93:1407—1416.

Zishka, K.M., and P.J. Smith. 1980. The climatology of cyclones and anticyclones over North America and surrounding ocean environs for January and July, 1950-77. Mon. Wea.. Rev. 108:387—401.

Zweidinger, R.B., J.E. Sigsby, Jr., S.B. Tejada, F.D. Stump, D.L. Dropkin, W.D. Ray, and J.W. Duncan. 1988. Detailed hydrocarbon and aldehyde mobile source emissions from roadway studies. Environ. Sci. Technol. 22:956—962.

Index

491